Seed Production Technologies of Horticultural Plants

园艺植物种子生产技术

Editor-in-Chief Wang Shan Tang Weihua
主　编　王　姗　汤伟华

中国财富出版社有限公司

图书在版编目（CIP）数据

园艺植物种子生产技术 = Seed Production Technologies of Horticultural Plants：英文 / 王姗，汤伟华主编 . — 北京：中国财富出版社有限公司，2023.12

ISBN 978-7-5047-8038-6

Ⅰ . ①园…　Ⅱ . ①王… ②汤…　Ⅲ . ①园艺作物—种子—教材—英文　Ⅳ . ①S604

中国国家版本馆 CIP 数据核字（2023）第 238065 号

策划编辑	刘静雯	**责任编辑**	刘静雯	**版权编辑**	李　洋
责任印制	尚立业	**责任校对**	张营营	**责任发行**	敬　东

出版发行	中国财富出版社有限公司		
社　址	北京市丰台区南四环西路188号5区20楼	**邮政编码**	100070
电　话	010-52227588 转 2098（发行部）		010-52227588 转 321（总编室）
	010-52227566（24小时读者服务）		010-52227588 转 305（质检部）
网　址	http: //www. cfpress. com. cn	**排　版**	宝蕾元
经　销	新华书店	**印　刷**	北京九州迅驰传媒文化有限公司
书　号	ISBN 978-7-5047-8038-6/S · 0061		
开　本	710mm × 1000mm　1 /16	**版　次**	2024年11月第 1 版
印　张	25.5	**印　次**	2024年11月第 1 次印刷
字　数	444千字	**定　价**	49.80 元

编写人员名单

主　编：

王　姗　汤伟华

副主编：

王　瑞　王媛花　刘　芳

参　编：

任俊辉　邵妍丽　陈　曦　张建祥　叶　丹

刘　洋　孙　卉　鲍华鹏　朱国兵　姜蓓蓓

张建红

Foreword

Seed Production Technologies of Horticultural Plants is based on the contents of seed science, vegetable production technologies, seed production technologies, and other disciplines, and is integrated and structured according to the requirements of vocational competencies of each vocational cluster in the horticultural plant seed industry chain, concentrating on the knowledge of seed production, processing, inspection, storage, and construction of seed production bases in the horticultural plant seed industry chain.

Based on the traditional textbook on horticultural plant seeds, this textbook adds vegetable seed production technologies, seed testing, seed marketing, seed management, and other content. It breaks the boundaries between the disciplines of agronomy and horticulture, techniques and management, breeding and cultivation, crops and vegetables, and realizes the organic integration of knowledge and technology from different disciplines, thus meeting the requirements of occupational positions (clusters) in related technical fields and the needs of cultivating corresponding senior technical and practical talents.

This textbook highlights the training requirements for students' vocational ability. The selection of theoretical knowledge is focused on the needs of work task completion. At the same time, the vocational needs for theoretical knowledge learning are fully considered, and the requirements of relevant vocational qualification certificates for knowledge, skills, and attitudes are integrated. The project design closely follows practical teaching, focusing on production practice skills according to the characteristics of horticultural plant growth, stressing the training of students' vocational abilities.

Contents

Module 1 Selection and Breeding of New Cultivars of Horticultural Plants

Learning Objectives

(1) Master the concept of cultivars and be able to classify cultivars according to the characteristics of different cultivars.

(2) Be able to classify germplasm resources according to the characteristics of horticultural plants.

(3) Understand the importance of conservation and utilization of germplasm resources of horticultural plants.

(4) Master the main methods of horticultural plant breeding.

(5) Acquire the ability to analyze and solve problems and abilities to accept and apply new technologies and skills in production practices.

Learning Scenario I Horticultural Plant Breeding Objectives and Germplasm Resources

I. The Main Objectives of Modern Horticultural Plant Breeding

1. High and Stable Yields

High and stable yields are the basic characteristic of superior cultivars and also the basic goal of horticultural plant breeding. The volume of yield is related to yield components. For example, the yield components of grapes include the total number of branches per plant or unit area, the percentage of bearing branches, the average number of fruit clusters of every bearing branch, and the average weight of a single cluster. The proper plant type is the basis of fertility and the morphological characteristic of high-yielding cultivars. The yield per unit area is basically the product of these components. Selection based on yield components is a better indicator of fertility potential among strains than selection based directly on the fruit yields of plants. However, the components are often mutually constrained and negatively correlated to different degrees, making it difficult to improve simultaneously. So, we should make different growth requirements according to the natural conditions of the region, the level of cultivation, and the developmental characteristics of the plants.

The harvest of many horticultural plants (such as a wide variety of vegetables) often takes place in batches, which can be divided into early-tirm, mid-term, and late-term yields based on the harvesting period. Since the prices of early products are significantly different from those of mid and late products, early production volume is sometimes a more important selection target than the total production volume. The comparison of yield among superior individual grape plants is shown in Table 1-1.

Table 1-1 Comparison of Yield among Superior Individual Grape Plants

Cultivar	A	B
Total Number of Branches	58	32
Bearing Branches	30	24
Fruit Clusters	55	40
Yield (g)	6,200	5,000
Percentage of Fruiting Shoots (%)	52	75
Fruiting Coefficient	0.95	1.25
Average Weight of a Single Cluster	113	125
Average Load of a Single Branch	107	156

2. Good Quality

The quality of horticultural products can be divided into sensory quality, nutritional quality, storage quality, and processing quality according to their use and utilization methods.

Sensory quality refers to the size, shape, color, flavor, aroma, and flesh quality of the plants or edible organs. Fruits and vegetables have high requirements for these qualities, while ornamental plants often have requirements for flower shape, flower color, leaf shape, leaf color, plant type, and fragrance. For the cultivars that need to be processed, stored, and transported, their sensory quality after being processed, stored, and transported should also be considered. The evaluation of sensory quality is often influenced by traditional habits, thus subjective to people's subjectivity. With the change in the use of products and people's consumption habits, people's evaluation of sensory quality also changes.

Nutritional quality mainly includes the composition and content of vitamins, mineral elements, organic acids, cellulose, protein, and other harmful components. The most important elements of the nutritional quality of vegetables are the contents of vitamins, mineral elements, organic acids, cellulose, protein, and amino acids, while for the fruit trees, they are the sugar content of fruits, the starch of dried fruits, and special chemical components (such as flavonoid for ginkgo). But when it comes to ornamental plants, except for the dual-purpose cultivars, nutritional quality won't be taken into consideration.

Storage quality refers to the storage tolerance of plants or edible organs, that is, whether fruits are easy to store and transported after ripening, including their appropriate storage temperature, storage or transportation duration, etc.

Processing quality refers to the relevant characteristics of the product suitable for processing, such as the lycopene and uniformity of fruit color for tomatoes, the freestone or clingstone and solute or insoluble substance for peaches, as well as the pigment content and sugar content for grapes. These qualities are particularly important during processing.

3. Strong Adaptability

Adaptability to environmental conditions is an important sign of superior horticultural plant cultivars, which refers to traits like low-temperature tolerance, low-light tolerance, heat tolerance, drought tolerance, and flood tolerance. Requirements for target traits of adaptability and stress resistance are often combined with requirements for yields and product quality. It is often required that the yields and product quality are relatively stable under a certain stress condition.

4. High Resistance to Diseases and Pests

Increasing the resistance and tolerance of horticultural plant cultivars to diseases and pests is one of the important breeding objectives. Since there are many kinds of diseases and pests, resistance breeding can only focus on the main kinds of diseases and pests. This objective can only be achieved in species where the damage is widespread and severe and where there are significant differences in resistance and tolerance within and between species. It is impractical to require new cultivars to have absolute resistance to diseases and pests in resistance breeding. Generally, it is only required that the number of pathogenic bacteria and the insect density can be reduced to a certain range when the disease is prevalent or when pests occur, thus there is no significant impact on yield and quality.

5. Suitable for Mechanized Production

To adapt horticultural plants to mechanized production, some of their traits must be improved to be suitable for mechanized tillage and harvesting. These traits include compact plant type, strong stalks without collapse, neat growth, uniform maturity, uniform size, uniform length, tough fruit skin, and moderate fruiting parts.

6. Different Maturity Stages

Having different maturity stages is the specific target trait in horticultural plant breeding. Among early, medium, and late maturity, early maturity, especially in vegetables and fruit trees, is of particular importance.

II. General Principles for Setting Breeding Objectives

1. Meet the Needs of Production and Market

A cultivar selected and bred must meet the needs of the production and the market at present and for a period of time. This requires consideration of the following aspects: Whether it meets the requirements of the modernization of agriculture, including national development plans for horticultural production and the requirements of agricultural mechanization, etc.; whether it suits the farming system of the area to be promoted, and whether the structure of crops and the matching cultivars are reasonable; whether it meets the requirement of the local and current large-scale production or meets some specific requirements (such as the requirement for foreign trade exports, processing, storage, and transportation).

2. Have High Economic and Social Benefits

The breeding objectives of horticultural plants should meet this requirement: The new cultivars must have higher economic and social benefits in production and supply than the original cultivars do. For example, if the new cultivar's price is similar to that of the original cultivar, its yield is higher than that of the original cultivar. If its yield is close to that of the original cultivar, its price is increased due to its excellent quality or earlier maturity. Alternatively, the cost of medicaments and labor for diseases and pest prevention and control are saved due to its improved resistance to diseases and pests.

In addition, some breeding objectives can produce greater social and ecological benefits. For instance, reducing pollution and sand wasteland, as well as other results with the same effect should receive more funding from the state or social groups.

3. Fully Consider the Possibility of Achieving the Breeding Objective (Objective Conditions)

The realization of the breeding objective is based on its realistic possibility. Things like the conditions of the breeder itself, the available germplasm resources, the natural environment and culture conditions in the selected and promotion areas, and the suitability of the trait index should be considered.

4. Balance the Relationship Between Realistic Needs and Long-term Interests

The breeding of horticultural plants takes a long time, usually as short as seven or eight years, or as long as a dozen years. Therefore, the development of breeding objectives should focus on both the realistic and near-term development needs, as well as the long-term development needs. It is important to consider what goals should be set after the immediate goal is achieved. We should develop phased breeding objectives for a longer-term and complex breeding project. For example, in a plan that requires 20 years to achieve, a number of transitional cultivars that may be acceptable to the market should be bred in 8–10 years.

5. Handle the Relationship Between Target and Non-target Traits

The breeding objectives should be as simple and clear as possible. In addition to highlighting the key points, the breeding objectives must be associated with specific component traits. The main problems of existing cultivars in production and development should be analyzed to clarify the target traits that need to be improved. As much as possible, quantitative objective indicators that can be tested should be proposed, so as to ensure the accuracy and clarity of the breeding objectives and provide objective and specific criteria for the final evaluation of the breeding objectives. For example, the target trait of yield should generally be shown in the biological yield and economic coefficient; the crop yield of available fruits can be translated into the number of fruits per plant, unit fruiting mother branch, or unit area, and the average quality of a single fruit; quality traits can be shown in the product size, shape, color, texture, flavor, and other sensory traits, and the content of sugar, acid, vitamin C, and other substances or other quality characteristics. For example, the breeding objective of *Clivia miniata* mainly is leaf length below 30 cm, leaf width above 10 cm, and leaf thickness above 0.2 cm. Although it is difficult to specify quantitative indicators for some traits, objective criteria that can reflect the degree of required traits should be put forward as possible according to the above examples.

There must be priorities in setting breeding objectives. No more than two traits should be targeted at a time. Moreover, it is necessary to clarify the main target trait and secondary target trait according to the importance and difficulty of the traits in the breeding, so that the main and secondary traits can be distinguished and improved in a coordinated manner. The relationship between target and non-target traits should also be handled. The major economic traits should meet the needs of production and consumers. Strong demand for one trait may sometimes have a negative impact on the existence of another trait. Such mutual restraint, for example, between early maturity and quality/yield, between maturity and storability, as well as between quality and stress/disease resistance, is manifested to varying degrees in different horticultural plants. If early maturity is the main target trait and quality is the secondary target trait, generally, the breeding process always improves quality after achieving earlier maturity. As there is a degree of negative correlation between early maturity and high yield/quality, the requirements for quality/yield indicators should usually be reduced appropriately.

III. The Concept and Types of Cultivars

(I) The Concept of Cultivars

Cultivars are horticultural plant groups created by human beings under certain ecological and economic conditions, based on the needs of human production and life, with relative genetic stability and relative consistency of major traits. In certain areas and under certain cultural conditions, a cultivar's yield, quality, fertility period, and adaptability can meet corresponding production needs, and it can maintain its persistence through ordinary reproduction methods.

1. Features of Cultivars

(1) They have relatively stable genetic traits;

(2) They have relative consistency in ecological, morphological, and economic traits;

(3) They have one or more morphological, physiological, and other characteristics that are different from those of other cultivars;

(4) Being grown in certain areas or conditions, their yields, qualities, and adaptability meet the needs of production;

(5) There is a time limit for its use.

2. The Role of Superior Varieties in the Production of Horticultural Plants

(1) Increase the yield per unit area: High yield is one of the basic characteristics of superior cultivars. Under the same conditions, the use of cultivars with high yield potential can increase the yield of horticultural plants by more than 20%.

(2) Improve product quality: Improving product quality, such as the quality of fruit color, flavor, and hardness, mainly relies on the selection and breeding of high-quality cultivars.

(3) Enhance the stress resistance and adaptability to ensure stable yield: The relatively stable yield performance of superior cultivars in different years is determined by their stress resistance and adaptability. For example, if a

horticultural plant cultivar has strong resistance to frequent diseases and pests, and environmental stresses; it can mitigate or avoid yield loss and quality deterioration in production; if the cultivar gets strong adaptability, its planting range can be expanded.

(4) Be conducive to the reform of the farming system and adapt to mechanization: These requirements can be met by selecting cultivars with proper fertility characteristics, growth habits, and plant types.

(II) Types of Cultivars

According to the different genetic compositions of the population, the horticultural plant cultivars can be divided into inbred line cultivars, population cultivars, hybrid cultivars, and asexual line cultivars.

1. Inbred Line Cultivars

The genetic composition of the population is basically homogeneous, and the individuals are basically homozygous. Selection and breeding of such cultivars are mainly done through selective breeding and sexual crossbreeding. In terms of selection methods, individual selection is the main method, while anther and pollen culture can also be used to induce haploid. Then, double the chromosome of the haploid plants, and after selection, breed the stable cultivar. For example, "Hai Hua No. 3" sweet pepper is bred in this way.

2. Population Cultivars

The population has a heterogeneous genetic composition, and individuals are heterozygous. Their cultivar populations can show differences, but one or more traits must be consistent and different from those of other cultivars. They are cultivars bred from cross-pollinated or often cross-pollinated horticultural plants, which are bred mainly through the mixed selection method. When reproducing this kind of cultivar, we must select and keep its traits. Some cultivars of flowers and vegetables (e.g. red onion, pointy spinach leaves) are population cultivars.

3. Hybrid Cultivars

The genetic composition of the population is homogeneous, and individuals are heterozygous. They're the F_1 hybrids produced by hybridization between two proper parents. Hybrid cultivars are mainly applied in annual and biennial vegetables and flowers. According to the combination of the parents, the cultivars can be divided into interspecific hybrids, hybrids between varieties, and interlineal hybrids. At present, mainly interlineal hybrids are used in production.

4. Asexual Line Cultivars

The genetic composition of the population is homogeneous, and individuals are heterozygous. It's a population consisting of one or several similar asexual lines that propagate asexually, such as the apple cultivar "Red Chief Delicious". In addition, populations that propagate with the seeds produced by absolute apomixis are also asexual cultivars. Asexual line cultivar selection mainly uses the mutations during the reproduction process as well as the genetic recombination mutations of the hybrid offspring. Mutant branches, tubers, root tubers, etc. are selected for strain comparison and selection, and after two or three generations of comparison and identification, a new cultivar can be selected.

From the definition and type of cultivars, we know that each cultivar has a certain specificity that distinguishes it from other cultivars. Their populations are uniform and adapt to the cultivation methods in a certain area and a certain period of time.

IV. Types, Collection, Documentation, Conservation, and Utilization of Germplasm Resources

Germplasm resources are the basis of breeding work. Without good germplasm resources, it is impossible to breed good cultivars. Germplasm, also called genetic material, is the genetic material that can be passed from parent to offspring. The carrier carrying germplasm can be a population, an individual, or part of an organ, tissue, or cell. With the development of biotechnology, carriers can also be individual chromosomes or even DNA fragments. Therefore, germplasm should include germplasm at different levels, such as population, individual, gametes, and molecules. A number of people call them variety resources, but this name obviously cannot include germplasm resources at the gamete level and at the molecular level and cannot include wild germplasm resources for which there are no cultivars. Therefore, the germplasm resources and the variety resources should not be mixed.

(I) Types of Germplasm Resources

Research and utilization of germplasm resources must be conducted based on survey and collection, and the types of germplasm resources must be mastered in order to conduct the survey, collection, and conservation of germplasm resources. Correct classification can reflect the historical origin of resources and the relationship of pedigree, reflect the connection and difference between different resources, and provide a basis for investigation, preservation, research, and utilization of resources. There are many ways to classify germplasm resources, such as the botany classification method, ecological classification method, cultivation classification method, and source classification method. According to the characteristics of horticultural plants, the following two classification methods are highlighted below.

1. Cultivation Classification Method

(1) Species: The species is the basic unit in plant taxonomy. A species has certain morphological characteristics and geographic distribution, often existing in the form of populations. Generally, different populations are reproductively isolated. However, in the fruit trees, different species in the same genus can often hybridize, for example, most *Malus, Pyrus, Castanea, Carya*, and also grape subgenus. In the stone fruits, different species of peach can hybridize with one another, but different species of peach, apricot, plum, and cherry are not easy to do so.

(2) Variety: Homogeneous plants differ in some major morphological aspects. For example, flesh-fingered citron is a variety of citron (Fig. 1-1), *Citrus reticulata* Blanco is a variety of pomelo, and bald lychee is a variety of lychee. Flesh-fingered citron and citron belong to the same species *Citrus medica* L., but the fruit morphology of flesh-fingered citron differs significantly from that of citron, so it is a variety of citron and its scientific name is *Citrus medica* var. Sarcodactylis Swingle.

Citron

Flesh-fingered Citron

Fig. 1-1 Citron and Flesh-fingered Citron

(3) Cultivar population: A cultivar population is a grouping of many cultivars of similar ecotypes or similar agrobiological characteristics. For example, the oriental cultivar population, the Black Sea cultivar population, and the western European cultivar population in European grapes; the central China cultivar population and the northern China cultivar population in peaches; as well as the universal sweet orange cultivar population, navel orange cultivar population, summer orange cultivar population, and blood orange cultivar population in sweet oranges.

(4) Cultivar: The cultivar is the basic unit of cultivated crops and the main object of breeding. It has stable economic traits. For vegetatively-propagated fruit trees or asexually-propagated flowers, it is actually a population of asexual lines from a mother plant or a population formed after the mutated shoot has been propagated; For sexually-propagated horticultural plants, it is actually the inbred lines of a superior plant or the F_1 hybrid population with heterosis.

(5) Strain: The strain has two meanings in practical application. The first one refers to different types within the cultivar. For example, *Citrus reticulata* Blanco cv. Succosa has large and small leaf cultivars. The other one refers to the variety that has performed well during the breeding process but has not become a cultivar. In this situation, the strain can also be called an individual plant, superior line, etc.

(6) Population Cultivar: It refers to certain horticultural plant cultivars propagated mainly by seed propagation. The individuals of them can maintain some similarity in major traits but are very different in minor traits.

2. Source Classification Method

(1) Local germplasm resources: It refers to the old local cultivars (or "farmholding cultivars") and the improved cultivars promoted locally. The old local cultivars are the products of long-term artificial selection and natural selection, profoundly reflecting the characteristics of local terroir, highly adaptable, and resistant to the climate and soil and cultivation conditions of the region, and also basically able to meet the production requirements for economic traits such as yield and quality. They can be directly used; another characteristic is that these resources are a complex group formed under the influence of long-term natural changes and cultivation conditions and contain rich genotypes, so they are a very valuable breeding material. However, due to the continuous improvement of basic agricultural conditions, the increasing variety of diseases and pests, and the increasing demand for yield and quality of cultivars, many qualities of local variety resources are incompatible with the requirements of breeding objectives, so these cultivars need to be gradually improved or replaced by new cultivars. In general, local germplasm resources are well adapted to local natural conditions and are the easiest material to collect. Therefore, local germplasm resources must be the primary and fundamental choice in research and utilization.

(2) Foreign germplasm resources: Resources introduced from outside regions or foreign countries have different biological and economic genetic traits compared with local germplasm resources, some of which are not seen in the local resources. Especially, many original cultivars produced in the origin centers and secondary origin centers can intensely reflect the genetic diversity of a certain plant, from which special germplasm that is not available in general cultivars can be screened. However, resources from outside the region and foreign countries are generally not fully adapted to nature, cultivation conditions, requirements for pest and disease resistance, and production of the region. In the breeding, we mainly select cultivars with certain favorable genes from the germplasm resources as breeding materials and introduce these favorable genes effectively into the cultivars to be improved. In addition, after being tested and proven to be able to adapt to the region's cultivation, some foreign germplasm resources can be directly promoted in the region. For

example, the Satsuma orange is now a globally cultivated fruit tree, which is achieved through a long period of introduction.

(3) Wild germplasm resources: Wild germplasm resources are natural, wild, and undomesticated horticultural plants. These wild or semi-wild plants are formed by long-term natural selection under harsh natural conditions, with strong adaptability and stress resistance (resistant to infertility, coldness, drought, salt and alkali, diseases and pests, etc.), rich in resistance genes. Their traits are most dominant, with strong heritability, and can be used in resistance breeding. Recent advances in plant distant hybridization and somatic cell hybridization have further expanded the scope of interspecific gene exchange, making the role of wild germplasm resources more prominent and thus receiving attention from various countries. Wild germplasm resources are also important resources for rootstocks and have an important role in expanding resistant cultivation and soil adaptation in cultivated areas of horticultural plants. Wild resources generally have small fruits, poor edible quality, and economic traits, and some resources have bitter and astringent fruits. Most of these undesirable traits have strong genetic power, so in breeding, we should pay attention to their advantages and disadvantages, and selectively use these wild resources.

(4) Artificially created germplasm resources: Artificially created germplasm resources are germplasm resources obtained by applying hybridization (including sexual and somatic cell hybridization) and mutagenesis. Since comprehensive traits that meet people's needs are not always available in the available resource types, the selection from natural germplasm resources alone is often not sufficient. Therefore, it is necessary to make genetic recombination or genetic mutation of germplasm resources through artificial creation, so as to produce excellent biological characteristics and economic traits to meet the special requirements of production and consumers for new cultivars, and further provide updated germplasm resources.

(II) Collection of Germplasm Resources

The collection is the primary link of germplasm resources work and is also the basis of this work. The collection of germplasm resources is generally carried out in two ways: exploration and solicitation. Exploration refers to the collection of local cultivars of fruits, vegetables, and flowers in rural areas or the collection of germplasm resources existing in nature. At the same time, we should understand first-hand information about the growth environment and sources of the materials. Therefore, exploration is the most basic method for the collection of horticultural plant variety resources. Solicitation is to request germplasm resources from germplasm resource centers, breeding institutions, or individuals by means of visit, communication, exchange, or acquisition.

(III) Documentation of Germplasm Resources

The purpose of germplasm resource documentation is to check and eliminate duplicate and erroneous materials and to give systematic numbers for scientific management and orderly preservation. Firstly, identify materials of the same breed with different names and namesake materials at the first planting observation. The similarities and differences can be compared by planting materials from different sources but with the same variety name next to each other. Secondly, develop a unified code according to the horticultural plant species, the age of collection, the serial number of the material, etc., and keep them according to the code number. In addition to the national code number, the germplasm resources conservation unit needs to have its own code number for management purposes.

(IV) Conservation and Utilization of Germplasm Resources

The principle of preserving germplasm resources of horticultural plants is to minimize the breeding period to avoid mutation and loss of certain unique and excellent traits. From a broad perspective, the conservation methods of germplasm resources can be divided into *in situ* conservation, off-site conservation, *in vitro* conservation, indoor conservation, conservation by utilization, gene library conservation, and germplasm archives. As per the material used for conservation, they can be classified as plants, vegetative bodies, seeds, pollen, etc. With the development of technology, they can now be preserved through cells, protoplasts, and DNA fragments (gene libraries). Fruit trees and woody ornamental plants are perennial heterogeneous individuals, so according to their characteristics in vegetative propagation, they should be preserved by being planted in a germplasm resource nursery. Vegetables and herbaceous flowers are usually propagated mainly by seeds, so they should mainly be preserved by seed storage, which is the simplest and most economical way to preserve germplasm resources. The wild horticultural plant germplasm resources can also be preserved by the method of *in situ* conservation combined with the establishment of protected areas.

1. Planting for Conservation (Off-site Conservation)

Planting for conservation is mostly used for fruit trees and asexually-propagating horticultural plants, where the collected resources are planted in nurseries for long-term conservation. The location of setting up a germplasm resources nursery should meet the following aspects: ① Represent climate, soil, and other ecological conditions within a certain area. ② Have a long history of cultivation, rich resources, and a good cultivation base, as well as equipment conditions. ③ Have relatively stable climate change so that the plants can grow and develop normally. ④ The local resources meet the needs of local development planning, and the transportation is convenient, which is conducive to material exchange.

The number of plants to be planted should be based on the requirements for breeding resources, i.e., it should be conducive to research and conservation without taking up too much space. After the resource materials are planted, all management measures should be combined with the specific local conditions to meet different requirements as much as possible. Make sure that the resource materials can grow well and reflect their traits and characteristics so that proper comparative studies can be done.

An observation and research system is also needed after the resource material has been planted. The recorded items should be based on the research and utilization needs, and in principle, the items should be few and precise to facilitate analysis and processing.

The botanical gardens, arboretums, medicinal botanical gardens, flower gardens, original seed plantations, and collection and determination nurseries of various germplasm resources are all off-site conservation methods.

2. *In Vitro* Conservation

(1) Seed storage and conservation: Seed conservation is necessary for germplasm resources that use seeds as reproduction material. This method is mainly used for annual and biennial vegetable and herbaceous flower resources, followed by wild fruit trees, rootstock materials, and their apomixis types.

For germplasm resources with seeds as reproduction materials, "Germplasm Resources Bank" (referred to as "Germplasm Bank") (Fig. 1-2) is commonly used in many countries around the world for long-term conservation. Seed conservation can be prolonged by controlling the temperature and moisture conditions. In recent years, some research centers have established double seed banks for the long-term conservation of crop variety resources. This method is to dry the seeds to reduce the moisture content to 6%–8%, then seal them in separate packages and store them at low temperatures. As for storage temperature, one is (-10±1)°C, which can keep the seeds for more than 30 years; The other is (-1±1)°C, which can keep the seeds for more than 10 years. In addition, FAO Genebank recommends two suitable seed storage conditions: ① Storage in sealed containers at or below -18°C with the seed moisture content of 5% ± 1% is the best method; ② Seeds could also be stored in sealed or unsealed containers at or below 5°C with relative humidity not exceeding 20%. For those germplasm resources that are kept in small quantities and do not require a long preservation time, they can generally be stored under dry and sealed conditions at room temperature; an even smaller amount of seeds can be placed in a glass desiccator, with self-indicating silica gel as a moisture absorber. Fruit tree seeds need to be stored at a low temperature of about 5°C. Under low-temperature storage conditions, the moisture content of the seeds of some fruit trees will decrease to between 12% and 31%, resulting in reduced reproductive activity, or even death. Therefore, further research is needed on the seed storage conditions of each fruit tree.

Fig. 1-2 Germplasm Resources Bank

(2) Pollen storage and conservation: Pollen conservation is also a simple and economical way to preserve germplasm resources, allowing for the conservation of larger amounts of germplasm on a smaller scale of plants. In breeding practice, pollen storage and transportation can be used to hybridize cultivars of different anthesis or geographically distant cultivars.

The main environmental factors related to pollen reproductive activity are temperature, moisture, light, and air pressure. Usually, lowering any of these factors can lead to a longer survival period.

Most pollens of fruit trees can maintain their reproductive activity for a long time at a low temperature of 0–20°C and relative humidity of 10%–30%. Moisture, temperature, light, and air pressure all have effects on pollen reproductive activity. Hence, multi-factorial tests related to these environmental parameters must be conducted to find the optimal conditions for the long-term storage of pollen while ensuring its reproductive activity. Pollen storage methods for some horticultural plants have been previously studied and now can be used.

(3) Storage and conservation of vegetative organs and *in vitro* test tubes: Cuttings and scions for asexually-propagating woody plants such as fruit trees and woody ornamental plants sometimes are only temporarily stored to prolong the dormant period. They cannot be kept in this way for a long time. This method aims to maintain the suitable physiological state of its regeneration so that the cuttings and scions can sprout normally after reproduction. It is usually required to preserve

them at a low temperature of -2–2°C with a relative humidity of above 90% so that the vegetative bodies will not be dry and lose water. To prolong the storage time, branches are generally wrapped in plastic films and buried in a moist heap of river sand, but they must be regularly inspected to make sure they do not dry out and lose water. For example, lignified branches of Rosaceae fruit trees with or without roots can be stored at 1°C for a year. Their survival rate can reach 50%–80% when transplanted into the field, and the survival rate will be even higher if preserved as scions. It has been reported that the branches of apple, pear, and other fruit trees can be preserved for one to seven years after being treated with liquid nitrogen and other low-temperature measures. When woody plant resources need to be preserved, their branches have more advantages over seeds or pollen in this regard because branches can grow into asexual lines with stable genotypes. In the germplasm resources area, the storage materials, the use of suitable storage conditions, and the techniques of propagation all need further study.

3. Indoor Conservation

The seeds of fruit trees, vegetables, and flowers that use seeds as reproduction material are mainly stored and preserved indoors. Except for tropical fruit trees, trees with large seeds, and aquatic flowers, dryness, and low temperatures are necessary to maintain long-term seed viability. The method is to dry and treat the seeds so that the moisture content is reduced to 6%–8%, then divide and seal the seeds and store them at low temperatures. Seeds can be stored for 10 years at -1±1°C and 30 years at -10±1°C. The National Crop Germplasm Repository of the Chinese Academy of Sciences (Beijing) and the National Crop Replicas Germplasm Repository of Qinghai (Xining) are both long-term low-temperature repositories.

4. Utilization

When germplasm resources are found to be useful, they can be used in breeding cultivars or intermediate breeding materials, which is an effective way to preserve them.

5. Gene Library Conservation

Gene file conservation techniques are to extract megabase-sized DNA from resource plants, cut them into many DNA fragments with restriction endonucleases, and then transfer the DNA fragments attached to the carrier to the fast-multiplying *Escherichia coli* through a series of steps to breed a large number of single-copy genes that can be preserved in the organism. This method can preserve the genetic resources of the species for a long time and obtain various desired genes through repeated culture multiplication and screening. The establishment and development of gene library techniques provide an effective way to save germplasm.

6. Germplasm Archives

The conservation of germplasm resources should include, in addition to the resources themselves, archives consisting of various information related to conservation, general including: ① Historical information on the resources, such as name, number, genealogy, source, distribution range, the number given by the original conservation unit, name of the donor, evaluation of the resources, etc.; ② Information on the accession of the resources, including the number given at the time of accession, date of accession, accession materials (seeds, branches, plants, tissue-cultured plant materials, etc.) and quantity, conservation methods, conservation sites, etc.; ③ Information on the evaluation of the accession, including methods, results, and year of the evaluation. The files should be stored in permanent numbers in order to facilitate the timely addition of new information. Archival data should be timely input into the computer to establish a database so that it can provide resources and information for breeders, data researchers, and society at any time.

V. Main Methods of Horticultural Plant Breeding

There are two ways for cultivar selection and breeding. The first one is the direct selection and utilization of the existing germplasm resources and their variant materials, which contains three methods: germplasm resources investigation (investigation), seed introduction (introduction), and selective breeding (selection). The second one is mutation breeding (breeding) of new cultivars by artificially creating mutations on the basis of existing resources. Mutation breeding includes methods like recombination breeding, heterosis breeding, radiation mutation breeding, *in vitro* culture breeding, polyploid induction breeding, and molecular breeding.

1. Germplasm Resources Investigation

Germplasm resources investigation is the basic method to directly select and use the existing germplasm resources. It discovers cultivars that have long existed locally but have not been well used. All breeding work can only be carried out on the basis of the mastery of a certain level of germplasm resources. So it can be seen as the foundation of breeding work.

2. Introduction

Introduction means introducing new cultivars or new crops and a variety of germplasm resources from other places (including foreign countries). It's a simple but efficient method. According to the trait performance of the horticultural plant cultivar in other regions and under their local conditions, they can be introduced to areas with similar conditions to identify their local adaptability and cultivation value. They can be used directly for production, domesticated, or used as germplasm materials for breeding methods.

3. Selective Breeding

Selective breeding is a method of breeding new cultivars via selection and elimination, using the mutation of existing cultivars or types in reproduction. It's an easy and effective method to improve existing cultivars and create new ones. Before humans carried out crossbreeding, all cultivated crop cultivars were created by selective breeding.

4. Recombination Breeding

Recombination Breeding, also known as sexual crossbreeding, is an important breeding method. It selects parents according to breeding objectives, combines the good traits dispersed in different parents into the hybrids via artificial hybridization, then selects and breeds their offspring, and obtains new cultivars with relatively stable genetic traits and cultivation and utilization values after comparing and identifying.

5. Heterosis Breeding

The first generation of hybrids between horticultural plant species and cultivars or inbred lines often show certain advantages, such as tall plants, large leaf area, increased growth potential, increased dry matter accumulation, earlier flowering, improved yield, and increased stress resistance. According to this principle, hybrid combinations with strong advantages are made. Using the F_1 hybrids as the new cultivars is called heterosis breeding, also known as heterosis utilization.

6. Mutation Breeding

Mutation breeding is a method that uses physical or chemical factors to induce genetic mutations or chromosomal structure mutations in crops and breeds new cultivars through artificial selection. Mutation breeding has the following characteristics: First, it can increase the frequency of mutation and expand the range of mutation; Second, it is more effective in improving a single trait in crops; Third, the trait induced is stabilized quickly, and the breeding year is short.

(1) Radiation mutation breeding: Radiation mutation breeding is a method that uses various kinds of radiation to treat seeds, plants, or other organs of plants to induce their mutations and then select new cultivars according to breeding objectives. Commonly used rays include X-rays, neutrons, etc. Space breeding also belongs to radiation mutation breeding, which is a method of carrying seeds of crops on satellites, using the space environment to provide mutagenic factors such as microgravity, high-energy particles, and high vacuum for mutagenesis. Seeds produced in space breeding will be used as breeding material to select new cultivars.

(2) Chemical-induced mutation breeding: It is a method of breeding new cultivars by using various chemicals to treat plant seeds, plants, or organs to induce their mutations and then select new cultivars according to breeding objectives. During this process, attention should be paid to the selection of appropriate mutagenic materials, mutagenic methods, and mutagenic doses, as well as the correct selection of mutant offspring in order to achieve the desired effect.

7. Ploidy Breeding

In this method, breeders artificially induce mutation in the number of plant chromosomes to produce new cultivars, mainly including haploid breeding and polyploid breeding.

Haploid breeding is a method of getting haploid material through artificial induction, doubling the chromosomes to make homozygous diploid plants, and breeding them into new cultivars after selection and identification. The advantages are as follows: ① Improve the efficiency of selection by overcoming the separation of hybrids. Haploids contain only one set of chromosomes. Each gene exists singly. Hence, recessive traits can be exhibited. Good genes can be screened early so that the selection efficiency is higher. In addition, when the chromosomes are doubled into homozygous diploids, the good traits can be inherited stably without multiple generations of separation and selection. As a result, the breeding time can be shortened. ② Overcome the infertility of distant hybridization. The hybrids obtained in the distant hybridization are often infertile due to the large differences between the parents. If F_1 pollen is cultured *in vitro*, a few haploid plants can be obtained, and then homozygous diploid plants can be obtained by staining. In the end, new cultivars are produced.

Polyploid breeding is a method of artificially inducing crops to get polyploids and selecting new cultivars from them. For example, octoploid triticale, diploid seedless watermelon, and triploid sugar beet are widely planted in production.

8. Distant Hybridization Breeding

Distant hybridization breeding refers to the hybridization between different species, subspecies, and genera, such as the hybridization between japonica subspecies and indica subspecies of rice (different subspecies of the same species), and the interspecific hybridization between wheat and rye (different species of the same genus). Distant hybridization breaks down species boundaries and combines various traits from different species at various degrees to form new cultivars.

Due to large parental differences in distant hybridization, compared with interspecific hybridization, distant hybridization is characterized by hybrid incompatibility, hybrid infertility, and strong and unstable separation of hybrid offspring. Some measures can be taken to overcome these difficulties. For example, appropriate parents, repeated pollination, mixed pollination, and stigma grafting technique can be used to overcome hybrid incompatibility; chromosome doubling, backcrossing, and *in vitro* culture of young hybrid embryos can be used to overcome hybrid infertility and separation.

Review and Reflection Questions

(1) What is a cultivar? What are the characteristics of the cultivars?

(2) How to summarize the importance of horticultural products in people's lives?

(3) What are the methods of conserving germplasm resources?

(4) What are the main objectives of modern horticultural plant breeding?

(5) What traits have been mentioned in the main objectives of modern horticultural plant breeding?

(6) What are the general principles for setting breeding objectives?

(7) What are the main methods of horticultural plant breeding?

(8) What are the advantages of haploid breeding?

Learning Scenario II Selection and Breeding Methods of New Cultivars of Horticultural Plants

I. Introduction

(I) The Concept and Significance of Introduction

Any plant species and cultivars have a certain distribution area in nature, and so do the horticultural plants. Introduction and domestication is the process of artificially relocating a plant from its existing distribution or cultivation area to other areas for cultivation, in other words, the introduction is the method to introduce new plant species, types, and cultivars that are not yet cultivated locally from other places. Some plants are so highly adapted that they can adapt to new environmental conditions without changing their genetic characteristics. When the difference between the natural conditions of the original distribution area and the introduced place is small, or the environmental conditions of the introduced place are more suitable for the growth of the plants, the plants would even grow better. The above-mentioned is called simple introduction; some plants have low adaptability or the ecological conditions of the introduced place are so different from the original place, the plant would grow abnormally or even die. However, after careful cultivation management or through hybridization, mutagenesis, selection, and other measures, the plant's genetic properties are gradually changed, and the plant can thus adapt to the new environment and grow normally. This kind of introduction is called domestication introduction.

(II) General Rules of Introduction

(1) The introduction of cultivars that grow in the same latitude area is easy to succeed. Plants distributed at the same latitude have the same photoperiod and basically the same growth and development period. However, due to differences in geographic location and climate, temperature conditions vary greatly from place to place. For example, China is in the Eurasian continental monsoon climate zone, which is the coldest region in the world in winter. Tianjin is at the same latitude as Lisbon, the capital of Portugal. It has an average temperature of -4.1°C in January and an extreme minimum temperature of -22.9°C, while Lisbon has an average temperature of 9.2°C in January and an extreme minimum temperature of -1.7°C. Guangzhou is at the same latitude as Havana, but the average temperature in Guangzhou is 8°C lower than that in Havana. Therefore, low temperatures may become a limiting factor in the introduction of seeds from European and American countries.

(2) The fertility period of southern species will be prolonged after being introduced to the North, and that of northern species will be shortened after being introduced to the South. Biennial herbs must go through a certain photoperiod before flowering and fruiting, so the introduction of seeds at different latitudes has a great impact on fertility. The perennial trees have overwintering problems in addition to the growth cycle, so their reaction to sunshine length is more complex than

that of herbaceous plants. When the trees in the South are introduced to the North, the sun exposure is prolonged during the growing season, and thus the growing period is prolonged, which promotes the sprouting of ramulus, reduces nutrient accumulation, hinders the lignification of tissues, and transformation of protective substances before winter, and at last, reduces the trees' cold resistance. For example, after Jiangxi's toona is sown in Tai'an, Shandong Province, the above-ground part often freezes to death because it cannot stop growing in due time. The situation of trees in the North being introduced to the South is just the opposite. Because of the shorter hours of sunshine, the trees' terminal buds stop growing in advance. This severely shortens the growing period, prevents normal life activities, and often makes the trees suffer from serious diseases and pests due to higher temperature and humidity.

(3) The introduction of cultivars that grow in regions with similar climatic ecotypes is easy to succeed. According to the plants' adaptation to environmental conditions, in production, horticultural plants are often divided into tropical plants, thermophilic plants, hardy plants, psammophytes, and aquatic plants. All types of plants have their own suitable climatic conditions and distribution areas. The introduction of the appropriate species or materials during the corresponding growing season or in corresponding artificially created protected environments is expected to be successful.

(4) The annual day-neutral plants can often be introduced without restrictions. Tomatoes, beans, cucumbers, and other horticultural plants can grow and develop normally as long as they have the right temperature and light conditions.

(5) Soil pH value restricts the introduction of some horticultural plants. Camellia and azalea prefer acidic soil, so it is difficult to introduce them to calcareous soil areas for open cultivation.

(III) Dominant Ecological Factors Affecting the Success of Introduction and Domestication

The environment of horticultural plants includes all conditions of the living space of horticultural plants, among which the factors that have an impact on the growth and development of horticultural plants are called ecological factors. Dominant ecological factors refer to the factors limiting plant growth and development, including climate, soil, and biological factors. These factors are mutually affected and constrained and work together on horticultural plants. The complex of such factors is called the ecological environment. Different plants and cultivars have different responses to the same ecological environment. The so-called ecotype is the plant's special need for or adaptability to certain ecological factors under the long-term influence of a specific environment. This ecotype or ecological habit is formed through genetic variation during long-term artificial selection and natural selection.

1. Temperature

Temperature is the most important factor influencing the acclimatization of horticultural plants. The effects of temperature on introduced plants can be broadly summarized into two categories. First, the temperature cannot adapt to or meet plants' basic needs of growth and development, resulting in the abnormal growth of introduced plants or fatal damage to different parts and organs of the plant due to high or low temperatures. Second, although the introduced plants can survive, due to unsuitable temperatures, flower buds cannot be formed, which may affect fruit yield and quality, making the introduced plants lose their cultivation and economic values.

The critical temperature is the limit of the lowest and highest temperature that a horticultural plant can tolerate, beyond which serious injury or death can be caused to plants. The critical minimum temperature in winter is critical to the success of the sorthern horticultural plants introduced to the North. For example, the critical low temperature that pineapple can tolerate is -1°C, and Guangzhou's temperature is generally above 0°C, so pineapple cultivars can generally adapt to it; Shaoguan often sees the low temperature of -2.3°C, so Shaoguan in northern Guangdong becomes the demarcation line of pineapple introduction. The resistance of citrus fruit trees to extremely low temperatures is roughly like this: sweet orange is -6–7°C, citrus is -8–9°C, and Satsuma orange is -9°C. In Wuxi, Jiangsu Province, the extremely low temperature in winter is about -12°C, so sweet oranges can't adapt to it. However, Satsuma orange and cold-resistant citrus introduced to the microclimate can overwinter safely. In addition to critical low temperature, the duration of low temperature should also be considered in the introduction. For instance, blue gum, a kind of garden tree, has a certain resistance to cold and can only tolerate a short period of the low temperature of -7.3°C. In addition, the low temperature and its long duration in winter are also the limiting factors for the adaptability of overwintering vegetables. Therefore, when introducing southern overwintering vegetables to the North, they are prone to failure due to low temperature and freezing damage. High temperature is the main limiting factor for the southward introduction of plants. In the growing period of horticultural plants, the physiological process is inhibited when the temperature is at 30–35°C, and the damage occurs when the temperature is above 35°C. For example, the critical high temperature for the growth of cabbage is 25°C. Once it's above 25°C, its vital activities are affected. In particular, high temperature and humidity conditions often cause the spread of certain diseases, seriously limiting the southward introduction of fruit trees, vegetables, and horticultural plants. For instance, some grape cultivars introduced to the south of the Yangtze River basin are exposed to high temperatures and humid ecological conditions, which often induce *Elsinoe* anthracnose and anthrax.

Effective accumulated temperature is also an important factor affecting the adaptability of horticultural plants during their introduction. For horticultural thermophilic crops, when they are introduced to areas where the difference of effective accumulated temperature above 10°C less than 200–300°C, their growth, development, and yield will not be significantly affected. The requirements for effective cumulative temperatures for grape cultivars at different maturity stages range from 2,000–2,400°C for very early-maturing cultivars, 2,400–2,800°C for early-maturing cultivars, 2,800–3,200°C for medium-maturing cultivars, 3,000–3,500°C for late-maturing cultivars, and 3,500°C and above for very late-maturing cultivars. In addition, the temperature of the growing season is a limiting factor affecting the adaptation of annual crops. For example, the length of the growing season of vegetables is mainly determined by the number of days between the initial growing temperature and the critical minimum and maximum temperature. For example, the low temperature at the start of cabbage growth is about 7°C, and the critical high temperature is about 25°C. The fertility period of "Beijing Daqingkou" cabbage is about 110 days in Beijing but about 80–90 days in Shenyang. Therefore, its head formation is not full after being introduced into Shenyang.

2. Illumination

Illumination (illumination intensity and illumination duration) also affects plant growth and development. Illumination is related to latitude, and the illumination duration is different at different latitudes. The higher the latitude is, the greater the difference between lengths of day and night. The longer the summer day, the shorter the winter day. At lower latitudes, daylight length is not much longer in summer than in winter. Plants growing in different latitudes for a long time have a certain response to the lengths of day and night, forming short-day plants (such as chrysanthemum) that need long photoperiod in summer and short photoperiod in autumn to bloom, long-day plants (such as *Gladiolus*) that only bloom when receiving short photoperiod in spring and long photoperiod in summer, and plants (such as peaches, apples, Chinese roses) insensitive to the length of sunlight. This is more obvious in annual vegetables (including overwintering ones). For example, beans and soybeans are short-day plants; onions, beets, carrots, lettuce, and spinach are long-day plants; tomatoes, eggplants, and peppers can flower and fruit with both long and short hours of sunlight. The length of sunshine is also a major factor affecting the development of certain vegetable organs. For instance, the bulb expansion of the onion introduced from the North to the South requires long hours of sunshine. But after being introduced to the South, it's usually planted in autumn and harvested in spring. When the bulb expands, it only gets a short winter sunshine period. Therefore, the onion only shoots, while the bulb does not expand.

3. Moisture

It is mainly the total amount of precipitation and its distribution during the year that has a great influence on the growth of horticultural plants. Water requirements of the introduced plants and their tolerance to drought and flooding should be taken into consideration when introducing them. Fruit trees like apricot, walnut, fig, and pineapple, as well as garden trees like acacia and juniper do not need so much water and have stronger drought resistance; they're followed by pear, peach, persimmon, chestnut, and grape; citrus, loquat, plum, cherry, glossy privet, laurel, southern magnolia, etc. need much water and have weak drought tolerance. The distribution of precipitation is often an important factor in determining whether the introduced cultivars can adapt to a place, especially for horticultural plants that have been introduced from the North to the South. For example, when the northern grape cultivars are introduced to the South, the rainfall in the South is unbalanced, mainly concentrated in April to June, which is the anthesis of the grape. It would affect the pollination and fertilization of the grape, and because of the high temperature and humidity, the grape branches and leaves are prone to excessive growth, resulting in the spread of diseases and insects. This will seriously affect the yield and quality. Similarly, excessive precipitation and high air humidity during the fertility period of vegetables often become the limiting factors for the southern introduction of northern plants. Therefore, these two should be considered when introducing vegetables, fruit trees, and garden flowers. According to the study on many hardy tree cultivars such as *Melia azedarach*, *Davidia involucrata*, and *Ficus carica* introduced in Beijing, the growth of these trees is affected by drought in early spring rather than the low temperature in winter. If the heat coefficient of moisture is large, the moisture can absorb heat at high temperatures and release heat at low temperatures, which can reduce the temperature difference and make the climate soft, favoring the growth of most plants. Therefore, some microclimatic environments in the North (e.g. near reservoirs, coastal areas) can often introduce southern horticultural plants successfully. For example, *Citrus unshiu* near Qingshan Reservoir in Lin'an, Zhejiang Province can overwinter normally. *Camellia japonica* from Laoshan Mountain of Qingdao can also overwinter in the open field. In addition, the height of the subterranean water level in a region will also affect the success of the introduction of a tree cultivar. For example, many species of Magnoliaceae have fleshy roots, which are not resistant to water and humidity, so they should not be introduced in places with high subterranean water levels.

4. Soil

Soil ecological factors include the water-holding capacity of the soil, permeability, salinity, pH value, and the level of the subterranean water level. The main factor affecting the success of the introduction is soil acidity. China's northern and northwestern belts are mostly alkaline soils, while the red soil hilly regions of southern China are mainly acidic, and the coastal waterlogged lowlands are mostly saline-alkaline or salty soils. The adaptability of different plant species to soil pH value varies widely. Berry trees, garden plants like *Gardenia jasminoides*, *Rhododendron simsii*, camellia, and so on grow in acidic and neutral soil; drupes, garden plants like *Amorpha fruticosa*, *Dianthus*, *Viola odorata*, and so on grow in neutral or slightly alkaline soil; *Robinia pseudoacacia*, *Tamarix chinensis*, horsetail beefwood, and so on are more salt tolerant.

5. Other Ecological Factors

Certain special limiting ecological factors should also be considered in the introduction, such as diseases and pests that are still difficult to control. For example, ulcer disease in Zhejiang, Guangdong, and certain citrus-producing areas limits the introduction of sweet oranges; Jujube blight in northern and northeastern China limits the introduction of date palms; European grape cultivars are susceptible to *Elsinoe* anthracnose in the Yangtze River basin, which affected the development of some high-yield and high-quality grape cultivars. Some areas have suffered severe wind damage. So the differences in wind resistance between cultivars must be taken into account when introducing seeds. In addition, plants will establish coordinated or symbiotic relationships with surrounding organisms during their long-term growth and evolution. For example, *Castanea mollissima,* pine, and *Pseudolarix amabilis* have mycorrhiza, so it is prone to failure if we only introduce plants but not mycorrhiza. Besides, in the introduction, attention should also be paid to the introduction of pollinating trees and special pollinators.

6. Historical Ecological Condition Factors

The adaptability of plants is not only related to the ecological conditions of the present distribution area, but also to historical ecological conditions. The natural distribution areas of modern plants are the results of the changes in a certain geological period, especially the most recent glacial period. During the glacial period, some of the plant species may have been wiped out because of the great changes in ecological history in the related areas, and only those plants that have adapted survive to this day. Therefore, extant plants, during their phylogeny, have experienced a variety of extreme living conditions and have a very rich and complex ecological history. Of course, there are also areas where the plants have not changed much during the climate change periods in geological history, so these areas' ecological histories are not so rich.

(IV) Introduction Techniques

On the basis of careful analysis and selection of introduced materials, the steps from a small number of trial introductions and intermediate propagation to large-scale promotion are adopted in the introduction experiments. The steps are to avoid blind introduction.

1. Selection of Introduced Materials

The selection of introduced materials should meet two conditions: First, the economic traits of the introduced materials should meet the requirements of the established introduction objectives; second, the possibility of adaptation of the introduced materials to local ecological conditions should be concluded on the basis of systematic comparative studies of the agro climate and soil conditions of the introduced areas and the requirements of the introduced materials for ecological conditions.

2. Collection and Numbering of Introduction Materials

Introduction materials can be collected through field surveys or mailing. Field surveys can help you check and verify the materials, prevent mixing, and collect reproduction material from an individual with typical cultivar characteristics and no pests and diseases. The collected materials must be registered and numbered in detail. The registered items include the species, cultivar name (scientific name, local name), species of reproduction material (seeds, scions, cuttings, seedlings, etc., and the rootstock name of graftings should also be indicated), source and amount of material, the date of receipt, and the treatment measures taken after receipt (including the temporary planting and final planting of seedlings). Each material received, as long as it is from different sources and collected at different times, should be numbered separately, and the relevant information of each material, such as botanical traits, economic traits, and ecological traits of source areas, should be recorded and described and then put respectively in the file bag with the same number for future reference.

3. Quarantine of Introduction Materials

To avoid the pests, diseases, and weeds introduced by the introduction materials, the materials introduced from outside the region, especially from abroad, must go through strict quarantine. The reproduction materials to be quarantined should be disinfected in time. Isolated planting should be carried out in a special quarantine nursery if necessary. Eradication measures should be taken if materials to be quarantined are found.

4. Introduction Test

Systematically compare and identify the introduced materials with local representative superior cultivars to determine their quality and adaptability. The soil conditions and management measures of the test site should be as consistent as possible. The introduction test generally includes the following steps.

(1) Introduction trial and observation. It means that the introduction materials will be planted in a small area to preliminarily identify their adaptability to the ecological conditions of the region and their utilization value in production. For perennial and large introduced fruit trees and the observation trees, three to five plants of each kind can be planted in either germplasm resources nurseries or cultivar gardens. While planting a small number of introductions, the top grafting method can be used to graft the introduced cultivars onto the canopies of the adult trees of local representative cultivars, promoting introduced cultivars' early flowering and fruiting, thereby accelerating the observation process of the introduced perennial plants.

(2) Cultivar comparison test and regional test. Conduct the cultivar comparison test on cultivars selected in the introduction observation to make a more accurate identification. Those with excellent performance in the cultivar comparison test should then conduct regional tests to determine their adaptation area and scope. Those perennial plants like fruit trees with excellent economic characteristics and adaptive performance in the introduction observation (or

top grafting) can also be adopted for productive intermediate reproduction with controlled numbers, during which process further inspection of their adaptability should be conducted.

(3) The requirements for the adaptability of the introduction plants should not be lower than those of local plants of the same variety or use. The identification period for adaptation should vary according to the plant species. The fruit trees' appropriate period is in the full bearing stage, while the appropriate period for the observation plants should be when the effect is the best.

5. Cultivation Test and Promotion

Those introduced cultivars, like the fruit trees, with good adaptability and excellent economic performance in the cultivar comparison test and regional test or the productive intermediate reproduction, can have the large-scale cultivation test. Based on a more comprehensive evaluation of their seed characteristics, it is necessary to delimit the most suitable, suitable, and unsuitable development areas, and then develop and promote corresponding cultivation technology schemes.

(V) Criteria for Success or Failure of Introduction

(1) It can grow normally without protection or with little protection.

(2) It can reproduce in its original reproduction way.

(3) The original quality and economic value of the plant being introduced are not reduced.

II. Selective Breeding

(I) Concept and Application of Selective Breeding

Selective breeding is a method of selecting qualified populations or individuals from existing germplasm resources or cultivars to cultivate new cultivars through comparison and identification. Selection is the main work of cultivars' breeding and the main method of maintaining seed characteristics in seed production. In the process of breeding new cultivars, the required genotype is selected according to the individual's phenotype, so that the individual's traits can be inherited stably by the new cultivars. There is a long history of selective breeding of horticultural plants in China. Chrysanthemums, orchids, ornamental gourds, etc. all have thousands of years of selective breeding history. Cultivated plants are made into different cultivars through selective breeding. C. Fideghelli counted 68 new peach cultivars and 258 new plum cultivars bred worldwide from 1990 to 1992, of which 48% and 25% were bred from crossbreeding respectively, 22% and 35% were bred by seedling selective breeding respectively, and 6% and 17% were bred by sport selection respectively. It is evident that selective breeding is an important breeding way for horticultural plants.

Selective breeding is usually applied to populations in which most of the major economic traits meet the requirements, a few economic traits are poor, and these poorly performing traits vary greatly among individuals. It is the complex selection of a variety with excellent and comprehensive traits from a population with multiple poorly performing traits that requires a long time to wait for a required variation with multiple qualified traits.

(II) Selective Breeding of Sexually-propagating Plants

1. Basic Selection Methods and Their Integrated Application

(1) Bulk-population selection. It means to select qualified excellent individuals from the mixed original populations to perform mixed seed collection. Sow the seeds in the same area for identification or reproduction in the next year. This kind of selection is simple, high-effective, and low-cost, so it can be widely used. Bulk-population selection is divided into the one-time bulk-population selection (Fig.1-3) and repeated bulk-population selection (Fig. 1-4). However, when the seeds of the individuals selected through the bulk-population selection are mixed together, the genotype of each individual cannot be identified according to the performance of the offspring, and it is impossible to distinguish whether the variant's traits are inherited. The bulk-population selection is generally used for the selection and purification of cross-pollinated plants to prevent the mixing and degradation of the cultivar, and also be used in the production process of original seeds.

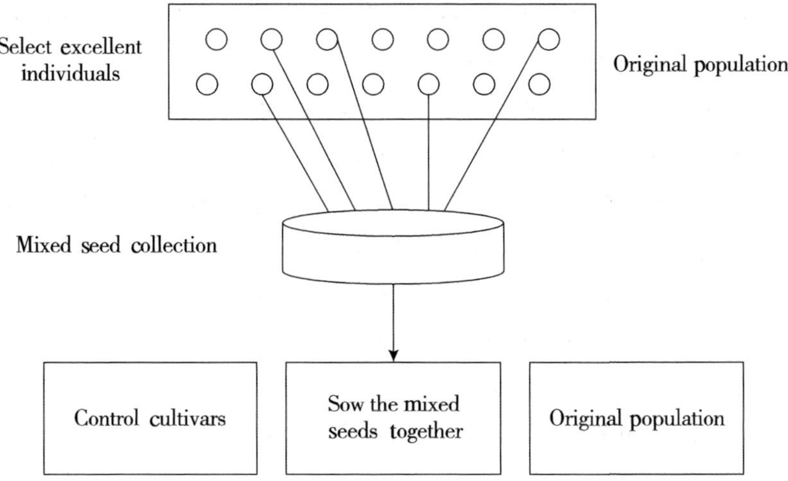

Fig. 1-3 One-time Bulk-population Selection

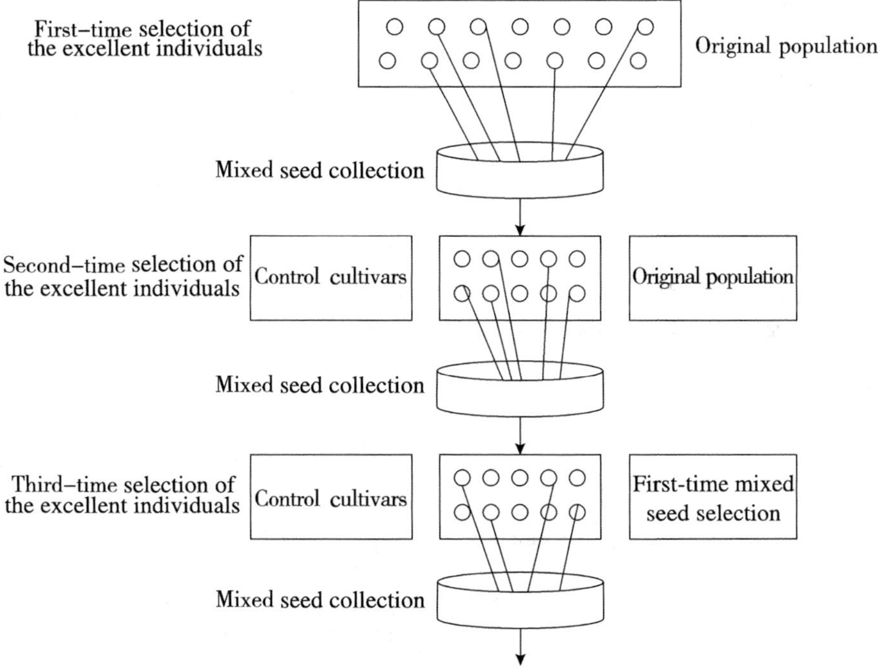

Fig. 1-4 Repeated Bulk-population Selection

(2) Individual selection. Select good individuals or fruits of each individual from the original populations, then collect and sow the seeds separately. The seeds

of each individual will be sown in a small plot. The plants in each plot are the offspring of one individual, which can be called a strain. Individual selection is to compare the quality of the strains and then select the individuals accordingly. It can identify the genotype according to the performance of the strain, and can quickly make the traits of the individuals reach a high consistency by reproducing them separately. However, individual selection is complicated and requires special experimental sites. Individual selection is also divided into one-time individual selection (Fig. 1-5) and repeated individual selection (Fig. 1-6). The one-time individual selection is sufficient for the purification of existing cultivars and the production of original seeds of self-pollinated plants, while the repeated individual selection is used for the hybrid offspring generated from cross-pollinated plants and artificial hybridization.

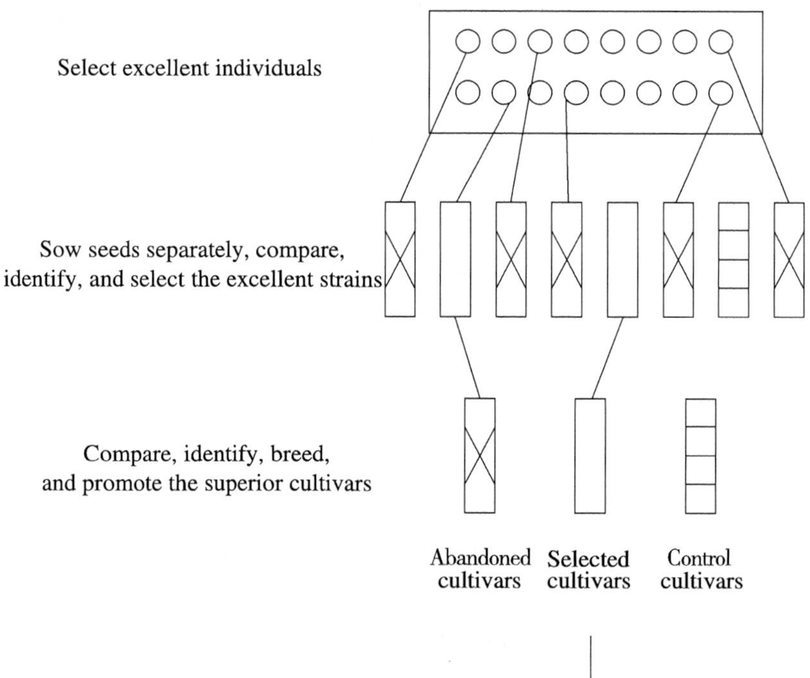

Fig. 1-5 One-time Individual Selection

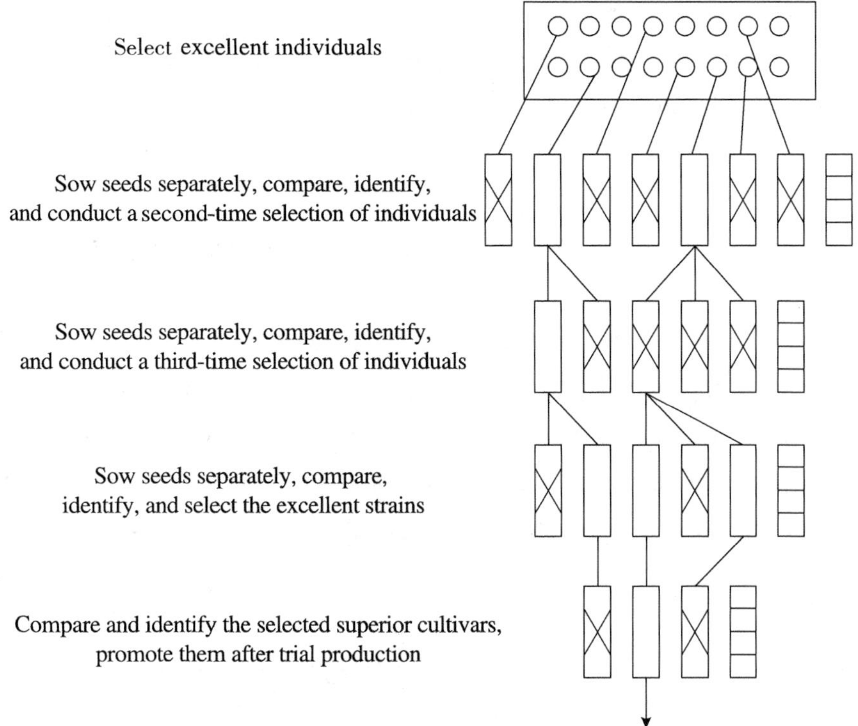

Fig. 1-6 Repeated Individual Selection

(3) Integrated application of the two selections. Both bulk-population selection and individual selection have their own advantages and disadvantages, so different combinations of selection methods are derived to complement each other in practice. The first is individual-bulk-population selection, which means to make an individual selection first to eliminate bad strains in the strain nursery, then eliminate bad plants in the strains left, and make the selected plants have free pollination. Collect the mixed seeds, and then conduct bulk-population selection on the first generation or generations afterward. The advantages are that strains with bad genetic traits can be eliminated after a strain comparison of the individuals' offspring, so the degradation of reproductive activity can be avoided in the bulk-population selection later, and that from the second generation, a large number of seeds can be produced in each generation. The disadvantage is that the effect of selection and purification is not as good as the repeated individual selection. The

second is bulk-population-individual selection, in which the one-time individual selection will be made after the bulk-population selection is performed on several generations. Its advantages and disadvantages are basically similar to those of the previous method. It's suitable for the original populations with obvious differences among plants.

2. Selective Breeding Procedures for Sexually Reproducing Horticultural Plants

The process of selective breeding, which includes collecting materials, selecting good individuals, and breeding new cultivars, composes a series of works like selection, elimination, and identification. Generally, an original material nursery, strain nursery, cultivar comparison pre-test nursery, cultivar comparison test nursery, production test, regional test nursery, etc. should be set up (Fig. 1-7).

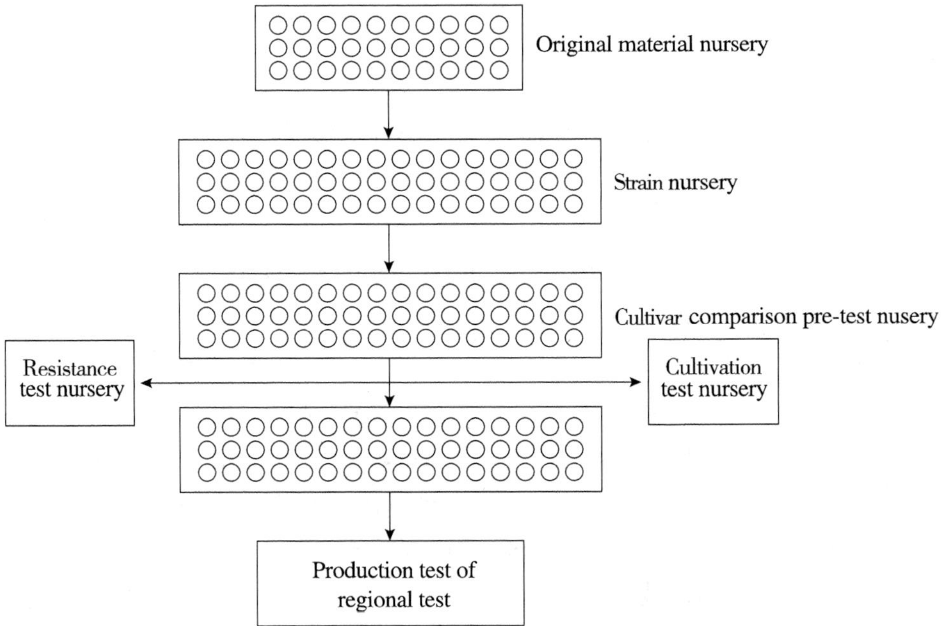

Fig. 1-7 Nursery Settings for Selective Breeding

(1) Original material nursery. Plant the original materials in an environment that represents the climatic conditions of the region, and set up a control group.

After being compared with the control group, the excellent individuals will be selected from the original materials for seed collection and then strain comparison. The seeds of the local varieties are often directly from the crop field instead of any special original material nursery.

(2) Strain nursery or selection nursery. Plant the superior strains or the offspring of bulk-population selection seed of excellent populations selected from the original material nursery or cultivars produced in a large area in the region, and then have the purposeful comparison, identification, and selection to select superior strains or populations for comparison and selection in cultivar comparison test nursery. Each strain or offspring of bulk-population selection should be planted in a small plot with the control group. This step should be done twice. The duration of the strain comparison depends on the consistency of the offspring populations of the selected plants. When the population is stable and consistent, the cultivar comparison pre-test can be carried out.

(3) Cultivar comparison pre-test nursery. Identify the consistency of superior or mixed strains selected through strain comparison and further eliminate some strains or mixed strains with poor economic traits. Expand the reproduction of the selected strain to secure the seeds used in the cultivar comparison test, which requires a large sowing amount. The pre-test period is generally a year.

(4) Cultivar comparison test nursery. Comprehensively compare and identify the superior strains or the offspring of mixed strains selected from cultivar comparison pre-test or strain comparison, thus to learn about their growth and developmental habits and finally select one or more new strains with yield, quality, maturity, and other economic traits better than those of the control cultivars. Cultivar comparison test must be carried out in accordance with the requirements of the formal field test, with the control group, repeated more than three times and controlled environmental differences. The test period is two to three years.

(5) Production test and regional test. The excellent strains from the cultivar comparison test are planted in the big production field, and will be directly judged and selected by the producers and consumers as new cultivars suitable for local producers and consumers. Production tests should be arranged in the main local

production areas, with a general area of not less than 1 mu (1 mu=667 m^2). Regional tests are conducted by the local agricultural authorities (regional agriculture bureau, provincial department of agriculture), and several representative test sites will be set up within the scope of the region. The production test and the regional test can be carried out simultaneously, and the schedule is two to three years.

(III) Selective Breeding of Asexually-propagating Horticultural Plants

The selective breeding of asexually-propagating plants includes sport selection, vegetative micro-mutation selection, and seedling selection. Only sport selection and seedling selection are introduced in the following.

1. Sport Selection

(1) Concept of sport and sport selection. Sport originates from the genetic variation that occurs naturally in somatic cells. When the variation of somatic cells occurs in the meristem of the bud or comes into the meristem of the bud after the dissociation and development, the sport bud is formed. Only when the mutated buds germinate into branches, or after flowering and fruiting, can they be found due to their obvious differences in traits from the original cultivars. Therefore, bud-sport always appears as branch-sport. Sometimes, such mutated buds are used accidentally for asexual reproduction and not discovered until they grow into new plants. In this scenario, we call the mutated plants plant-sports. Sport selection refers to the selective breeding for the variants of bud-sport.

For the vegetative cultivar of horticultural plants, in addition to the bud-sport caused by genetic variation, there is also a mutation called fluctuation or modification, which is not inheritable and is caused by various environmental factors. An important issue in sport selection is to correctly distinguish these two varieties of variation so as to select the real superior sport.

(2) Objective and selection criteria of sport selection. The sport selection usually takes the original superior cultivar as the object to improve some of its disadvantages through selection while maintaining the excellent traits of the original cultivar. So, its breeding objective is more targeted. For example, in the sport selection of citrus,

when selecting and breeding the cultivars suitable for processed orange cans with sweet syrup, the main traits of the cultivars should be seedless or with few seeds, given the existing *Citrus reticulata* Blanco cv. Succosa processed already have great color, fragrance, taste, and shape; while the main objective traits of the seedless Satsuma orange should focus on the fitness of fruit shape, petal shape, juice sac, etc. for processing. After the objective is determined, corresponding selection criteria should be formulated. For example, in the 1970s, the criteria in the preliminary sport selection of *Citrus reticulata* Blanco cv. Succosa with few seeds in Taizhou, Zhejiang was that the single fruit should contain less than four seeds.

(3) Procedures for sport selection (Fig. 1-8). The sport selection is divided into two sections. The first section is to select the superior lines from the production garden; the second section is to compare and identify the asexually-propagating offspring of each superior line, including repeated selection and final selection.

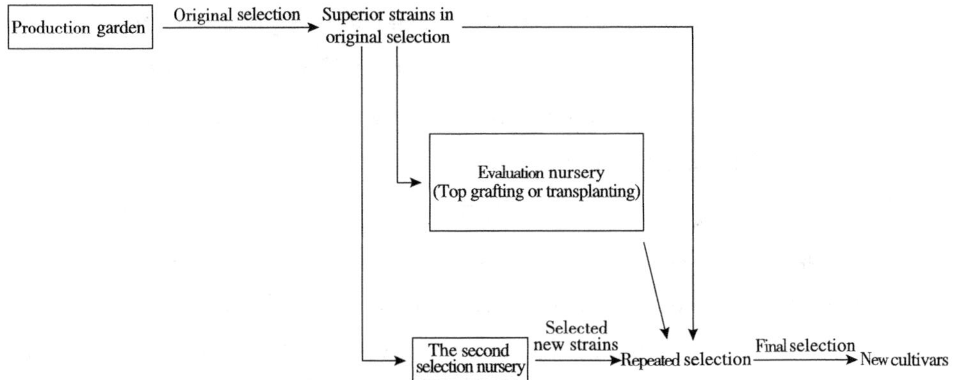

Fig. 1-8 Procedures for Sport Selection

① Original selection: It includes the discovery of superior lines, screening, identification, isolation, purification, etc. The masses should be widely mobilized to select and report the superior lines as well as explore the variation. The primarily selected superior ones should be registered and numbered with obvious marks and written on the record form. Then set up control groups for analysis and comparison after getting their fruits or seeds respectively. During the analysis and comparison, methods like pruning and tissue culture are usually used to purify the chimeras and stabilize the variation traits. Varieties of different traits can be treated separately. The varieties that can be clearly determined as the modification should be eliminated immediately. For the varieties with unclear or unstable variation traits, please continue the observation. If necessary, measures like grafting and pruning can be adopted to promote the proliferation of the variants. Then, proceed to analyze them. For the varieties with excellent traits, please take the following treatment: Those that can not be determined as the sport can be identified in the top grafting evaluation nursery; those that can be determined as the sport but whose traits have not been fully understood can be sent directly into the selection nursery; those with excellent sport performance and without poor traits can directly participate in the repeated selection.

② Repeated selection: It involves the top grafting evaluation nursery and selection nursery. Plants in the top grafting evaluation nursery have an early fruiting period, so they can provide the fruits and scions needed for identification in a short period of time. Still, the influence of rootstock should be considered and minimized as much as possible during the identification process. The main role of the selection nursery is the comprehensive and accurate identification of the bud-sport lines, not only examining the main economic traits but also the quantitative traits controlled by micro-effect multiple genes. The nursery requires consistent fertility with no less than 10 plants on each line and five plants per row in a small plot of singular rows. There should be two repetitions, control plants and protection rows. Every kind of line in the nursery should have a file and be observed continuously for more than three years. Evaluation should be carried out annually. According to the evaluation results of more than three years, the submitting unit will submit a second selection report to select the best single line to enter the final selection.

③ Final selection: The competent department will organize relevant personnel to conduct a final selection and evaluation of the strains. The following materials are required from the submitting units of the strains for the final selection: A report on the selection process, evaluation, and development prospect of the strains; evaluation data of the strains in the seed nursery for more than three consecutive years; experimental results and evaluation opinions of the strains in different ecological areas; the physical materials of the strains and the control groups (if fruit, not less than 25 kg; if the individual plant, not less than 50 plants). The strains will be examined and evaluated based on the above information and physical materials. After the final selection, the submitting unit may name the strains and promote them as new cultivars.

2. Seedling Selection

(1) Concept and significance of seedling selection. For perennial horticultural plants that can bear fruit normally, different places often adopt vegetative propagation or seed propagation according to different production and cultivation habits. The propagation of seeds is often called seed propagation. Selective breeding aiming at improving the economic traits and quality of the seed reproductive population is called seedling selective breeding, or seedling selection for short.

Compared with the vegetative population, the seedling population is usually characterized by common variation, many variation traits, and a larger variation difference. It has great potential for breeding new cultivars. As their variants are formed in local conditions, generally speaking, they have good adaptability to the local environment. The selected new varieties are easy to be promoted locally, with less investment and faster returns.

(2) Procedures and methods of seedling selection.

① Seedling selection for seedling populations. In the production areas of walnut and Chinese chestnut in North China, mass seedling selection is performed based on the common variation of the seedling population. With the grafting propagation method, selected trees can be cultivated into asexual line cultivars, thereby quickly obtaining superior varieties. The general procedure is as follows.

Seed reporting and pre-Selection: First, organize fruit farmers to discuss and clarify the significance, specific methods, requirements, and standards of seed selection, and then encourage more people to participate in seed selection and reporting. Second, organize professionals to investigate and verify the high-quality trees reported by fruit farmers on the spot, and mark, number, and register the remaining trees as pre-selected trees after excluding the individual plants that evidently do not meet the requirements for seed selection.

Original selection: Professionals will sample and identify the pre-selected trees, and after two to three years of investigation, trees with stable traits can be selected as original plants. Graft and breed 30–50 original plants as experimental trees of the selection nursery and for identification at multiple production sites. The mother plant can be observed while watching the performance of its vegetative offspring. Please eliminate environmental differences to improve the identification effect. In addition, some scions can be cut and grafted onto nearby trees without compromising the growth of the mother plant. This way, they can bear fruit earlier and be identified. Both the addition selection and the original selection shall be conducted in the seedling nursery.

Final selection: An objective evaluation of the plants selected in the original selection should be done by professionals and experienced local farmers based on the comparison and identification of the fruits produced by the asexual progeny of the primary plants for three consecutive years, as well as their mother plants, grafted trees, and the trees growing in multiple identification sites. The selected plants with good performance can be recommended to variety approval committees at all levels for promotion. At the same time, it is necessary to cultivate some mother tree gardens that can provide a large number of good scions.

② Seedling selection of the new population. Asexually-propagating horticultural plants have a strong heterozygous genetic basis. Once they undergo the sexual process, even if they are self-pollinated, they will have character separation. Therefore, you can select high-quality individual plants from the sexual offspring of asexually-propagating horticultural plants with the single plant selection method, and fix their characters by the asexually-propagating method to make them become new horticultural cultivars. The procedure for seed selection is as follows: After the selected plants bloom and bear seeds, sow them in the seed nursery for selection, select the high-quality individual plants to determine the superior line among them, harvest their vegetative organs, number the organs as asexual strains, and compare and identify these asexual strains, among which the best performers will become vegetative cultivars.

III. Sexual Hybridization Techniques

(I) Concepts and Applications of Sexual Crossbreeding

Hybridization is the process by which different gametes combine to produce a hybrid. The genetic basis of hybridization is genetic recombination, which is an important source of genetic variation. Sexual crossbreeding, also known as combination breeding, is the breeding selection of new cultivars with relatively stable heritability and cultivation and utilization values by integrating the excellent traits scattered in different parents into hybrids through artificial hybridization and selection and evaluation. Sexual crossbreeding can be divided into close hybridization and distant hybridization according to the distance of the parental relationship. Close hybridization is the hybridization between types or cultivars of the same species, featuring high compatibility and a high success rate. It is the most commonly used method for the sexual crossbreeding of all kinds of plants. Hybridization between different plants within the same species is sometimes called mating. Distant hybridization is the interspecific or intergeneric hybridization between different species. It's the hybridization between geographically distant ecotypes, featuring low compatibility and a low success rate. It mainly uses the superior genes of a certain trait of the relative species or wild species to synthesize new germplasm through artificial hybridization. Then, it provides parents for sexual crossbreeding and heterosis use.

(II) Methods of Sexual Hybridization

There are many kinds of methods of sexual hybridization. The most common is the pairwise cross between two parent cultivars. When a single cross does not achieve the breeding objective, a backcross or polycross can be performed.

1. Single Cross

A single cross refers to a cross involving only two parents, also known as the pairwise cross or two-parent cross. If the cross between two parents, A and B (expressed as A×B), the mother (♀) A should come first and the father (♂) B should come after in writing. For a cross between two parents, if A×B means direct cross, B×A means reciprocal cross. Direct cross and reciprocal cross are relative terms. If there is only one combination, A×B or B×A, there is only a direct cross and no reciprocal cross. According to the principle of genetics, if the traits are controlled by the nuclear genome, the progenies produced by the direct cross and reciprocal cross are the same; the progenies produced by the direct cross and reciprocal cross are different only when the traits are controlled by the cytoplasmic genome. Therefore, the purpose of carrying out direct and reciprocal crosses is to exploit traits controlled by the cytoplasmic genome. For example, Nanjing Forestry University found some differences in the progenies of the direct cross and reciprocal cross in the crossbreeding process of liriodendron.

The single crossing is simple and the variation of the hybrid is easy to control. However, because it is only affected by the genotypes of its two parents, the variation range of the offspring's traits is small, and the range of selection will be limited to some extent.

2. Backcross

A cross between one parent and a hybrid or its offspring is called a backcross. Backcross can be expressed as (A×B)×A (or B) and is designed to enhance the trait of one of the parents in the hybrid. Backcross can be performed only once or multiple times. Parents used for backcrossing are called recurrent parents. The purpose of backcrossing is to change the genetic composition of the backcross offspring so that the genes of the recurrent parents are continuously increased and the genes of non-recurrent parents are continuously decreased. In the selection of backcross offspring, it is necessary to select individuals with special traits introduced from non-recurrent parents. For example, if a cultivated species crosses with a wild plant to introduce disease-resistant genes into its offspring, a backcross is needed to restore the agronomic traits of the offspring.

3. Polycross

A polycross is one in which three or more parents participate in the cross, also known as a composite cross. According to the order of the third and subsequent parents participating in the cross, it can be divided into the additive cross and synthetic cross.

(1) Additive cross: A single cross is carried out first to obtain a single cross hybrid. Cross the single cross hybrid or individuals selected from its offspring with the good traits of both parents with the third parent. The resulting hybrid or offspring can then be crossed with the fourth or more, or with the fifth or even more parents. Additive cross can be expressed as [(A×B)×C]×D. One parental trait can be added per cross. The more parents are added, the more excellent traits the hybrids will obtain, but the longer the breeding years will be. For sexually reproduced vegetable plants, three parents are the most common when they are subjected to additive crosses. After these vegetable plants are crossed, it is necessary to purify the main target traits through multiple generations of self-cross selection in order to develop stable cultivars. An additive cross involving three parents is also called a triple cross. For example, the Shenyang Agricultural University bred Shen Nong No. 2 tomato characterized by early maturity, abundant yield, limited growth, and large fruit with this method, which combines the early maturity, erection, and limited growth characteristics of Klotkstaki, the rapid fruit development, limited growth, consistent fruit color, good fruit shape of Ai Hong Jin, and the early flowering, dwarf nature of Bi Song.

(2) Synthetic cross: Synthetic crossing refers to crossing four parents in pairs, producing two single cross hybrids, and then crossing the two hybrids. This hybridization can be expressed as (A×B)×(C×D). The "Record Yellow Bellflower" bred by Soviet Union horticultural breeder Michurin was produced via this hybridization method. According to the theory of this type of mating, the nuclear genetic composition of the parents A, B, C, and D in the resulting hybrid each accounts for a quarter of the total. Sometimes, in order to strengthen the traits of a parent in the hybrid, you can repeat the parent in the crossing, such as (A×B)×(A×C), in which case the nuclear genetic composition of parent A accounts for half of the final hybrid. Synthetic crossing, compared with additive crossing, can combine the good traits of most parents in a short period of time. If the target trait is recessive, it is also necessary to allow the single-cross hybrids to self-cross. After that, select individuals with comprehensive superior traits from the separated F_2 to

cross with different single cross hybrids (F_2).

Compared with single crossing, the biggest advantage of polycross is that the good traits scattered in many parents can be integrated into one hybrid, which greatly enriches the heritability of the hybrid and makes it possible to breed excellent cultivars with good comprehensive traits, wide adaptability, and versatility.

The variation range of polycross hybrid offspring is large, so the sowing population of its offspring should be large as well. Generally speaking, it is necessary to sow more than 500 F_1 plants to increase the number of individuals combining most of the good traits of the parents.

(III) Selection and Matching for Hybrid Parents

Parental selection refers to the selection of cultivars with superior traits as hybrid parents according to the breeding objectives. Sometimes, the target trait genes are difficult to find in cultivated cultivars. However, some resistance and quality genes, such as the CMV resistance gene of tomato, the downy mildew resistance gene of luffa, and the high lycopene content gene, can be found in related wild species, semi-wild species, or mutagenic materials. Thus, the material of hybrid parents may include cultivated species, semi-cultivated species, wild and semi-wild species, mutagenic materials, etc. Parental matching refers to the grouping of two (or several) parents from the selected parents and the grouping method (such as the choice of parents in a single cross and the grouping method of the two parents in polycross). Proper selection and matching of parents can enable you to obtain more variation types in line with the breeding target, thus improving breeding efficiency. If the parents are not selected or mated properly, even if a large number of cross combinations are made, the variation types that meet the breeding target may not be obtained, resulting in unnecessary waste of manpower, material resources, and time.

1. Principles of Parental Selection

(1) Specify the target traits of the selected parents.

According to the breeding objective, it is necessary to determine the requirements for selecting the traits of the parents and prioritize these traits. In the parents, the desired traits should be at a higher level, and the necessary traits should not be lower than the general level. For example, if the breeding objective is disease resistance and high quality, the parents should have a high level of these two desired traits, and the level of necessary traits such as maturity stages, plant type, and product organ morphology should not be lower than the general level and can be accepted by producers and consumers. For another example, in the breeding of canned peaches, the trait of insoluble flesh is more important than the trait of yellow flesh, so when selecting parent materials, priority should be given to the parents of insoluble flesh.

It is important to note that some important economic traits, such as high yield, high quality, and maturity stages, are complex traits composed of multiple unit traits. For example, tomato yielding ability is determined by unit traits such as plant width (which determines the reasonable density of planting), number of flowers per plant, fruit setting rate, and fruit weight. In the breeding of tomato cultivars with high yield, if the selection of parents is directly based on the trait of large fruit per plant, the parents may all have heavy but few fruits, and it is difficult to get the offspring with high yield by crossing these parents. If some of the selected parents are of the trait of small plant width that is suitable for dense planting, some bear extensive fruits, and some bear heavy fruits, then it is possible to mate the parents with high levels of different unit traits, synthesize the good unit traits of different parents and obtain hybrid progenies with a high level of complex traits. Therefore, the unit traits that constitute the target traits should be analyzed, studied, and clarified in the selection of parents.

(2) Master a large number of original materials required for the breeding objectives, study and understand the genetic rules of the target traits.

According to the requirements of the breeding objectives, the more the original materials are collected, the easier it is to select suitable hybrid parents. A better understanding of the genetic rules of the target trait will lead to fewer mistakes in parental selection. For example, in canned peach breeding, the two kinds of peaches, Wuyun Peach and Mount Gang 500, both have white solute flesh. The self-crossed offspring of Wuyun Peaches all have white solute flesh, while some of

Mount Gang 500's self-crossed offspring have yellow non-solute flesh. Obviously, this is because the two pairs of traits of the Wuyun Peach are homozygous (YYMM), while those of Mount Gang 500 are heterozygous (YyMm).

(3) The parents should have as many superior traits as possible.

If the parents have more good traits and fewer bad traits, it is easy to select and mate two parents who can complement each other; otherwise, it is necessary to take polycross, which increases the complexity of work and breeding duration. While focusing on parents with more excellent traits, you should also pay attention to some undesirable traits with high heritability. If the parents have this kind of trait, it will add a lot of difficulties to the transformation of the offspring. In particular, it is necessary to pay attention to the undesirable traits of some perennial fruit crops. For example, you'll notice the trait of the bitter taste of trifoliate orange when crossing orange with trifoliate orange, and the trait of the small and poor quality fruit of Siberian crabapple when crossing big apple with Siberian crabapple. Therefore, during parent selection, the materials with strong heritability of bad traits should be avoided as much as possible.

(4) Prioritize the rare favorable traits and valuable types.

In terms of single traits, some favorable traits appear more often. For example, there are many cultivars of big apples weighing more than 200 g, and cultivars that can be stored for four to five months are also common, but there are few compact cultivars controlled by a single gene. Only Weisaixu and a few others have been found so far. In addition, in terms of trait combination, there are many quality orange cultivars with weak cold resistance, but almost no sweet orange cultivars with both good cold resistance and good quality; there are many cultivars of peach with high yield, but few with both insoluble flesh and high yield. Pear cultivars resistant to venturia, black spot and ring rot, and large seedless grape cultivars with high temperature and moisture resistance are all precious breeding resources. Special attention should be paid to these cultivars in fruit tree breeding to select new cultivars with high quality or strong stress resistance in hybrid offspring.

(5) The heritability of good parental traits should be strong.

The breeding practice has proved that the good traits in hybrid offspring tend to come from the parent with strong heritability. For example, if a large-fruit tomato is crossed with a small-fruit tomato, the fruit size of the offspring tends to be small; if a cucumber fruit with a blunt tail is crossed with a cucumber fruit with a sharp tail, the phenotype of the offspring fruit tends to be the sharp tail. This indicates that the heritability of small fruit and sharp tail traits is greater. Usually, the heritability of the wild or primitive trait is greater than the cultivated trait; the heritability of the purebred trait is greater than the hybrid trait; the heritability of the locally stable trait is greater than the unstable trait; the heritability of the maternal trait is greater than the paternal trait; the heritability of traits of adult plants and self-rooted plants is greater than that of young plants and grafted plants; the heritability of the oligogenic controlled trait is greater than polygenic controlled traits. In the selection of parents, the parents with strong heritability of good traits and weak heritability of bad traits should be selected to increase the number of individuals with good traits in the hybrid offspring population.

(6) Prioritize local cultivars.

Local cultivars are products singled out from long-term local natural selection and artificial selection that have better adaptability to local natural conditions and cultivation conditions. Local customers are accustomed to these local cultivars, and their shortcomings are well understood. The cultivars coming out of these local cultivars are also highly adaptable to local conditions. A number of horticultural plant cultivars are descended from local cultivars. For example, one of the parents of the Jinyan series cucumbers bred by the Tianjin Institute of Agricultural Science is *Neoalsomitra integrifoliola*, a local cultivar. One of the parents of the Ningqing Cucumber bred by the South China Agricultural University and Guangdong Academy of Agricultural Sciences is the local cultivar Guangzhou Erqing. The grape variety BeiChun is the offspring of Ousha Grape and local *Vitis amurensis*. The pear cultivar *Pyrus pyrifolia* is a hybrid of the local cultivar Huangmi and Sanhua.

2. Principles of Parental Selection

(1) Parental trait complementation.

Trait complementation refers to the combination of the good traits of both parents into one individual. The complementation of good traits has two implications. One is the complementation of different traits, and the other is the complementation of different unit traits that constitute the same trait. Let's take a look at the complementation of different traits first. For example, if a cucumber cultivar is to be bred with the early maturity and disease resistance, one parent should be selected with the early maturity trait and the other with the disease resistance trait. In breeding apple cultivars with strong resistance to disease, if the Golden Delicious apple is used as the mother to cross with the Hongtaiping apple, the father Hongtaiping apple can make up for the weak cold resistance of the Golden Delicious apple. Similarly, the Golden Delicious apple can make up for the shortcomings of the Hongtaiping apple, such as small fruit, sour taste, and short storage duration. The new cultivar Jinhong apple, which is crossbred from this combination, has the traits of strong cold resistance, stable high yield, medium fruit size, and long storage duration, basically integrating the good traits of its parents. As for the complementation of different unit traits of the same trait, let's take early maturity as an example. The early maturity of some fruit vegetable cultivars mainly refers to early bud flowering, while the early maturity of some other fruit vegetable cultivars mainly refers to fast fruit growth. If parents with these two different types of early maturity unit traits are selected and mated, there is a possibility that variants that mature earlier than the parents will appear in the progeny.

Trait inheritance is complex. Even if the traits of the parents are complementary, the traits of the offspring often do not show the simple mechanical combination of the advantages and disadvantages of the parents, especially the quantitative traits. For example, yield and soluble solids are quantitative traits. In the breeding of pumpkin cultivars with high yield and high soluble solids content, if the cultivar with high yield and low soluble solids content is crossed with the cultivar with high soluble solids content and low yield, the hybrid progeny will generally not have variants with high yield and high soluble solids content. If parents with relatively high levels of both traits, that is, one with high yield and low soluble solids (not very low) and the other with low yield (not very low) and high soluble solids, are crossed, it is possible that the progeny will match or even surpass the parent in both traits.

(2) Parental mating of different types or geographical origins.

Different types refer to parents with significantly different growth and development habits, different cultivation seasons, or differences in traits, for example, pole bean and dwarf bean, growth-determinate tomato and growth-indeterminate tomato, spring cucumber and autumn cucumber, spring cabbage and summer cabbage, South China cucumber and North China cucumber, large-topped bitter melon and long-bodied bitter melon, pepper and sweet pepper, erect eggplant and spreading eggplant, etc. The affinities of different types of parents are mostly more divergent than the genotypes of the same type. As for parents of different geographical origins, although there may be little difference in general traits, their genotypes may be more differentiated than those of cultivars from the same area, and their adaptability to the natural environment may also be quite different. For example, one parent of the cucumber cultivar Xiaqing 3, which was developed by the Guangdong Academy of Agricultural Sciences, is Female Line 75, which was selected from Japanese cucumber cultivars, and the other parent is Guiqing, a cultivar from Guangzhou. The geographical origins of the two parents are far apart, and the hybrid Xiaqing 3 is widely adaptable. As for fruit trees, the Beijing Botanical Garden crossed Muscat Hamburg (18% sugar content) from European grapes with local *Vitis amurensis* (1.5% sugar content) from northeast China. The average sugar content of the hybrids is 20%, and the highest is 24.9%. This is not only significantly higher than the mid-value of the two parents (16.5%) but also significantly higher than the parent with the higher sugar content, showing transgressive inheritance. By comparing the progenies produced through interspecific and interspecific hybridizations of 89 apples from Tashkent, the Uzbek Schroeder Institute of Fruit and Viticulture of the Soviet Union concluded that the greater the difference between parents in terms of ecological geographical conditions, the greater the ratio of superior types in offspring. When parents of different types or different geographical origins are matched, the segregation of the progeny is often quite obvious, and thus it is easy to select the desired recombinant traits. Of course, matching parents of different types or different geographical origins is not always better than matching parents of the same type or within the same region, because the key lies in the degree and nature of parental genotypic differences.

(3) Use the parent with the most excellent traits as the mother.

In some cases of the inheritance of traits controlled by the cytoplasmic genome, the offspring's traits are more inclined to the mother. Therefore, if the parent with more excellent economic traits is used as the mother and the parent with traits needing improvement is used as the father, there will be more individuals with comprehensive good traits in the hybrid offspring. For example, when the breeding objective is to improve the disease resistance of early-maturing high-quality cultivars, disease resistance is a trait that needs to be improved. When selecting parents, you need to select cultivars without disease resistance but with the required quality, early maturity, and other economic traits as the mothers, as well as cultivars with disease resistance as the fathers. When a cultivated cultivar is crossed with a wild cultivar, the cultivated cultivar is usually used as the mother and the wild one as the father. When a local cultivar is crossed with a foreign cultivar, the local cultivar is often used as the mother.

(4) Match parents according to the genetic rules of traits.

If the objective trait is a qualitative trait, one of the two parents must have the trait. Otherwise, the hybrid offspring cannot have the trait. Genetics clarifies that it is impossible to segregate individuals with the dominant trait from the hybrids of parents with the recessive trait, so when the target trait is dominant, one of the parents should have the dominant trait, not necessarily both parents. When the target trait is recessive, it is still possible for the offspring to segregate the desired recessive trait, even though neither parent shows the trait. But the condition is that one parent is heterozygous, which can't always be done. Therefore, at least one parent should have the target recessive trait during the match of parents. For example, the consistency of maturity in all parts of tomato fruit is often positively correlated with the absence of green fruit shoulder in young fruit, which is a recessive trait, and green fruit shoulder is the dominant trait. In order to breed cultivars with consistent maturity in all parts of the fruit, at least one of the parents lacking green fruit shoulder in young fruit should be selected.

Genetics clarifies that the cytoplasm also has the capacity for genetic transmission. The direct cross and reciprocal cross will yield different results for traits controlled or influenced by cytoplasmic genes. After studying the performance of apple hybrids in the direct and reciprocal crosses at fruit tree institutes in several provinces, including Heilongjiang, Liaoning, Jilin, and Shaanxi, it was found that maternal prepotency was obviously shown in two traits, namely, fruit maturity stages and the cold resistance of trees. The heredity of plant types of some fruits and vegetables also tends to favor the maternal side. Therefore, when matching parents, attention should be paid to the choice of the direct or reciprocal cross.

(5) Match parents with high general combining ability.

General combining ability refers to the average performance of all combinations of a parent cultivar or strain crossed with other cultivars. The level of general combining ability is determined by the gene accumulation effect of quantitative inheritance, and there may be transgressive variants in the hybrid offspring due to the traits controlled by the gene accumulation effect, which can be stabilized into fine cultivars by selection. It is possible to breed stable cultivars with better traits than the parents by selecting parents with high general combining ability. However, the general combining ability cannot be estimated according to the performance of parental traits now. It can only be identified according to the performance of hybrids. Therefore, it is necessary to design a test of combining ability or a test of combining ability in combination with the selection of F_1 hybrids to analyze the level of the general combining ability of parental cultivars or strains. In addition, you can also learn about the commonly used parental cultivars from crossbreeding records. These cultivars usually have high general combining ability.

(IV) Hybridization Techniques

1. Preparation

(1) Develop a breeding program.

All aspects of the hybridization work should be fully considered before the hybridization in order to achieve the purpose of hybridization, so it is necessary to develop a crossbreeding program. The program should include the breeding objective, the selection and matching of hybrid parents, estimation of hybrid progeny, hybridization tasks (including the number of combinations and hybrid flowers), hybridization process (such as pollen collection and hybridization dates), operation protocols (selection criteria of flowering branches and flowers for hybridization, emasculation, pollen collection and handling, pollination techniques and management requirements after pollination, etc.), etc.

(2) Equipment preparation.

The main tools for hybridization include forceps for emasculation or special emasculation scissors, powder storage bottles and desiccators, pollinators, plastic plates, magnifiers, pencils, 70% alcohol, isolation bags, covering materials, tying materials, record books, etc.

2. Sexual Hybridization Technical Procedures

The specific hybridization techniques for various horticultural plants are different, but the general requirements and technical procedures are basically the same. The technical procedures are described as follows.

(1) Breeding of parent seed plants and selection of flowers for hybridization.

Select typical, healthy, disease-free plants with strong growth vigor from selected parent types as the parental seed plants of hybrids. Generally, 10 plants are selected.

From the selected parent seed plants, you need to select those with strong flowers and buds and remove excess or unhybridized buds, flowers, fruits, and branches to ensure full growth of the hybrids' flowers, fruits, and seeds. Generally, a peach plant needs to keep three to four flowers, and the rest need to be removed; an apple and pear plant should have two to three flowers per inflorescence; the flower number of grape plants varies with cultivar; a poplar plant needs three to five flowers per branch; a *Gladiolus gandavensis* plant needs four to six flowers per branch; for cruciferous and umbelliferous vegetables, you need to choose flowers on the main and primary branches; for Liliaceae vegetables, you need to choose the upper and middle flowers of the inflorescences; for tomatoes, you need to choose the first to third flowers on the second inflorescence; for eggplants and peppers, you need to choose the first flower and the first pair flowers; for Cucurbitaceae vegetables, you need to choose the second to third pistillate flowers; for legumes, you need to choose flowers on the middle and lower inflorescences.

Parent seed plants of hybrids should be cultivated under strict management, and it is necessary to pay attention to the control of pests and diseases to ensure the robust growth of the plants.

(2) Pollen techniques.

① Pollen collection.

From father plants with typical traits, collect the well-developed buds or branches that are about to flower, take out the anthers indoors and place them in a petri dish with paper on the bottom, and then place the dish in a desiccator. Generally, at room temperature, after a certain period of time, the anthers will dehisce. Collect dried pollen from the anthers in a vial, label the cultivar on it and store it in a desiccator for later use.

For many horticultural plants, especially vegetable plants, it is common to pick the father's flowers and pollinate the mother when the anthers mature, but it is necessary to isolate the father's flowers before they bloom.

② Pollen storage.

Sometimes, because the anthesis of the father and the mother do not meet or are far apart, it is necessary to properly handle the pollen to maintain its viability for a certain period of time.

The length of pollen life varies with plant species. The viability of Chinese tulip tree pollen seriously declines after five days at 4°C; pear pollen and citrus pollen can be preserved for two to three weeks at room temperature under dry conditions; grape pollen and loquat pollen can be preserved for two months, while persimmon pollen only two days under the same condition; pine pollen and fir pollen can still be active for several years if preserved at low temperatures and under dry conditions. Generally, under natural conditions, the pollen life of self-pollinated plants is shorter than that of cross-pollinated plants. Pollen life is closely related to temperature, humidity, and light, in addition to genetic factors. For example, according to an experiment by Adams, apple pollen can be preserved for three months in dry conditions, but if it is placed in a 2–8°C and 80% relative humidity environment, it will lose its vitality in only five weeks. The pollen of lily still has a high germination rate after 194 days at 0.5°C and 35% relative humidity. The pollen of peach and pear can be preserved for one to two years at 0–2°C and 25% relative humidity. The pollen of some tropical and subtropical fruit trees, such as litchi and pineapple, has a short life and storage period, so it generally can't be stored for a long time. You'd better collect it right before you use it.

Pollen storage is collecting pollen and drying it in the shade until it is not sticky, removing inert matter, and then putting it in small bottles, preferably to one-fifth of the bottle's capacity. Seal the mouth of each bottle with double-ply gauze, put a label on each bottle, and then put them in a desiccator with an absorbent such as anhydrous calcium chloride at the bottom. The desiccator should be placed in a cool, dry, and dark place, preferably in a refrigerator at about 4°C.

③ Determination of pollen viability.

There are many methods to measure pollen viability. Here we'll mainly introduce the following four methods.

A. Direct pollination method: Pollinate the pollen directly on the pistil stigma of the mother and then count the number of fruits and seeds. The disadvantage of this method is that it takes a long time and is susceptible to climatic conditions.

B. Morphological identification method: Observe the morphology of pollen grains with a microscope and determine the viability of pollen grains according to their morphology. In general, deformed, crumpled, or void pollen is not viable.

C. Medium method: It is a commonly used method. Sow the pollen on a solid medium containing 1%–2% (mass ratio) agar and 5%–15% (mass ratio) sucrose, or an aqueous medium of 10%–20% (mass ratio) sucrose. Keep the medium in an incubator of about 20°C. After a period of time, examine the pollen germination rate and pollen tube growth microscopically, and then determine pollen viability accordingly. Various horticultural plants have different requirements for medium formulations in terms of, sucrose concentration, pH Value, microelement (boron), vitamin dosage, etc.

D. Staining method: Staining methods include the iodine reaction method, tetrazolium reaction method, magenta acetate method, hydrogen peroxide, benzidine, α-naphthol reaction method, etc. The iodine reaction method refers to the use of iodine-potassium iodide to stain the pollen. It is only applicable to the identification of sterile pollen without starch and not to sterile pollen with starch. The tetrazolium reaction method is used to identify the histochemical reaction of dehydrogenase activity. Fertile fresh pollen has dehydrogenase activity, while sterile or senescent pollen loses dehydrogenase activity. The iodine reaction method, tetrazolium reaction method, and magenta acetate method are commonly used to determine the fertility of flowers and vegetable plants, while the hydrogen peroxide, benzidine salt, and α-naphthol reaction methods are commonly used for fruit trees and ornamental trees. With these reagents, those with viability are stained, and those without viability are not. The staining methods have the advantage of being fast, but they're more indirect.

(3) Emasculation and isolation.

Emasculation is the removal of male organs from the mother flowers of hermaphroditic plants to prevent the absence of hybrid seeds due to self-pollination. Emasculation, in a broad sense, should also include the physical and chemical killing of stamens or pollen, as well as the removal of staminate flowers on dioecious plants and the removal of male plants in the dioecious test plot. Manual emasculation is usually used in the sexual crossbreeding of horticultural plants, especially vegetable plants. Emasculation is usually done one day before flowering, that is, the stamen is completely removed before the anther of the father is dehiscent. Please don't damage the pistil during emasculation.

Isolation is to prevent the mother from receiving pollination from non-target pollen and to prevent pollen in the flowers of the father from being contaminated by pollen from other closely related plants in case of unwanted crosses. Therefore, the flowers of the mother and father plants should be isolated.

In artificial hybridization, mechanical isolation, such as bagging or net room isolation, is often used. The bag is mostly made of light, transparent, waterproof, flexible sulfite paper, sulfuric acid paper, or cellophane. The size and dimensions of the bags depend on the species of plant and the size of the flower or inflorescence. For some woody plants (such as pine and fir), the anthesis is synchronized with the growth spurt of twigs, and the flowers are on the twig tips. The bag should be appropriately lengthened to prevent the twigs from breaking the paper bag. Mother flowers should be bagged and isolated during the pistil validity period before flowering and after pollination (i.e. from the beginning to the end of pollen fertilization). Father flowers should be bagged and isolated from the day before flowering to the time of pollen collection; as for some crops with large flowers, such as melons and morning glory, wire code (an article used by electricians to fix wires), thin iron wire or thick wire can be used to bundle and clamp their corolla for isolation; as for some entomophilous flower crops, they can be isolated in the net room. It is necessary to plant hybrid parents in the net room to prevent pollinators from entering and causing unwanted hybridization. In order to ensure the isolation effect of the net room, it is necessary to build the net room in a place sheltered from the wind; appropriately expand the planting distance of seed plants to prevent the contact of flowers and branches of the father and the mother; prevent pollinators from entering, and kill them immediately when they are found indoors.

(4) Pollination.

Pollination is the pollination of the pollen from the father's anthers onto the stigma of the pistillate flower of the mother. The anthers from the staminate flowers of the father with petals removed can be directly applied to the stigma of the pistil of the mother. You can also use a brush, sponge ball, cotton ball or rubber head, styrofoam head, and other fine soft objects to stick the pre-collected pollen held in the vessel and apply it to the stigma. The best time for pollination is usually on the day when the staminate and pistillate flowers bloom because this is the period when pistil and stamen pollen are most vigorous. Pollination during this period can increase the hybrid's fruiting rate and the number of seeds. However, due to the influence of various factors, sometimes some horticultural plants can be pollinated one day earlier or later, and a certain number of hybrid seeds can still be obtained. Before changing father anthers, it is necessary to disinfect pollination tools, fingers, etc. with 70% (volume fraction) alcohol to avoid pollen contamination.

(5) Marking and registration.

To prevent misplacement of hybrid seeds when harvesting, hybrid flowering branches and flowers must be marked. After emasculating the mother flower, you need to hang a tag on its base, on which the combination name, plant number, emasculation date, and the number of flowers should be recorded. After pollination, the date of pollination and the number of flowers should be recorded on the tag. After the fruit is mature, it should be removed together with the tag, and the date of harvest should be recorded on the tag. It is best to use plastic tags in case of falling off after being blown by wind and rain. A pencil should be used to write the content on the tags, and for those tags hanging on plants with long maturity stages, water-resistant paint pens should be used to write to ensure legible handwriting at harvest time.

In addition, it is necessary to prepare a hybrid journal for later analysis and summary, and this will prevent the absence of reference information if the tags on the mother plant fall off or are lost.

(6) Management after hybridization.

During the first few days after hybridization, the paper bags and other items used for isolation should be checked. If they fall off or break, accidental hybridization may have occurred and this hybridization will be invalid and a new hybridization should be made. After the expiration date of pistil fertilization, accidental hybridization is unlikely to occur, at which point the items used for isolation can be removed, usually about a week after hybridization. When removing bags of fruit crops, the hybrid fruiting rate can be checked for the first time. After its physiological fruit dropping, the second check, namely, the check of effective fruiting rate, can be carried out. Put gauze bags on the fruit before it is about to mature to prevent the pre-harvest fruit drop.

In order to create favorable conditions for the development of hybrid seed plants, it is necessary to strengthen the management of mother plants, apply more phosphorus and potassium fertilizer, pay attention to the prevention and control of diseases, pests, rodents, and birds, timely remove the flowers and fruits without hybridization, conduct topping and remove the side vine (branch) if necessary.

(7) Collection and storage of hybrid seeds (fruits).

The hybrid fruit should be collected promptly when the fruit reaches physiological maturity, especially for some fruits whose seeds are easy to fall off after maturity, such as the fruits of *Brassica* in Cruciferous, legumes, Liliaceae, and Compositae vegetables, as well as flowers like peonies and *Impatiens*. In fruit trees, excessive ripening of some species can affect the germination rate, such as early-maturing peach cultivars and cherries. The fruits of these crops should be collected promptly when they are ripe or nearly ripe. Generally, hybrid fruits should be placed in a dry place sheltered from the wind after collection, and then threshed after several days of after-ripening.

During the collection process, you need to prevent misalignment and mixing of different cross combinations. If the tag of the hybrid fruit is missing or the handwriting is fuzzy and cannot be checked, normally, the fruit should be eliminated.

According to the characteristics of the cultivars, you need to dry the seeds in the sun or shade after threshing them, and put them into bags in time, with the name of the combination, harvest date, and registration number marked outside the bag, put the corresponding label inside the bag, and then put these hybrid seeds in low-temperature, dry, rodent-proof storage conditions. Some seeds are susceptible to pests, so pesticide treatment should be conducted before storage.

When some horticultural plants such as *Paeonia suffruticosa*, *Rosa chinensis*, *Litchi chinensis,* and citrus fruits are in a state of exsiccosis, their seed germination will be affected. So after collection, the fruits should be promptly threshed, washed, and preserved in the sand, or the seeds need to be sown immediately.

(V) Handling of Hybrid Offspring

Professional breeding institutions often need to make a large number of hybrid combinations and prepare a large number of hybrid materials for each crossbreeding. In order to accurately select the cultivars that meet the requirements from the multifarious materials, corresponding selection methods and technical measures should be taken according to the species and material characteristics of crops. The common selection methods include pedigree breeding, mixed-individual selection method, and single seed passage method.

1. Pedigree Breeding

(1) First filial generation (F_1). Sow the seeds separately according to the cross combination, and sow or plant about dozens of plants of each combination in a row. Sow the paternal and maternal plants in rows on both sides to identify false hybrids. Since F_1 of self-crossing horticultural plants and F_1 of the inbred line of outcrossing plants have the same performance, only the unsatisfactory combinations need to be eliminated according to the performance of the combinations. In the selected combinations, individual plant selection is generally not needed. Only the false hybrids and significantly inferior plants need to be eliminated, and the seeds of the remaining plants are collected according to the combinations. Since the recessive good traits and the recombination types of various genes aren't shown in F_1, the selection of combinations should not be too strict.

(2) Second filial generation (F_2). Sow F_1 seeds separately according to the combinations, with a control group every 5 or 10 plots. F_2 is the generation with the most intense segregation of traits, with great variation among individuals and a wide variety of variant types, and is the most important selection generation. In order to get the desired variant type, the population of this generation should be larger, with no fewer than a few hundred plants per combination; due to the quantitative traits controlled by the genes, there should be no less than 1,000 plants.

For this generation, it is necessary to make a comparison between combinations first, eliminate the combinations without outstanding individual plants, and select the excellent individual plants from the selected excellent combinations. The selection criteria should not be too strict in case of losing good genotypes. For F_2, the selection is mainly based on strong genetic traits such as the growth habit of the plant, the shape and color of the product organ, and the date of the ripening time; for the traits easily affected by the environment, such as yield, nutrient content, disease resistance, etc., it is not yet possible to determine their merits based on individuals, so it is important to select more excellent individual plants. In principle, the number of plants per strain in the next generation can be smaller (no less than dozens of plants) and the number of strains can be larger.

Generally, the number of selected plants in the superior combination is about 5%–10% of the total population of the combination, while the selection ratio in the suboptimal combination can be lower.

(3) Third filial generation (F_3). Sow F_2 selected fine individual plants separately. The plants in each plot should only be the offspring of each individual plant. These plants are a strain. Plant dozens of plants for each strain and set up a control group every 5 or 10 plots.

Starting from F_3, the comparison of the advantages and disadvantages of strains should be emphasized. Select superior strains according to main economic traits and consistency, and then select individual plants with traits that are still in segregation from the selected strains. The number of selected strains should be large, and the number of selected individual plants per strain can be small, about 10 plants, to prevent the good strains from being missed. If there are a few fine individual plants in the strains that are about to be eliminated, the fine individual plants can be retained, but not too many.

If a relatively uniform strain is found in F_3, then when it is self-crossing, you can collect the mixed seeds after roguing. When it is outcrossing, you can artificially control the pollination in the strain after roguing and then collect seeds.

(4) Fourth filial generation (F_4) and subsequent generations. Sow selected F_3 superior individual plants separately, each plot only containing the offspring of each individual plant. The offspring then become a strain. F_4 strains from the same F_3 strain (i.e. the offspring of the same F_2 individual plant) are called strain groups. The strains within a strain group are called sister strains. The differences among different strain groups are greater than those of sister strains within the same strain group. The comprehensive traits of the sister strains are often similar. Therefore, it is necessary to compare the advantages and disadvantages among the strain groups and select the superior strains from the superior strain groups. Then, select superior individual plants from the superior strains.

The plot area of F_4 should be larger than that of F_3, with about 100 plants in each plot. Only after the second repetition can the yield, quality, and disease resistance be compared more accurately.

Starting with F_4, qualities like genetic stability and uniformity emerge. The plants with these qualities can be moved into the evaluation nursery for yield tests. The plants participating in the yield test can be renamed strains. However, for plants with not yet stable traits, individual plant selection should be continued, in which process the seeds should be systematically numbered.

With the increase of the generation, the number of excellent and consistent strains is gradually increasing, and the focus of work has shifted from selecting individual plants to selecting excellent strains for upgrading. If the strain groups behave neatly and relatively uniformly, you can collect mixed seeds according to the strain groups to maintain relative heterogeneity. This way, you can also obtain more seeds that can be distributed to multiple sites for experiments. If no good material appears in F_5 or F_6, the strains can be eliminated. However, the selection generations of often cross-pollinated plants are slightly longer than those of self-pollinated plants.

2. Mixed-Individual Selection Method

The method is an improved mixed selection method. In the breeding of self-pollinated horticultural plants, this method is simple, practical, and effective. The basic procedure is as follows: Starting from F_1, combine (or don't) plants and carry out mixed planting. For early generations, the selection is generally not needed. You only need to pick the plants with qualitative traits and high heritability from the mixed plants. When hybrid traits tend to be stable, that is, about 80% of that generation are homozygous individuals (F_4 or F_5), you need to plant all the seeds collected from the previous generation to accelerate the expansion of the population, which can reach several thousand to ten thousand plants according to different plant characteristics. Conduct an individual plant selection in F_4 or F_5. If there are 2,000–3,000 plants, at least 400–500 plants should be selected, with as many types as possible. In F_5 or F_6, you need to plant the plants by strain. The number of plants of each strain should be small, about 10–20 plants per strain, and it is best to set up a second repeat. The ratio of selected superior strains must be small, about 5%–10%, for later evaluation.

3. Single Seed Passage Method

This is a derivative of the mixed-individual selection method and is suitable for self-crossing crops. The procedure is as follows: In general, starting with F_2, take one seed from each individual plant in each generation and carry out mixed sowing the next year (In practice, three seeds are often taken from each plant to form three seeding materials. Two are sown and one is kept). For each generation, no selection is carried out. Only the individuals with simple inherited undesirable traits are eliminated until F_4 or F_5, when the genetic traits are stable and no longer separated. In F_4 or F_5, collect the seeds of each individual plant separately and sow them separately for the next generation's strain comparison and evaluation. Neat and consistent strains that meet the requirements of the breeding objectives are selected and upgraded for testing.

When asexually-propagating plants are crossbred, segregation may occur in F_1, and good individual plants can be selected in both F_1 and F_2. The selected good individual plants should have asexual propagation, which forms the asexual line.

In the process of breeding new cultivars through sexual hybridization, the control of environmental conditions should be stressed. This is because the selection of individual plants or strains in the hybrid generation is based on trait performance resulting from the interaction between genotype and environment. Therefore, the environmental conditions for breeding hybrid progeny should correspond to the breeding objectives. For example, in order to cultivate a high-yield cultivar, the hybrid offspring should be cultivated under good fertility and water conditions; in order to develop a disease-resistant cultivar, the hybrid offspring should be cultivated in areas and seasons with serious diseases, and sometimes the hybrid offspring should be tested with adversary cultivation or artificial inoculation of pathogens.

IV. Heterosis Utilization

(I) Concept of Heterosis

Heterosis refers to the phenomenon that F_1 plants produced by the crossing of two parents of different genotypes are superior to their parents in terms of growth potential, viability, reproductive capacity, stress resistance, yield, quality, and other traits. The performance degree of heterosis differs, and so does the performance of plant combinatios. It must be noted that not all F_1 will show heterosis; on the other hand, if the same combination (F_1) is planted under different environmental conditions, the heterosis performance will be different because it is the result of the interaction between environmental factors and genes.

The opposite of heterosis is inbreeding depression, which is also a common biological phenomenon. For cross-pollinated plants, different degrees of depression generally occur after self-fertilization, and the degree of depression varies with different plant species. For example, cruciferous vegetables decline faster, while watermelon, melon, and other melon crops decline more slowly. In the early stage of self-fertilization, the decline is faster and tends to slow down as the number of self-fertilization generations increases.

(II) Measures of Heterosis

According to the practical application of heterosis, we can see that heterosis is decided by how much better the F_1 is than the better parent or the average of the two parents in terms of quantitative traits. The strength of heterosis is usually measured in the following ways.

(1) Mid-parent heterosis (H_m): The ratio of the difference between the mean value of a quantitative trait in F_1 and the mean value (mid-parent value) of the same trait in both parents (P_1, P_2). The formula is as follows:

$$H_m = \frac{F_1 - (P_1 + P_2)/2}{(P_1 + P_2/2)} \times 100\%$$

(2) Over-parent heterosis (H_p): The ratio of the difference between the mean value of F_1 and the mean value of the better of the two parents, using the mean value of the better parent (P_h) as a metric. The formula is as follows:

$$H_p = \frac{F_1 - P_h}{P_h} \times 100\%$$

(3) Over-standard heterosis (H_s): The comparison of the difference between the mean value of a quantitative trait of F_1 and the mean value of the same trait of

the standard cultivar CK (the cultivar currently being promoted in production). The formula is as follows:

$$H_s = \frac{F_1 - CK}{CK} \times 100\%$$

(4) Heterosis: It has been widely used in the production of horticultural plants, such as tomato, onion, eggplant, cucumber, zucchini, kale, broccoli, cabbage, radish, carrot, lettuce, pepper, celery, and so on. The F_1 hybrids of ornamental plants, such as coconut, begonia, snapdragon, pansy, corn poppy, carnation, petunia, marigold, and scarlet sage, are also produced by the crossbreeding of parents and used directly in production. From the perspective of application value, the heterosis of plants is manifested not only in the increase of yield and yield factors in the F_1 hybrids, but also in the increase of the proportion of yield in the early stage. Therefore, their F_1 hybrids have higher economic value, stress resistance, population consistency, and ornamental value.

(5) Divergence heterosis (mean dominance $\sqrt{\frac{H}{D}}$): A measure of the difference between F_1 and mid-parent values using the mean difference between the two parents as the unit. The formula is as follows:

$$H = \frac{F_1 - \frac{1}{2}(P_1 + P_2)}{\frac{1}{2}(P_1 - P_2)}$$

In this formula, the divergence value is used as the unit of the measure of heterosis. With this measure, the level of heterosis intensity can be read directly from the H value, making it easy to compare combinations and traits individually or collectively. This measure reflects the negative correlation between the H value and the difference between the two parental values. It has been proved by breeding practice that the greater the difference between the two parental values, the less likely the combination to have over-parent heterosis; otherwise, when $P_1=P_2$, the H value in this formula would go to infinity, which means the assumption is completely unrealistic that the smaller the difference between the two parental values is, the stronger the heterosis will be.

(III) Comparison between Dominant Crossbreeding and Conventional Crossbreeding

1. Genetic Effects

Theoretically, conventional crossbreeding mainly utilizes additive effect and partial epistatic effect, which can be passed down to the next generation stably; dominant crossbreeding utilizes additive and non-additive effects that can't be passed down to the next generation stably.

2. Crossbreeding Procedure

In conventional crossbreeding, plants cross first, then self-cross, and finally, get stable cultivars of homozygous genotypes. In dominant crossbreeding, strains cross first to get excellent inbred lines of homozygous genotype, then through combining ability analysis and selection, and finally, get hybrid cultivars of heterozygous genotype.

3. Seed Production

Conventional crossbreeding is simple. You only need to collect the seeds from the crop field or seed production field every year. Dominant crossbreeding is complicated as the seeds cannot be kept in the crop field. As a result, the parent breeding area and hybrid seed production area must be set up every year.

(IV) Procedures for Heterosis Utilization

Heterosis utilization mainly refers to the utilization of F_1 hybrids, which can be roughly divided into three types in terms of their parental properties: F_1 hybrids between cultivars, F_1 hybrids between cultivars and inbred lines, and F_1 hybrids between inbred lines. Interspecific F_1 hybrids are mainly used in self-pollinated crops. For cross-pollinated crops, the work of F_1 hybrid selection should start from the selection and breeding of inbred lines.

1. Selection and Breeding of Superior Inbred Lines

The inbred line is a system of inbred progeny with consistent traits and relatively stable heritability produced by consecutive generations of self-crossing of an individual plant of a cultivar combining the selection method. A good inbred line should have characteristics such as high combining ability, strong disease resistance, high yield (including a high yield advantage of the selected hybrid combination and strong growth, high yield, and high seed yield of the inbred line), and most good traits inheritable.

(1) General methods and steps for selection and breeding of inbred lines.

① Basic materials for selecting plants.

Generally, good cultivars or hybrids should be selected as the basic materials for selecting plants. In particular, the latest promoted cultivars or excellent local cultivars should be prioritized as the basic materials. Cultivars with poor agronomic traits should normally not be used except when their special genes are needed. If the combining ability of the existing cultivars or hybrids is known, materials with high combining ability should be selected. Please don't select too many basic materials, generally only 10. Otherwise, the workload is too much to bear.

The self-crossing plants selected from the basic materials are called "basic plants", which constitute the S_0 generation; the next generation obtained by the self-crossing of the good individuals selected from the S_0 generation is the S_1 generation; S_2, S_3,··· and subsequent generations are obtained in the same way.

② Superior plant selection and self-crossing.

Select good individual plants within the selected good cultivars or hybrids for respective self-crossing. The number of plants that should be selected for self-crossing within each cultivar or hybrid, and the number of the progeny plants that should be planted of the strain to be self-crossed are determined by the specific conditions of the test materials, usually from a few to dozens of plants. Generally, for cultivar materials, more individual plants should be selected for self-crossing, and not so many of their offspring need to be planted; for hybrids, relatively fewer plants should be selected for self-crossing, and many of their offspring should be planted. For purer cultivars, fewer individual plants can be selected for self-crossing. If otherwise, more representative individual plants can be selected. Usually, 50–200 plants are planted per S_1 strain, and the number of S_1 strains is usually between a few to two hundred.

③ Selective elimination generation by generation.

Compare and evaluate S_1 generation strains according to the breeding target traits, first eliminate some undesirable inbred lines, and then select several to dozens of plants from each selected inbred line to continue self-crossing. The number of plants is about several hundred to several thousand. Plant 20–100 plants per S_2 generation strain. Thereafter, the selection and elimination continue, and eventually, about a few dozen remain. The number of plants planted for each strain may increase slightly as the number of strains decreases. For inbred lines, 4–6 generations of selection are generally carried out until the lines with high purity, no separation of main traits, and no obvious decline in viability are obtained. Then, isolate the inbred lines, let them propagate separately, leave them free to pollinate, and collect seeds within the lines. But you need to strictly prevent them from hybridizing with pollen of other lines or cultivars.

Generally, the following method is used for numbering throughout the process of inbred line selection. For example, A1→10-3-8 represents the 1st plant of S_0 generation, the 10th plant of S_1 generation, the 3rd plant of S_2 generation, and the 8th plant of S_3 generation of cultivar A.

④ Determination of combining ability.

After different inbred lines through the above self-crossing and selection process are obtained, the combining ability should be measured to screen out the superior inbred lines with high combining ability. The details of the method will be introduced later.

(2) Recurrent selection method for breeding inbred lines.

The recurrent selection method is a breeding method in which the desired genes are gathered together through repeated selection and hybridization. The specific practices are as follows: Select the good individuals from the heterozygous population and carry out polycross among the progeny strains after one self-crossing to obtain the population for the next selection. This method is different from the general method of inbred line breeding, which selects good individuals from the heterozygous population and then repeats the self-crossing and selection. The use of this method allows for the pooling of genes that are scattered across individuals and chromosomes in heterozygous populations, maximizing the chances of selection and genetic recombination. In the early generations, the plants are selected for their high degree of heterozygosity, and the selection effect is better. The original materials used for recurrent selection can be naturally pollinated cultivars, mixed cultivars, offspring of a considerable number of inbred lines crossing each other, double cross hybrids and single cross hybrids, etc. Materials with a narrow genetic base are not suitable for recurrent selection because they are not easily improved. According to the existence of test crosses, the different types of test-cross species, and the objective of genetic content improvement, the recurrent selection method can be divided into three types, which are briefly described as follows.

① Single-round recurrent selection method.

Select good individual plants for self-crossing. Stock the seeds of individual plants and sow them according to their strains, then cross the strains with each other and retain their seeds. This is the whole process of the first cycle (round). According to the breeding objective, identify the merit of the candidate population. You may conduct a second or even more round of selection. This method carries out selection based on phenotypes, so it is suitable for selecting traits with high heritability (such as maturity stages, plant height, disease resistance, insect resistance, etc.).

② Combining ability recurrent selection method.

Select good individual plants from the original population and carry out two kinds of crosses. One is the self-crossing of the individual plants selected, and the other is to use the selected plants as the mothers and let them perform test cross with heterozygous populations or compound hybrids and other test species. The purpose of self-crossing is to retain the progeny of good individual plants and progressively make the genotype homozygous; the test cross aims to assess the general combining ability of the selected individual plants. Compare the production performance and other horticultural traits of F_1 obtained by the test cross, select a suitable number of good combinations (about 10%), and let the progeny of the self-crossing mother plants of these good combinations randomly mate in the isolation area to form an improved population. Then, the first selection round is finished. A number of recurrent selections can be made according to the strengths and weaknesses of the selected population. This selection method uses a heterozygous population or a compound hybrid as a test species to perform a test cross with good individual plants. Since the genotypes of the test species are heterozygous, the identification results of the test cross combination provide a measure of additive effects and reflect the general combining ability of the selected inbred line. This method is suitable for the inbred lines with high general combining ability and can also be used to improve the production performance of the original population.

③ Interactive recurrent selection method.

In this method, it is necessary to prepare two populations, A and B, as the original populations, and then use a considerable number of randomly selected individual plants from population B as the test parents to improve the combining ability of population A. A considerable number of randomly selected plants from population A are used as the test parents of population B. This can increase the combining ability of A and B and improve both A and B populations. This method was developed by R. E. Comstock, H. F. Robinson, and P. H. Harvey (1949). The improved A and B populations can be directly crossed, inbred lines can be selected from each population for type $(A_1 \times A_2) \times (B_1 \times B_2)$ cross or various types of hybrids between A and B can be bred.

2. Determination of Combining Ability

(1) Concept of combining ability.

The breeding practice has proved that the yield of hybrid offspring of parents with good appearance and high yield is not necessarily higher; some parents are not very good, but when they mate with other parents, the offspring are excellent. This phenomenon is related to the combining ability of the parents. Combining ability is a concept put forward by Sprague and Tatum in 1942. According to the research results of maize crossbreeding, they divided combining abilities into general and special combining abilities.

① General combining ability.

It refers to the average yield (or other economic traits) performance of the inbred line or cultivar (homozygote) of a parent in a series of cross combinations. To be exact, the general combining ability of a parent is the difference between the average yield (or other trait values) of the F_1 of the parent crossed with other parents and the total average yield of all the F_1 of all parents crossed with each other. In heterosis utilization, measuring the general combining ability can reduce the blindness of hybrid combination selection and matching. Besides, the combination with high special combining ability can only be realized based on the parents with high general combining ability. In this way, the most ideal hybrid combination can be obtained.

② Special combining ability.

It is the difference between the actual yield (or other trait values) of a particular combination and the predicted average yield (or other trait values) based on the general combining ability of the two parents. Or, it is the result that the yield (or other trait values) displayed in a particular cross combination is better or worse than the average of all cross combinations.

When chance errors are not considered, the special combining ability can be expressed by the following formula:

$$S_{ij}=X_{ij}-\mu-g_i-g_j$$

In this formula, S_{ij} is the special combining ability effect of the cross combination of i parent and j parent; X_{ij} is the average effect of the F_1 plot of the cross combination of i parent and j parent; μ is the total average effect of the F_1 population of all cross combinations; g_i (g_j) is the general combining ability effect of i (j) parent.

③ The relationship between combining ability and breeding.

General combining ability is mainly determined by the additive effect of parental genes, while special combining ability is controlled by the dominant effect of genes and the non-allelic interaction effect. Heterosis is mainly caused by dominant effect and non-allelic interaction, which means that the strength of heterosis of a given combination depends mainly on the specific combining ability of the combination. For traits that mainly depend on general combining ability, combination breeding can be used to breed stable cultivars to avoid the trouble of annual seed production. For traits that mainly depend on special combining ability, or for traits in which both general combining ability and special combining ability have a great influence, the heterosis of F_1 hybrids should be fully utilized.

Since there is no direct correlation between the additive and dominant and epistatic effects of genes, it is possible that two parents with low general combining ability may have very high F_1 trait values due to high special combining ability after crossing, while the other parent pair may have high general combining ability and low special combining ability, and thus also have high F_1 trait values. This means that although high-yielding combinations may be obtained within combinations with either high general combining ability or high special combining ability, the highest combinations can only be obtained by selecting the parents with high general combining ability and then the combinations with high special combining ability.

(2) Determination method of combining ability.

① Top cross method.

The top cross method refers to the mating of common cultivars (including hybrids) as test species with each test inbred line (or cultivar) and the comparing of the yield (or vegetable trait value) of each test cross cultivar in the next generation. For the combinations with high yields of the test cross cultivars, the combining ability of the tested inbred lines (or cultivars) is high; conversely, the combining ability of the tested inbred lines (or cultivars) is low. The advantage of the top cross method is that the number of combinations needed to be prepared and compared is small, and the test results can be easily compared between the tested inbred lines (or cultivars). The disadvantage of this method is that the general combining ability and the special combining ability cannot be measured separately but mixed together; in addition, the data obtained are not symbolic as the results may be different with a different tester. Therefore, this method is suitable for testing and comparing the combining ability of early generations (e.g. S_0 or S_1), and timely eliminating some strains with relatively low combining ability. It is also suitable for cases where the tester is one of the parental strains of the F_1 hybrids in the final formulation. For example, by using a male sterile line or a self-incompatible line as a tester, it is possible to select from a large number of inbred lines the inbred line with the highest combining ability and obtain a good cross combination.

② Unequal matching method.

The unequal matching method is also called the irregular matching method or simple matching method. It means matching the inbred lines into several combinations according to the principle of parental selection and matching. Generally, the superior inbred lines are matched into many combinations, and the undistinguished inbred lines are matched into fewer combinations so that the actual number of combinations formed by each inbred line is not equal. That's why it's called the unequal matching method. This method is simpler. As long as each parent is matched into more than two combinations, the general combining ability of each parent and the special combining ability of each combination can be calculated according to the equation of definition. However, some inbred lines have too few combinations, which makes the calculation of combining ability unreliable. Therefore, this method is suitable for the situation where there are many parent materials and the best combinations are expected, but only a few combinations can be compared due to the limitation of conditions; in addition, when the original breeding materials are collected and the selection and breeding of inbred lines are started, this method is often used to test the combining ability of the breeding materials, providing a basis for the selection of inbred lines of key breeding materials.

③ Half-round matching method.

The half-round matching method is also called the half-diallel cross method, which refers to the matching of each inbred line (or cultivar) with other inbred lines (or cultivars) one by one, but excluding self-crossing and reciprocal crossing combinations. The number of mating combinations can be calculated by the formula: $n = P(P-1)/2$, where n is the number of combinations and P is the number of parental inbred lines (or cultivars). If there are 10 inbred lines, 45 combinations need to be made. Conduct field tests on the combinations prepared according to the experimental design, record, analyze, and verify the repeated results for each combination, calculate the average value, and then calculate the general combining ability and special combining ability according to the following formula.

$$gca_i = \frac{X_i}{P-2} - \frac{\sum X..}{P(P-2)}$$

$$sca_{ij} = \overline{X}_{ij} - \frac{X_{i.} + X_{j.}}{P-2} + \frac{\sum X..}{(P-1)(P-2)}$$

In this formula, X_i. is the sum of the values of a trait in all combinations with inbred line i as the parent; X_j. is the sum of the values of a trait in all combinations with inbred line j as the parent; $\sum X..$ is the sum of the values of a trait in all combinations in the test; P is the number of parents; $\overline{X}_{ij}$ is the average value of a trait of the F_1 produced by i as the mother crossed with j as the father.

The advantages of this method: It can help people understand whether the combining ability of a certain trait is mainly determined by general or special combining ability; it can also enable breeders to select superior combinations more accurately. The main disadvantage is that the number of combinations matched is relatively large and the workload is heavy. The differences in individual traits between the direct and reciprocal crosses of some combinations could not be measured by this method.

(3) Determination of mating method between inbred lines.

After the selection of good hybrid combinations and their parents' inbred lines by the combining ability test, it is necessary to further determine the optimal combination mode of each inbred line in order to obtain the most productive hybrids. According to the number of parental inbred lines used for the F_1 hybrids, the hybrids can be divided into single, double, and triple cross hybrids. Both double and triple crosses are seed production methods used to reduce the cost of hybrid seed production. At present, most vegetable F_1 hybrids are single cross hybrids.

① Selection and breeding of single cross hybrids.

The F_1 hybrids made from two inbred lines are called single cross hybrids. The advantages of single cross hybrids are their strong heterosis, uniform plants, and simple seed production procedures. In addition, their parents can be maintained as a stable strain and can produce hybrids with the same genes every year. The disadvantage is that some inbred lines have poor fertility, less seed quantity, and high cost.

Single cross hybrids can be selected and bred in the following methods:

A. Use good inbred lines for diallel crossing to produce single cross hybrids: Match the good inbred lines that have been tested for general combining ability to produce possible single cross combinations with bag pollination. After identification and comparison, regional tests, and so on, those with superior performance can be popularized and applied in production.

B. Use elite inbred lines as test species: In the preparation of single cross hybrids, if the number of inbred lines available is large, the mechanical diallel crossing could be made, but the workload would be too heavy. Therefore, excellent inbred lines with high general combining ability can be selected as elite inbred lines and crossed with other lines for yield evaluation. The hybrids that meet the requirements of breeding objectives can be selected for further demonstration and promotion. In this way, not only can new single cross hybrids be bred, but the combining ability of inbred lines can also be measured.

C. Improve the existing single cross hybrids: After several years of cultivation, the original single cross hybrid may be overshadowed by the new inbred line, so it must be improved. The method is to replace the defective inbred line with a new one, making the F_1 more suitable for production. In recent years, the United States also adopted the method of sister mating to form improved single cross hybrids. For example, A×B can be changed into $(A\times A_1)\times B$ or $(B\times B_1)\times A$ or $(A\times A_1)\times(B\times B_1)$, which are all improved single cross hybrids of A×B.

② Selection and breeding of double cross hybrids.

Double cross hybrid refers to the F_1 hybrid produced by two single cross hybrids of four inbred lines. The advantage of double cross hybrid is that the amount of seed used in the parental inbred lines can be significantly reduced and the yield of hybrid seed can be significantly increased, thus reducing the cost of seed production. In addition, the genetic composition of double cross hybrid is not as simple as that of the single cross hybrid, and although the plant is less uniform, the double cross hybrid is more adaptable. Although the yield is slightly lower than that of the single cross hybrid, it is more stable in production. The disadvantage is that its seed production is more complex, so it is less used in production.

③ Selection and breeding of triple cross hybrids.

The F_1 hybrids made from three inbred lines are called triple cross hybrids. In a three-line cross, a single cross line is usually used as the mother and another inbred line as the father. Because of the strong viability and high fruiting rate of single cross lines, heterosis can be effectively utilized to reduce seed production costs. The same principle and procedure of double cross hybrids can be used to predict the productivity of triple cross hybrids. This type of grouping is rarely applied to vegetables.

(4) Cultivar comparison tests, production tests, and regional tests.

After these processes, one or more good cross combinations may be developed, but they are not yet ready for production. It is necessary to carry out cultivar comparison tests and production tests seriously to determine whether it has the value of promotion according to its performance in all aspects, where it is suitable for promotion, and whether it has the value of being reported to the national regional test. The standard for measuring the merit is decided by the control species (including the unified control species and local control species). Generally speaking, the cultivar whose yield is more than 15% higher than that of the control species, or whose yield increase is not obvious, but there are 1–2 main economic traits significantly better than that of the control species, can be recognized as valuable for promotion. Otherwise, it should be rejected.

Review and Reflection Questions

(1) What is the significance of seed introduction?

(2) What are the general rules for seed introduction?

(3) What are the dominant ecological factors affecting the success or failure of introduction and domestication?

(4) What are the characteristics of the floral organ structure and pollination and fertilization of self-pollinated plants?

(5) What principles should be grasped in setting selection criteria for selective

breeding?

(6) What are the selective breeding procedures for sexually-propagating horticultural plants?

(7) What is the sport selection procedure?

(8) What are the sexual hybridization methods?

(9) What are the principles of hybrid parent selection?

(10) What are the genetic variation and breeding characteristics of self-pollinated plants?

(11) What is the procedure for heterosis utilization?

(12) What are the principles of breeding hybrid offspring?

Learning Scenario III Variety Purity Protection of Horticultural Plants

I. Purity and Relativity of Horticultural Plant Cultivars

(I) Concept of Variety Purity

Variety purity can be understood from both production practices and genetics. The so-called purity in production refers to the relative consistency and stability of biological, agronomic, and economic traits of a cultivar population, which can be passed on to the next generation. Only when the biological, agronomic, and economic traits of a cultivar population are relatively stable and consistent, can it be easier to cultivate, manage, and achieve high quality and high yield.

Genetically, even relatively stable and consistent populations of self-pollinated crops with low natural outcrossing rates may differ genetically from individual to individual. However, after several generations of random mating, the genetic composition of the genetically inconsistent populations will be in equilibrium. That is, genetic equilibrium will be achieved. Therefore, the gene frequency and genotype frequency of this cultivar population reach a relatively stable state, and the characteristics of the cultivar also show relative stability. Generally, we refer to cultivars with relatively stable and consistent biological, agronomic, and economic traits as pure breed, and the opposite as impure.

(II) Relativity of Variety Purity

Pure and impure breeds are relative concepts. Absolute pure breeds do not exist. We can understand this from the following aspects.

1. Genetic Factors

At present, the cultivars used in production are mostly bred by the crossbreeding method, and they are the new cultivar groups that meet the breeding objectives after successive generations or even more than 10 generations of selection from the separated generations of hybrid offspring. From the perspective of genetic rules, regardless of self-pollinated crops or often cross-pollinated crops, the population will not reach complete homozygosity after several or more than a dozen generations of self-pollination and selection, and there is always "residual variation". For example, for a trait controlled by 10 pairs of genes, after seven generations of complete self-crossing, only 85.42% of the population is homozygous, leaving 14.58% of individuals heterozygous. In addition, most of the economic traits closely related to yield and quality are quantitative traits, controlled by micro-effect multiple genes, and easily affected by environmental conditions. It is difficult to distinguish them according to phenotypes, even if one or two pairs of genes are in the heterozygous state during selection. These individuals in the heterozygous state will inevitably show trait separation in the subsequent generations. Therefore, the purity of a cultivar is relative in terms of genetic composition analysis. Even if individuals within a newly bred cultivar have very similar phenotypes, the genotypes of these individuals cannot be identical.

2. External Factors

After a cultivar is applied to production, due to the interference of some external factors (such as mechanical mixing, biological mixing, incorrect artificial selection, etc.), its genetic composition will change, which will make its original gene frequency and genotype frequency change, and then make its phenotypes change, such as biological and morphological changes, economic characteristics and trait changes. This will make the original cultivar lose its typicality and relative consistency, resulting in lower yield and lower quality. Therefore, the purity of the cultivars used in production will gradually decrease until the use value is lost as the planting area expands and the breeding generations increase. Therefore, we should do well in superior variety breeding while focusing on cultivar updating to ensure that superior varieties of good quality and high purity are used in production.

3. Ex-situ Effect

Any superior variety is bred under certain ecological conditions and cultivation conditions. Some individuals with different genotypes may show little difference in phenotype under the environmental conditions of the breeding land, but this difference may become larger or even show very clearly after being introduced to another area with ecological and production conditions. This phenomenon of some traits of cultivars that have changed greatly due to changes in environmental conditions is called the ex-situ effect. The ex-situ effect also illustrates the problem that what we normally think a high-purity cultivar, or even a newly bred cultivar, is only very close or similar in phenotype, while the genotypes are not identical between individuals. This shows that variety purity is only a relative concept, and that is the truth of "differences existing in high purity cultivars".

II. Reasons for Mixing and Degradation of Horticultural Cultivars

1. Mixing and Degradation of Cultivars

Mixing and degradation of cultivars refer to the phenomenon in which the loss of the original good morphological characteristics of cultivars, the decline of stress resistance adaptability, yield, and quality caused by low purity and seed degeneration in the production and cultivation process. Mixing and degradation of cultivars are common phenomenon in plant production. The phenomenon of various variation types, uneven plant height, different maturity stages, nonuniform growth vigor, segregation phenomenon in disease resistance, and stress resistance will occur in the population when a cultivar is mixed and degraded, and it will seriously affect the yield and quality. Such as some early-maturing spring common head cabbage cultivars in the process of seed production, the seeds made often lose the original trait of strong winterness due to the lax original seed selection, or lax isolation in hybridization. Large-area bolting in spring often occurs when cultivating spring commercial vegetables with such seeds, thus losing their commerciality.

Mixing and degradation of cultivars are two different concepts. Mixing refers to the phenomenon that different types or cultivars of seeds are mixed within a cultivar population, or that natural hybridization and genetic mutations have occurred in the previous generation, so variation types are separated from the offspring population, causing the low purity of the cultivar. If the upright cylindrical cabbage seeds are mixed in the head cabbage seed population and after using this batch of seeds for commercial vegetable production, head and upright cylindrical cabbage will both grow on the planting field, resulting in irregular products, inconsistent harvest period, and non-conformance to the requirements of product commerciality.

Degradation refers to the phenomenon where the genetic changes in a cultivar lead to the loss of production and utilization value, resulting in inferior economic traits, stress resistance decline, and fruit quality reduction. Cultivar degradation is frequent and widespread in the production of horticultural plants. For example, the excellent traits of tall plants, large flowers, pure and bright flowers, etc. can be found on the tulip, *Gladiolus gandavensis*, and other flowering bulbs in the first and second years of introduction. With the increase of propagation generations, these flowers will gradually show the phenomenon of plants becoming shorter, flowers becoming smaller, inflorescences becoming shorter, flowers becoming darker, and so on.

Mixing and degradation of cultivars are closely related. The degradation of cultivars is caused by the cultivar mixing, and the degraded cultivars are bound to aggravate the mixing. Therefore, although the mixing and degradation of cultivars belong to different concepts, the two are often intertwined and difficult to separate completely. Generally speaking, when the phenomena of low purity, seed deterioration, stress resistance decline, and product quality reduction appear, we call them mixing and degradation of cultivars.

2. Reasons for Mixing and Degradation of Cultivars

The root cause of the mixing and degradation of cultivars is the lack of a complete superior variety breeding system. The specific reasons are mainly the following.

(1) Biological mixing: Biological mixing refers to the phenomenon in which natural hybridization occurs between different types or cultivars due to insufficient isolation during cultivar propagation, thus changing the genetic composition of the superior variety. Biological mixing is often the most important cause of the mixing and degradation of cross-pollinated crop cultivars. For example, the common head cabbage will stop head formation if there is a natural hybridization between the common head cabbage and cauliflower or turnip cabbage; the cabbage will not form a head after a natural hybridization between cabbage and pakchoi or flowering Chinese cabbage. In the superior variety breeding process of often cross-pollinated crops and self-pollinated crops, subspecies, varieties, and cultivars that can hybridize should also be properly isolated. Otherwise, natural hybridization may occur to some extent. Among the factors causing biological mixing, in addition to pollination habit and isolation distance, there are also weather conditions such as wind speed, pollinating insect species, the size of the seed collection field, and its surrounding geographical environment (e.g. terrain height, obstacles nearby).

(2) Mechanical mixing: Mechanical mixing refers to the phenomenon in which breeding cultivars are mixed with different cultivars, crop seeds, or plants without operating in accordance with the technical regulations during superior variety breeding. Once a cultivar has been mechanically mixed, it will show inconsistent height and fertility period, thus affecting the yield and quality. There are many possibilities for mechanical mixing, which can be caused by human negligence in the process of seed treatment (seed drying, seed soaking, seed dressing, seed coating, etc.), sowing, planting, harvesting, threshing, drying, transportation, packaging, and other processes; volunteers of the previous crops, grafted own-rooted seedlings and the undecomposed organic fertilizer used as fertilizer being mixed with other living seeds may also cause mechanical mixing. Mechanical mixing is one of the main causes of the mixing and degradation of cultivars.

Failing to take timely purification, strict roguing, and other effective measures after mechanical mixing may result in biological mixing, thus aggravating the degree of mixing and degradation of cultivars.

(3) Mutation of the cultivars: At present, most of the cultivars promoted in production are made by crossbreeding. Some cultivars seem to be stable and consistent in terms of major traits, but there are still some traits, especially the quantitative trait controlled by multi-gene, which are not completely stable. The phenomenon of character separation will occur in the process of self-propagation of cultivars due to gene recombination, which decreases the typicality and consistency of cultivars, leading to the mixing and degradation of cultivars. Under natural conditions, cultivars sometimes undergo genetic mutations due to some special environmental influences, and most of these mutations are unfavorable. If there is no manual selection and elimination, the increasing number of unfavorable mutation types and individuals will cause a decline in variety purity.

(4) Incorrect selection and seed retention: Artificial selection is an important tool to prevent mixing and preserve purity during seed production. However, if an incorrect artificial selection is used, it can also artificially cause the mixing and degradation of cultivars. For example, people often mistakenly reserve plants with large fruit for pursuing high yields in the production process of export-oriented cultivars of cluster red peppers, and the produced peppers will exceed the size of export requirements after several generations of selection, resulting in difficult exports, low prices, and high yield without high efficiency. For another example, people often mistake the hybrid seedlings with good performance and heterosis as the strong seedlings of the cultivar for selection and reproduction during the execution of seedling thinning, resulting in mixing and degradation.

(5) Genetic mutations: Different genetic mutations may occur due to various natural conditions after a new cultivar promotion. Among the mutations that occurred, most of them have a negative impact on human needs, and if these mutants continue to propagate, it will inevitably increase the individual variation rate in the superior variety population. For example, there will be some easy bolting individuals when propagating superior varieties of cabbage or radish with

bolting resistance. If these mutants cannot be eliminated and the propagated seeds continue to be produced, premature bolting individuals will appear and affect the yield of commercial vegetables.

The bud sport caused by gene mutation is the main reason for the mixing and degradation of asexually-propagating cultivars of fruit trees, flowers, vegetables, and other horticultural plants. Although bud sport sometimes result in beneficial variants for human needs, most of them are deteriorations. If the bud sport shoots are used for propagation, it may lead to cultivar degradation. This is why there are some differences among individuals in the offspring population of the same individual plant propagation, and even different performances of the offspring population of the same individual plant propagated by taking shoots in different years.

(6) Adverse ecological and cultivation conditions: The cultivars cultivated in production are selected under excellent cultivation conditions, especially those selected by means of recombination breeding and heterosis in recent times, which are synthesized by the outstanding traits of different ecological zones. In the process of superior variety breeding, if the growth and development conditions are not met for a long time, and the dominant traits are not fully expressed for a long time, it will cause degradation. For example, if the water sprouts are used for propagation in the propagation of citrus, delayed fruiting and few fruits will be caused to their offspring, showing cultivar degradation.

(7) Virus infection: For some horticultural plants of vegetative propagation, the accumulation of virus increases from generation to generation due to virus infestation, and the damage will become more and more serious until the good traits of the cultivars are lost and cause degradation.

III. Methods of Horticultural Plant Cultivars to Prevent Mixing and Preserve Purity

Preventing the mixing and degradation of cultivars is a relatively complex work, involving various working links of seed production, which is technical and lasts for a long time. To carry it out well, it is necessary to establish a sound superior variety breeding system, strengthen organizational leadership, develop rules and regulations, conduct planting planning, strengthen inspection and supervision and the construction of superior variety breeding teams, and formulate the measures to prevent mixing and preserve purity according to the causes of mixing and degradation of cultivars.

(I) Establish Strict Technical Operating Procedures to Avoid Mechanical Mixing

During the whole process of sowing, harvesting, storage, and packaging, the superior variety breeding procedures must be carefully observed, the rotation of breeding fields must be reasonably arranged, the receiving and distribution procedures of seeds must be strictly implemented, the treatment and sowing of seeds must be done well, and the occurrence of mechanical mixing must be eliminated from various working links such as sowing, harvesting, transportation, threshing, drying, storage, and packaging.

Continuous cropping should be avoided in seed production fields to prevent them from mixing with the seeds remaining in the soil from the last season after seedling emergence. Do not mistake the cultivar or cultivar grade in the process of receiving and distributing seeds, and strictly check the purity, cleanliness, germinability, and pest and disease of the seeds. In the case of doubt, it must be thoroughly solved before sowing. Seed treatment before sowing, such as seed selection, seed soaking, seed dressing, and other measures, must be done separately for different cultivars. The tools should be cleaned, with specialized personnel in charge assigned. When sowing, the same crop of different cultivars of propagation plots should be separated by a certain distance, and the sidewalk should be left in the breeding field for roguing. The operation specifications of the harvesting, transportation, threshing, drying, storage, and packaging should be carefully implemented, and the harvesting, threshing, drying, and storage of the breeding field must be arranged separately. Specialized personnel should be arranged to maintain the cleanliness of each operation's appliances and site, and conduct frequent inspections. The labels should be attached inside and outside containers and packaging materials.

(II) Take Strict Isolation Measures to Prevent Biological Mixing

In order to ensure the purity of good-quality seeds, when propagating seeds, strict isolation measures must be taken between varieties, cultivars, or types that are easy to cross with each other. Specific isolation methods include mechanical isolation, anthesis isolation, and spatial isolation.

1. Mechanical Isolation

Mechanical isolation is mainly applied to propagate small amounts of original seeds and seeds with high propagation factors and small usage of seeds. The method is to take the bagging, net cover, net room, etc. for isolation during the anthesis. Bagging is used for individual flower or inflorescence isolation. The bagging can be made of newsprint, sulfuric acid paper, and other tough, rain-resistant paper, depending on the size of the inflorescence. The net cover is generally used to cover individual plants and is mostly made of nylon mesh. Net room is used for population isolation and can be made of plastic tunnel skeletons covered with nylon mesh. When mechanical isolation is used, pollination problems must be solved for cross-pollinated vegetables. After the paper bag isolation, only artificially assisted pollination can be used. After the net cover isolation or net room isolation, both artificially assisted pollination and insect (e.g. bees, flies, etc.) assisted pollination can be used.

2. Anthesis Isolation

Anthesis isolation is also known as time isolation. Take certain cultivation measures and treatments to stagger the anthesis of different cultivars that are prone to hybridization in order to avoid natural hybridization. Generally, the methods of sowing and planting at different stages, vernalization and light treatment, topping, and pruning combined with the use of growth regulators are used. However, most horticultural crops have long anthesis, which makes it difficult to take this measure in production.

3. Spatial Isolation

Spatial isolation means that cultivars, varieties, subspecies, and species that are easy to have natural hybridization are separated from each other by a certain distance for reserving seeds in the open field. This approach is the most commonly used in the production of good-quality seeds. The distance of isolation should be determined according to the factors affecting natural hybridization and the size of the impact on the economic value of the product after hybridization.

(III) Adopt Correct Selection Methods and Seed Retention to Prevent Cultivar Deterioration

(1) Master the correct selection criteria and carry out regular roguing. Be familiar with the standard traits of the cultivar, and formulate the correct selection criteria before roguing. Only in this way can we achieve the purpose of selection and purification. Roguing is to remove plants that do not conform to the typical traits of the cultivar due to biological mixing, mechanical mixing, and genetic recombination or genetic mutation, as well as plants with pests and diseases and inferior plants with weak growth vigor. Roguing should be carried out during each fertility period, such as the seedling stage, vegetative growth vigorous stage, anthesis, fruiting stage, etc. The identification is made according to the characteristics of each cultivar in different fertility periods, among which the selection of the main economic traits (such as product organs) formation stage is the most important and strict.

(2) The original seeds used to propagate should have a larger population. Generally, it is required that at least 50 plants should be collected, and slight differences should be allowed between selected seed plants on the premise that the main economic traits are the same, especially those of product organs, to enrich their genetic basis to prevent too few plants from causing genetic drift and inbreeding.

(3) Implement a reasonable selection and seed retention system and adhere to the continuous directed selection. A reasonable selection system should be formulated and implemented, and the seed plants should be selected and eliminated successively and directionally every year according to the fertility period for the phenomenon that some units only care about seed retention, regardless of seed selection, only carry out rough piece selection, not strain

selection, or they may have inconsistent selection criteria, etc.; roguing should be carried out before the piece selection during the propagation of seeds used for production; use the original seeds to propagate seeds used for production; big plant seed collection method is used to produce the original seeds; the seed-to-seed method is used only to reproduce seeds used for production.

(IV) Improve Seed Collection Conditions to Prevent Cultivar Degradation

According to the principle of evolution of biological adaptation, in addition to preventing genetic degradation caused by mechanical mixing, biological mixing, genetic mutation, and genetic drift caused by small populations in the process of superior cultivar breeding, creating good conditions that are similar to the production of large fields for the growth and development of seed plants is a fundamental measure to prevent the degradation of cultivars.

In addition, some regions or seasons are capable of commercial production of horticultural crops, but the conditions for breeding good-quality seeds are not met. For example, when propagating the cultivars of cabbage and radish with bolting resistance in the southern warm winter areas, the phenomenon of not bolting, flowering normally will occur because the temperature is high in the winter and not have the conditions to make the seed plants pass vernalization. If the seeds are barely retained, the phenomenon of premature bolting (Fig. 1-9) will occur when the seeds obtained are used for the production of commercial vegetables. For another example, when potatoes and some asexually-propagating flowers are retained in warm areas, cultivar degradation often occurs due to infection with virus diseases.

The purpose of seed collection and cultivation is to harvest high-quality seeds, which requires more strict requirements for the seeding time, cultivation environment, and field management. The determination of the seeding time is to ensure that the development, flowering, and fruiting of the seed plants can occur in the most suitable season. For example, vegetables and flowers can be produced annually with the help of facilities, but the requirements for the propagation of good-quality seeds are more

Fig. 1-9 Premature Bolting of the Plant

stringent for the season. The purpose of strengthening field management is to enable the full expression of the main economic traits of each individual plant to facilitate selection based on the phenotype. When propagating original seeds, the cultivation measures favoring the enhancement of the full expression of the characteristics of the cultivar should be used, which sometimes are the opposite of propagating seeds used for production or producing commercial fruits and vegetables. For example, when propagating the original seeds of some disease-resistant cultivars, cultivation conditions for the identification of disease resistance should be provided to eliminate the susceptible strains, while when propagating seeds used for production or producing commercial fruits and vegetables, the occurrence of the diseases should be avoided so as not to affect the yield.

Review and Reflection Questions

(1) What aspects should we know about the relativity of variety purity?

(2) What are the specific reasons for the mixing and degradation of horticultural cultivars?

(3) What are the methods for preventing mixing and preserving the purity of plant cultivars?

Learning Scenario IV New Cultivar Protection of Horticultural Plants and Cultivar Certification

I. New Cultivar Protection

(I) The Significance of New Cultivar Protection

Excellent cultivars are one of the basic factors for obtaining a high yield, high quality, and high efficiency in agriculture and forestry production. On the one hand, protecting the rights and interests of breeders of new cultivars is of great significance in encouraging their enthusiasm. On the other hand, plant breeding is a project that takes a long time and requires more capital investment. At the present stage, the financial investment of China's scientific research system in the breeding business is still far from meeting the needs of breeding high-quality cultivars, and the implementation of the protection of new botanical cultivars will open up a compensatory way for the source of breeding funds. These are conducive to breeding more new cultivars of high quality, thus promoting the development of agriculture and forestry production. New botanical cultivars are in the category of intellectual property rights, and the formulation and implementation of protection regulations for new botanical cultivars is not only the respect for the labor of breeders and the protection of their rights and interests but also one of the measures to connect the relevant national economic regulations with international standards.

(II) International Measures for the New Botanical Cultivar Protection

Many countries around the world attach importance to the protection of new botanical cultivars and legally protect the interests of breeders by adopting legislative means. Only the breeder who has obtained the cultivar protection right has the right to breed, sell or transfer the cultivar. The name of the legislation varies according to different countries. The countries that adopted the cultivar protection law are Britain, the Netherlands, etc., and the countries that adopted the license protection law are Italy, Korea, etc. Some countries, such as the United States and France, use both laws.

For example, in Japan, the revised *Plant Variety Protection and Seed Act* implemented in 1978 stipulates that "new cultivars cultivated by individuals or public authorities in fruit tree breeding must be approved by the breeder for the sale of these seedlings after the registration of the new cultivars." "The seedling dealer has to pay the breeder a commitment fee to obtain the right of sales and undertake the production and sale of seedlings." "The new cultivars bred by the state while conducting seedling registration can be implemented through the window of the Fruit Tree Seedling & Clonal Association to accept fee business, and the compensated transfer fee for seedling and scion can be collected according to the production volume and handed over to the treasury."

(III) The Main Contents of the *Regulation of the People' s Republic of China on the Protection of New Varieties of Plants*

In 1997, China issued the *Regulation of the People's Republic of China on the Protection of New Varieties of Plants*, with eight chapters and 46 articles. The content includes the conditions for granting new plant variety rights; the application, acceptance, examination, and approval of plant variety rights; the interests and ownership of authorized cultivars; protection duration of plant variety rights and punishment for infringement; etc., all of which are stipulated and scheduled to come into force on October 1, 1997. The main contents are as follows.

1. Conditions for Granting New Plant Variety Rights

The new cultivars applied for authorization should be the genus or species of plants listed in the *National List of Plant Varities Protection*. Authorized new cultivars refer to plant cultivars that have been artificially cultivated or developed from the found wild plants, with novelty, specificity, uniformity, and stability, and are suitable for naming.

(1) Novelty: It refers to the reproduction material of the cultivar applied for that has not been sold before the application, or has not been sold in China for more than one year with the permission of the breeder; The vines, fruit trees, and

ornamental trees have not been sold outside China for more than six years, and other plants have not been sold for more than four years.

(2) Specificity: It refers to the plant cultivars that are clearly distinct from the known plant cultivars before filing the application.

(3) Consistency: It refers to the main trait characteristics that are consistent after propagation, except for allowable variation.

(4) Stability: It refers to the characteristic features that remain relatively stable after the repetitive propagation or at the end of a specific propagation cycle.

2. Application, Acceptance, Examination, and Approval of Plant Variety Rights

The administrative departments of agriculture and forestry of the State Council are responsible for the acceptance and examination of applications for new plant variety rights, granting plant variety rights, issuing plant variety right certificates, and registering announcements for new cultivars that meet the conditions.

(1) Application and acceptance: Chinese units or individuals can apply directly or entrust the agency. Foreign individuals or units applying for plant variety rights in China shall be handled in accordance with the relevant agreements between China and their countries. When applying new cultivars bred in China to foreign countries, Chinese units or individuals should register with the approval authority.

The application form, specification, and photos of the cultivar should be submitted to the approval authority in accordance with the prescribed format. The approval authority shall accept the application that meets the requirements, specify the application date, give the application number, and notify the applicant to pay the application fee within one month from the date of receipt of the application.

(2) Examination and approval: The examination and approval authority shall complete the preliminary examination according to the new cultivar authorization conditions within six months from the date of accepting the application, and make a public announcement to those who pass the preliminary examination and notify the applicant to pay the examination fee within three months. The application shall be rejected if the unqualified applicants are informed to present their opinions or make amendments within three months and still fail after the deadline reply or

amendment.

The approval authority will conduct a substantive examination after the applicant pays the examination fee, and those who meet the conditions will be granted the plant variety right, a certificate will be issued, and a registration announcement will be made. If the applicant is unsatisfied with the unqualified examination, he/she can request a review within three months.

3. Interests and Ownership of Authorized Cultivars

The regulations clearly specify that the units and individuals who complete the breeding enjoy exclusive rights for their authorized cultivars. Any unit or individual shall not produce or sell the propagation material of the authorized cultivars for commercial purposes without the permission of the owner who owns the plant variety right and shall not reuse the reproduction material of the authorized cultivar for commercial purposes in the production of the reproduction material of another cultivar.

The application right for new cultivars belongs to the unit when performing the unit's task with the use of the unit's material conditions to complete the job-related breeding; the application right for non-job-related breeding belongs to the individual; the plant variety right is stipulated by contract when finishing the breeding by entrustment or cooperation, and when there's no contract, the plant variety right belongs to the unit or individual who is entrusted or jointly finishes the breeding.

4. Protection Duration of Plant Variety Rights and Punishment for Infringement

Plant variety rights are protected for a certain period of time, and the regulations provide for 20 years for vines, forest trees, fruit trees, and ornamental trees, and 15 years for other plants from the date of authorization. During the protection period, if the owner of the plant variety right declares in writing to give up the plant variety right, fails to pay the annual fee as required, fails to provide testing materials as required, or the cultivar no longer meets the characteristics and features at the time of authorization, the approval authority may declare the termination of the plant variety right and register the announcement.

During the protection period of the authorized cultivar, if reproduction materials are produced or sold for commercial purposes without the permission of the cultivar owner, the cultivar owner or the interested party may request the administrative departments of agriculture and forestry of the government at or above the provincial level to handle the matter according to their respective authority, or may file a lawsuit directly to the people's court. Counterfeit authorization of the cultivar shall be handled by the administrative departments of agriculture and forestry of the government at or above the county level.

(IV) The Relationship between the New Botanical Cultivar Protection and Cultivar Certification

New botanical cultivar protection and cultivar certification are two regulations (ordinances) of different natures. For the crop species that have been stipulated to be qualified by cultivar certification before promotion, the new cultivars bred should pass cultivar certification before production, sale, and promotion even though they have been approved to grant plant variety rights.

This is the first time in China to protect the rights and interests of breeders by promulgating regulations on the new botanical varieties, and the practical implementation of which requires the formulation of corresponding rules and a certain period of time for publicity and education. However, the new botanical cultivars applied for variety rights are limited to those species listed in the National *List of Plant Varities Protection*, and not many species of horticultural plants are included in the list. Therefore, it is necessary to supplement and improve the list in the future.

II. Cultivar Certification

(I) The Significance of Cultivar Certification

Article 12 of Chapter 2 of the *Seed Law of the People's Republic of China*, adopted at the 16th meeting of the Standing Committee of the 9th National People's Congress on July 8, 2000, states: "The State implements a system for the new botanical cultivar protection, and grants new plant variety rights to those cultivars cultivated artificially or developed from discovered wild plants that possess novelty, specificity, consistency, and stability to protect the lawful rights and interests of the owners of new botanical cultivars. Specific methods shall be implemented in accordance with the relevant state regulations. The breeder shall receive the corresponding economic benefits as per the law if the selected breed is promoted and applied. "For a particular plant, before forming a new cultivar, it must be evaluated whether it has novelty, specificity, consistency, and stability, which involves the work of cultivar certification.

Cultivar certification refers to the examination of newly selected or newly introduced cultivars by authoritative specialized agencies and making a decision on whether and to what extent they can be promoted. "Major crop and forest tree cultivars should pass national or provincial certification before promoting and applying, and the applicant can directly apply for provincial certification or national certification. The major crop and forest tree cultivars determined by the administrative departments of agriculture and forestry of the people's governments of provinces, autonomous regions, and municipalities directly under the Central Government shall be examined and approved at the provincial level." "The major crop and forest tree superior varieties that pass the national certification shall be announced by the administrative department of agriculture and forestry under the State Council, and then they can be promoted to suitable ecological regions in the country. The major crop and forest tree superior varieties that pass the provincial certification shall be announced by the administrative departments of agriculture and forestry of the people's governments of provinces, autonomous regions, and municipalities directly under the Central Government, and then they can be promoted in suitable ecological areas within the administrative region; the introductions can be conducted in adjacent provinces, autonomous regions, municipalities directly under the Central Government of the same suitable ecological zone after the consent of the administrative departments of agriculture and forestry of the people's governments of provinces, autonomous regions, and municipalities directly under the Central Government." (Articles 15 and 19 of Chapter 3 of the *Seed Law of the People's Republic of China*). It can be seen that the cultivars must be qualified and published before they can be officially propagated and promoted. The vegetable category in China's horticultural plants was the first to implement the cultivar certification system. Fruit plants were also initially implemented in the past few years, while ornamental plants started relatively late due to the various species and complex situations, as well as the weak foundation of the original work.

The implementation of the cultivar certification system is conducive to strengthening the management of cultivars, promoting superior varieties according to local conditions, giving full play to the role of superior varieties, and realizing regionalization of cultivar layout, thus avoiding blindness in cultivar breeding and promotion and promoting production development.

The basis of cultivar certification is the cultivar test. New cultivars shall be subject to 2–3 years of multi-point regional tests and production tests to master their characteristic features, select the best ones that meet the requirements, and promote them in the adaptive area after certification. Therefore, cultivar tests are the essential intermediate links from breeding to the production of new cultivars, while cultivar certification is the decision of whether the tested cultivars meet the promotion requirements.

(II) Certification Agency and Its Work Content

Currently, China has set up the National Crop Variety Approval Committee and Forest Variety Approval Committee at both the national and provincial (municipalities directly under the Central Government and autonomous regions) levels (referred to as the variety approval committee) and set up Crop Variety Approval Teams at the local (city) level (referred to as the variety approval teams). Approval institutions are usually composed of representatives of agricultural and forestry administrative departments, seed departments, scientific research units, agricultural and forestry colleges, and other relevant units. Under the National Crop Variety Approval Committee, there are professional variety approval committees for each crop, including vegetables and fruit trees, and under the provincial variety approval committees, there are professional groups for each crop. The daily work of the variety approval committee is handled by a special agency set up by the agricultural administrative department at the same level.

The main work tasks of the variety approval agencies are as follows: ① leadership and organization of regional tests and production tests; ② a comprehensive review of the cultivars submitted for approval, and the decision on whether to promote and in what range to ensure that the new cultivars through the approval can play a greater role in production; ③ implementation of the *Seed Law of the People's Republic of China* and *Regulation of the People's Republic of China on the Protection of New Varieties of Plants* and provide opinions on superior variety breeding and promotion.

(III) Application Requirements and Procedures

1. Application Requirements

(1) After 2–3 years of continuous regional tests and 1–2 years of production tests, it showed stable traits and excellent overall traits in the tests. The cultivar applied for national cultivar approval should participate in the regional test and production test of agricultural crops and forest tree cultivars with excellent performance, passing the approval of one provincial cultivar approval committee; or the cultivar can be approved by two provincial cultivar approval committees.

(2) The cultivars submitted for approval are required to be more than 10% higher than the main promoted cultivars of the same type locally in terms of yield, or the yield increase is significant by statistical analysis, and other traits are equivalent to the control cultivars. The yield is similar to that of the main promoted cultivars of the same type locally, but one or even more traits such as quality, maturity stage, and resistance are significantly better than those of the control cultivars.

2. Application Materials

Fill in the application form carefully according to the requirements of the declaration of approval, which usually requires the attachment of the following materials: ① year-end summary report of regional and production tests each year; ② disease (insect) resistance identification report of designated professional units; ③ quality analysis report of designated professional units; ④ standard mapping photos and physical specimens of cultivar characteristics; ⑤ cultivation techniques and technical points of propagation (seed production); ⑥ copies of the qualification certificate of the cultivar approved by the next level of cultivar approval committee (group); ⑦ a sufficient number of original seeds.

3. Application Procedures

Application procedures usually include: ① application and signature by the breeding (introduction) unit or individual; ② review and identification seal by the breeder unit; ③ recommendation and signature from the unit hosting the regional tests and production tests; ④ approval and signature by the cultivar approval committee (group) in the region of the breeder.

(IV) Cultivar Certification, Naming, and Registration

1. Certification

Each professional committee (group) will hold a meeting to discuss and seriously review the cultivars submitted for approval and decide whether to pass the approval by secret ballot. When the number of votes the cultivars gain exceeds more than half of the total number of legal members (the members present must account for more than 2/3 of the members in the committee), these cultivars will pass the approval, and the committee shall organize the comments and submit them to the official meeting of the director and deputy director of the cultivar approval committee for examination and approval, and then issue the approval qualification certificate.

The certification of disputed cultivars shall be submitted to the next professional committee for review after the field visit. For example, if the breeding units or individuals have objections when failing the certification, they can further provide relevant information to apply for review, and the second review will not be conducted if the review is not passed.

2. Naming and Registration

The cultivars approved by the National Crop Variety Approval Committee and Forest Variety Approval Committee are registered and published by the State Ministry of Agriculture and the State Forestry Administration with uniform numbers; the cultivars approved by provincial certification are registered and published by the provincial (municipalities directly under the Central Government and autonomous regions) agricultural and forestry administrative departments with uniform numbers and reported to the National Crop Variety Approval Committee and Forest Variety Approval Committee for the record. The name of the new cultivar can be proposed by the breeding unit or individual, and the name will be designated by the variety approval committee. For the introduced cultivar, the original name or the exact translation will generally be adopted.

Review and Reflection Questions

(1) What is the significance of the new botanical cultivar protection?

(2) What are the main contents of the *Regulation of the People's Republic of China on the Protection of New Varieties of Plants*?

(3) What is the significance of cultivar certification?

(4) What are the application requirements and procedures?

Module 2　Production Principles of Horticultural Plant Seed

Learning Objectives

(1) Propose the correct specific methods to prevent and control mixing and degradation according to the causes of mixing and degradation of cultivars.

(2) Master the basic procedures and technical essentials for the production of original and good-quality seeds of conventional cultivars of horticultural plants.

(3) Master the application conditions and seed production techniques of horticultural plant hybrids.

(4) Master the main methods and techniques of asexual line seed production of horticultural plants.

(5) Have complex and systematic thinking and the ability to complete typical work tasks.

Learning Scenario I　Procedures and Methods of Seed Production

I. Classification of Horticultural Plant Seeds

In botany, a seed is a reproductive organ that develops from an ovule. In agricultural production, as production materials, seeds have a much broader meaning than in botany and generally refer to plant organs that can be used directly as seeding material. The *Seed Law of the People's Republic of China* states: "The seeds mentioned in this seed law are the planting material or reproduction material of crops and forest trees, including grains, fruits, roots, stems, seedlings, buds, leaves, flowers, etc." Generally, agricultural seeds, including seeds of horticultural plants, can be classified into the following types.

(I) True Seeds

True seeds, which are botanically referred to as seeds, develop from ovules, such as legumes (with a few exceptions), various cruciferous vegetables, melons, eggplants, tomatoes, peppers, tea, citrus, pears, apples, ginkgoes, etc.

(II) Seed-like Fruits

It refers to the dried fruits of some horticultural plants that are not dehiscent when ripe and can be used directly as seeding material, such as sunflowers, large achenes, schizocarps of the Apiaceae (e.g. carrots and celery), nuts of the beech family (e.g. chestnuts) and the Chenopodiaceae (e.g. beets and spinach), and drupes with lignified endocarp of the Rosaceae.

(III) Vegetative Organs Used for Propagation

Many rhizome crops have organs for natural asexual propagation, such as the root tubers of sweet potatoes and Chinese yam (*Dioscorea polystachya*), the tubers of potato and *Jerusalem artichoke*, the bulbs of taro and *Sagittaria Trifolia*, the bulbs of shallot, garlic and onion, the terrestrial stem of sugar cane and cassava, and the rhizomes (lotus roots) of lotus. Most of the above-mentioned crops can also flower and bear fruit, but except for a few cases, such as crossbreeding, good-quality seed purification, and rejuvenation using seed propagation, they are generally planted with their vegetative organ to give full play to their special advantages.

(IV) Artificial Plant Seeds

Artificial plant seeds are the embryos (mainly referred to as somatic embryos) generated by culturing plants *in vitro*, wrapping them in substances containing nutrients and with protective functions to form granules that can germinate and grow into normal plants under suitable conditions. They are also known as synthetic seeds, artificial seeds, or asexual seeds. Artificial seeds are very similar to natural seeds, both of which consist of viable embryos and external structures with nutritional and protective functions (equivalent to endosperm and testa) to form a granule suitable for sowing or reproduction, and that's why they are called artificial seeds. Artificial seeds are essentially in asexual reproduction and have many advantages over natural seeds: Firstly, they can be used to reproduce special plants that do not bear fruit under natural conditions or have very expensive seeds; secondly, they can be propagated quickly, for example, with a fermenter of 12 L volume, the somatic embryos of carrots produced in more than 20 days can make ten million artificial seeds, which can be planted on dozens of hectares; thirdly, they can fix the heterosis, so that F_1 hybrids can be used for multiple generations, etc.

II. Original Seed Production Procedure for Conventional Cultivars

(I) Seed Grading

Currently, there are three levels of seed classification in China: breeder seeds, original seeds, and good-quality seeds.

1. Breeder Seeds

Breeder seeds are the first batch of parent seeds or cultivars with stable genetic traits bred by breeders, which are used for further propagation of original seeds. The typicality for the cultivar of breeder seed is the strongest. The main characteristics and features of its plants performed under good growth conditions are the cultivar's or parent seed's trait criteria.

2. Original Seeds

Original seeds refer to the first to the third generation of seeds propagated by the breeder seeds according to the technical operating procedures, or the seeds produced according to the original seed production technical procedures stipulated by China to meet the original seed quality standards. In China, original seeds are divided into the original seed of the first generation (equivalent to the foundation seeds in other countries) and the original seed of the second generation. Original seeds are used to propagate good-quality seeds that play a vital role in seed production, so there are strict requirements for their propagation generations and commercial quality. The quality standards of the original seeds of various crops in China are mainly determined by four indicators: purity, cleanliness, germination rate, and moisture.

3. Good-quality Seeds

Good-quality seeds refer to the first and second generation of seeds that can meet the quality standards for good-quality seeds, and the F_1 hybrid seeds that meet the quality standards for hybrid good seeds. These are propagated from conventional original seeds according to the technical operation procedures. Good-quality seeds are used for large-scale production and seed market trading and are the main commercialized seeds, so they are also called seeds for production.

(II) Original Seed Production Procedures of Horticultural Plants

The formation process of a cultivar according to the sequence of breeding stages and different generations is called the superior variety breeding procedure. In the current production of original seeds, there are two different procedures or technical routes: one is the repeated propagation method of original seeds, and the other is the circular selection method.

1. Repeated Propagation Method

The repeated propagation method refers to the direct production and supply of breeder seeds by breeding units or breeders every year, which can fundamentally guarantee the quality and typicality of the seed source. Breeding units or breeders should pay attention to the production and preservation of the original seeds, which can adopt the method of producing in one year, storing for many years, and using in several years to maintain the characteristics of the cultivars.

(1) Basic procedure of the repeated propagation method. The repeated propagation method, also known as the purity preservation propagation method, implements graded propagation starting from the breeder seeds to the production

seeds. Seeds of each grade can only be planted once. After that, the next grade will be planted, and each grade cannot reserve seeds for planting. In this way, it is possible to propagate up to four generations from the breeder seed to the production seed, and the next round of seed production still follows this procedure.

(2) Characteristics of the repeated propagation method. The repeated propagation method requires strict measures and testing systems for preventing mixing and preserving the purity during the whole process of producing the original seeds to minimize the probability of mechanical and biological mixing.

When using the repeated propagation method in producing original seeds, except for the necessary roguing, there is no need to carry out artificial selection due to the good quality of the seed source, which will not cause gene loss. Further production of production seeds from this can maintain the purity and characteristics of the cultivar. Due to the limited number of original seeds, this method of producing original seeds for some crops with small propagation coefficients has to undergo multiple generations of propagation before being put into production, which is time-consuming and will increase the chance of mixing and degradation.

The repeated propagation method can not only apply to the production of seeds of conventional cultivars of self-pollinated crops and often cross-pollinated crops but can also be used for the production of purity preservation of inbred line and "three-line" parental seeds.

2. Circular Selection Method

The circular selection method is actually an improved mixed selection method, which is more effective than other methods for the production of original seeds of cultivars with more serious mixing and degradation.

(1) Basic procedure of the circular selection method. First produce original seeds through "individual selection, plant-to-row identification, interbreed" after the mixing and degradation of a cultivar or the promotion and application of new cultivars, and then expand the propagation of production seeds, thus circling and purifying the production of original seeds. This original seed production procedure is commonly used for self-pollinated crops or often cross-pollinated crops.

(2) Basic method of the circular selection method. There are three methods for the circular selection method, such as the three-nursery system, the two-nursery system, and the one-nursery system.

① Three-nursery system. This method consists of four steps: individual selection, plant progeny row identification, strain identification, and interbreeding.

A. Individual selection: The material for individual selection must be a cultivar with a promising use that is widely promoted in production, or a cultivar that performs well in tests and demonstrations, and is ready to be promoted. Choose to perform this in the field of superior variety with a high variety purity, preferably in the first and second generation of the original seed. To facilitate selection, the selection nursery planting population should be large, sowing should be low-density, and excellent cultivation techniques should be adopted to facilitate the full expression of plant traits. In addition, selectors must be familiar with the characteristics, features, and typicality of the cultivar. Based on the characteristics of the cultivars, selectors should select the typical plants with robust plants, good productivity, strong disease resistance, proper fertility period, and full grains for harvesting, numbering, and seed testing. The individual plants are stored separately after the final selection for the next year's plant progeny row identification. To ensure that the selected population does not deviate from the original cultivar's typicality and produce enough original seeds, the number of selections should be large.

B. Plant progeny row identification: Select a field with flat terrain and uniform fertility, and plant the individual plants selected in the previous year in the nursery at low density, or carry out single-seed punch planting, planting one or several rows per plant. The field management is uniform and consistent, and observation and comparison are carried out in each vital period of growth and development, keeping the superior ones and eliminating the inferior ones according to the typicality and neatness of the plant progeny rows. Carry out strictly roguing, mixed harvest, respectively threshing, and seed testing in the selected plant progeny rows and store separately for the following year's strain comparison test.

C. Strain identification: The plant progeny rows selected in the previous year should form a single line and be planted separately at low density in the strain nursery in the form of one plot per line. Further comparative trials on its typicality, productivity, and adaptability in the field, as well as the cultivation management, observation, and evaluation are the same as in the nursery. Select the superior and remove the inferior based on the overall performance of the strain. Perform the mixed harvest and storage after strict roguing of the selected strains. If there are differences between the strains, they can also be harvested in separate strains, and after the indoor identification and final selection, the selected strains are mixed.

D. Interbreeding: Plant the seeds from the previous year's mixed harvest in the foundation seed nursery to expand propagation. The foundation seed nursery requires safety isolation, good fertilization conditions with low-density planting, and other technologies to improve the propagation coefficient. Strictly roguing in the field, and the threshing, drying, and storage after harvesting must be arranged separately, from which the original seeds are produced.

As a basic method of original seed production, the "three-nursery system" also has some drawbacks. This method has a long period for producing the original seeds; the amount of seeds obtained is small, its costs and seed costs are high. If the number of individual selections is too small, it will easily lead to genetic drift in the population and destroy the genetic stability of the cultivar. In addition, the purifier's knowledge of the typicality of the cultivar is generally less than that of the breeder, so deviation is inevitable in selection. Special attention should be paid to these factors that may affect the selection effect in practice.

② Two-nursery system. The "two-nursery system" is the same as the "three-nursery system" except for the omission of the strain nursery. It is a common method for producing original seeds of self-pollinated crop purification. After roguing and mixed harvest of plant progeny rows, the original seeds are directly mixed and propagated in the foundation seed nursery for production.

③ One-nursery system. The theoretical basis of the one-nursery system is that self-pollinated crop cultivars are a relatively genetically stable population. The genetic balance of cultivars can be maintained by removing the impure and inferior plants caused by mechanical mixing and genetic mutations during the seed propagation process. The one-nursery system is a method for the rapid propagation of original seeds. Its production procedure can be summarized as "single-seed punch planting, ramet identification, whole plant roguing, mixed harvest". To a certain extent, the adoption of the one-nursery system overcomes the disadvantages of the three-nursery system, such as long production period, high cost, narrow genetic base caused by the selection of a few individual plants and genetic drift caused by artificial directional selection, and it is simple, easy to implement, efficient and practical.

III. Good-quality Seed Production Procedure for Conventional Cultivars

Compared with hybrids, the good-quality seed of conventional cultivars is propagated stably and can be reused, so there is no need to change seeds every year. The principle and technology of good-quality seed production are similar to that of the original seed production. Still, its production procedure is much simpler than the latter. It will be generally performed under appropriate isolation conditions to prevent mixing and preserve purity, expand reproduction, and provide field production seeds.

(I) Seed Production Field Selection

1. Seed Production Field Selection

In order to obtain high-yielding, high-quality seeds, seed production fields should have the following conditions: Natural climate, soil conditions, and others suitable for the growth and development of the crop and cultivar; flat terrain, fertile soil, convenient drainage and irrigation, stable yields; no serious diseases and pests, weeds, and other hazards; no quarantine pests and diseases; the crop rotation can be used for crops that are not suitable for continuous cropping; concentrated in a row, convenient transportation, with good isolation conditions.

2. Types of Seed Production Fields

There are two types of seed production fields for conventional cultivars: primary seed production fields and secondary seed production fields.

The primary seed production field procedure is simple and suitable for small grain crops with high propagation coefficients. In addition to regular renewal with the original seeds, only the mixing selection of individuals is made in the seed production fields each year, and the selected plants are used as seeds for the next year's seed production field, while the rest are mixed and threshed as seeds for the field after strict roguing. The primary seed production field has the characteristics of less use of land, fewer propagation generations, less producing seed, and a lower probability of cultivar mixing and degradation.

The secondary seed production fields are suitable for crops with small propagation coefficients. In the first year, a mixed selection of individuals shall be made in the primary seed production field. The selected plants are mixed as seeds for the primary seed production field in the following year, while the rest are used as seeds for the secondary seed production field in the following year after strict roguing and mixed harvest. In the second year, the mixed selection of individuals should be continued in the primary seed production field, and the previous year's process is repeated. Perform the mixed harvest after strict roguing of the secondary seed production field and use them for sowing in the crop field. In contrast to the primary seed production field, the secondary seed production field has the disadvantage of more use of land, increased propagation generations, and a higher probability of mixing and degradation. However, it provides more seeds used for production, and since two rounds of expansion and propagation are carried out, the number of individual plants selected can be appropriately reduced compared to the primary seed production field, thereby ensuring excellence in selection quality.

When producing good-quality seeds, remember that the propagation generations of original seeds in the seed production field should not exceed three. Otherwise, the quality of good seeds cannot be guaranteed.

3. Area of the Seed Production Field

The area of the seed production field should be determined mainly according to the seed production plan and the propagation coefficient of the cultivar. In order to fully ensure the seed supply plans, a certain margin should be left when performing the specific arrangements. The cultivar of crop seed production field surplus area can refer to the following ratio: rape (0.3%–2%), rice (5%–10%), wheat (7%–10%), corn (3%–5%), cotton (15%–20%), millet (1%–2%), and potatoes (8%–10%).

(II) Prevent Mixing and Preserv Purity

Preventing mixing and preserving purity are the most basic requirements of superior variety production. In the production of superior variety, in order to prevent mechanical mixing and fulfill reasonable isolation of seed production fields to prevent out-pollination mixing, the following links should be carefully noted.

1. Conduct a Mixed Selection of Individuals

The method of the mixed selection of individuals is similar to the original seed production, mainly based on the characteristics and features of the original cultivars, taking into account the robust growth, consistent maturity, full grains, no pests and diseases, etc., and selecting the best on the basis of purification. However, different from the selection in original seed production, it has a larger number of selections. Therefore, it is important to adhere to the criteria to ensure selection quality.

2. Strictly Roguing

Roguing is very important in seed production fields, usually performed several times during the seedling, flowering, and maturity stages. Full morphological characteristics mainly characterize the maturity stage of self-crossing crops. The roguing of cross-pollinated and often cross-pollinated crops must be made in time before flowering and dispersal of pollen to avoid biological mixing.

3. Regular Seed Renewal

Regular seed renewal is a fundamental measure to ensure seed purity. Although the combination of the above measures can effectively maintain the cultivar's characteristics and purity to a certain extent, mixing and degradation are inevitable as the number of propagation generations increases. Therefore, the seeds in the seed production field should be renewed with their original seeds after 10 generations (generally three generations) to ensure the yield potential of the cultivar and extend the utilization life of the cultivar.

IV. Accelerated Propagation

The original seed number of newly bred, newly introduced, or purified superior cultivars is very small, and it isn't easy to rapidly promote and apply. Propagation using conventional methods is slow and inefficient. Therefore, it is very important to adopt special methods to accelerate reproduction. There are two types of methods to accelerate propagation: one is to expand the propagation coefficient, such as low-density sowing propagation, tillering planting, tissue culture, and other methods; the other is to use natural conditions for off-site and off-season propagation to increase the number of propagation.

(I) Low-density Sowing Propagation

On the one hand, low-density sowing propagation can save the amount of seed used. Under the same number of seeds, using low-density sowing with careful management, seedling transplantation, and individual plant planting can plant a larger area and improve the utilization rate of seeds. On the other hand, it can expand the individual growing space and nutrient area to improve the productivity of an individual plant, thereby increasing the propagation coefficient and speeding up the promotion of good-quality seeds.

(II) Tillering Planting

Crops with the tillering habit can be tillered once or more times to improve propagation coefficient by sowing earlier and promoting tillering. Potatoes, sweet potatoes, and other asexually-propagating crops can be planted by buds, seed pieces, clumping, cuttings, or multiple-time branching.

(III) Tissue Culture

The tissue culture technique is based on the characteristics of cell totipotency, and the plant roots, stems, leaves, flowers, fruits, and even endosperm of seeds are cultured into a complete plant under sterile conditions. Tissue culture techniques are currently used to rapidly propagate many plants. Tissue culture is used for rapid propagation in the following three situations.

(1) The direct differentiation of adventitious buds from epidermal and mesophyll cells' roots, stems, and leaves. Then, the rooting is induced, and finally, forming complete plants.

(2) To cultivate terminal buds, axillary buds, or lateral buds to differentiate bud clusters and proliferate mass buds through subculture. Then, remove these buds and put them into the rooting medium for root induction, and finally, complete plants are formed.

(3) To take young tissues of the plant body for culture *in vitro*. Carry out the dedifferentiation culture to make them produce callus. Then, put them into a differential medium to induce the callus to produce buds and roots or to produce embryos for further growth into small plants.

The main advantages of tissue culture technology are higher propagation coefficient, less material used, better maintenance of the original cultivar characteristics and features, and the effect of removing viruses and improving yields without the limitation of seasons and regions.

(IV) Adding-generation Propagation

Using the characteristics of vast territory, complex terrain, and large differences of various regions in China, conduct off-site propagation, off-season propagation, and multiple generations of propagation in one year to speed up the superior variety breeding process. Artificial climate chambers can also be used for rapid adding-generation propagation of precious materials in different seasons.

V. Seed Production of Asexual Line Cultivars

(I) Genetic Characteristics of Asexually-propagating Crops

The offspring system produced by an individual plant through asexual propagation is called the asexual propagation line (referred to as the asexual line). The main genetic characteristics include the following aspects.

1. Complex Genetic Basis

The asexual lines of potatoes, sweet potatoes, and other crops are generally hybrid offspring, and even if they are self-crossed crops, they are mostly not self-crossing homozygous. Therefore, their genetic characteristics are as follows: the genotype of asexually-propagating crops is heterozygous, and the genetic basis is quite complex.

2. Stable Traits and Not Easy to Separate

No matter whether homozygous or heterozygous the mother's genetic base is, the offspring produced by asexual propagation is similar to the mother in phenotype without segregation. In this way, all plants within an asexual line are identical in their genetic basis and have the genotype and characteristics of the original parent, i.e. the mother. The asexually-propagating crops can adopt the same selection methods as self-pollinated crops to preserve pure cultivars in the superior variety breeding.

3. Crossbreeding by Sexual Propagation

Asexually-propagating crops can also be sexually-propagating for crossbreeding under suitable natural and artificially controlled conditions. At this time, whether self-pollinated potatoes or cross-pollinated sweet potatoes, there is a large separation in the F_1 hybrid because their parents turn out to be heterozygotes with a complex genetic base, thus providing rich variable materials for breeding selection.

In the segregation of offspring, the superior traits can be rapidly stabilized and become a new asexual line by selection and asexual propagation once the superior variation appears.

(II) Characteristics of Asexual Line Seed Production

Asexual propagation methods and asexually-propagating crops are the results of long-term natural and artificial selection during biological evolution. Its existence indicates that it is progressive and adaptive, therefore of great significance in evolutionary, genetics, and productive terms.

1. Strong Stress Resistance and Adaptability

In the production of asexual line seeds, there is no need to undergo the process of flowering, pollination, fertilization, formation of fruits and seeds, as well as seed development and maturation, so they have greater stress resistance and adaptability.

2. High Seed Purity

Asexual line seeds can fundamentally maintain the excellent characteristics of the mother, and the offspring can achieve the ideal state of "keeping the tradition intact". Because the genotype of the offspring and the mother is theoretically the same. The genotypes of individuals within the same asexual line are also the same, with neat consistency and purity up to 100%. This is also the reason why modern clone technology is extremely active.

In general, the asexual line seeds have a larger volume, lower propagation coefficient, higher sowing amount, and higher cost in production, and their seeds are not easy to save (because they are fresher and livelier and have higher moisture content). However, the roots, stems, and buds of asexually-propagating crops are strong in differentiation and can often be used for cutting propagation by cuttings, split stems, and other materials, which can be produced annually in a factory manner.

(III) Categories and Characteristics of Seed Production of Asexually-propagating Plants

In the production of asexually propagated horticultural plants, there are two types of reproduction material: one is the direct sowing of specialized roots and stems as seeds in the crop field; the other is the planting of seedlings propagated by grafting, cuttings, layering, and other nursery methods in the crop field. According to the definition of the *Seed Law of the People's Republic of China*, these seedlings should be included in the category of seeds.

Horticultural plants with specialized roots and stems as sowing materials exhibit the production characteristics of annual and biennial plants, i.e. sowing in the same year and harvesting the product in the same year or the following year, such as many vegetable and flowering plants. Seeds for propagation need to be produced and sown annually. It is easy to produce degradation of the cultivar as the number of propagation generations increases.

Horticultural plants that use cultivated seedlings as sowing materials exhibit the production characteristics of perennial plants, with one year of planting and multiple years of harvesting; the majority of fruit trees, some woody flowers, and woody condimental vegetables belong to this category. In production, it often shows degradation due to virus infection.

No matter for those species with specialized roots and stems as reproduction materials, horticultural plants with cultured seedlings as reproduction materials, or even those horticultural plants with true seeds or fruits as sowing materials, micropropagation (tissue culture) can be used to accelerate propagation or to remove the virus for overcoming the degradation caused by virus infection.

(IV) Seed Production Methods for Specialized Root and Stem Propagation Plants

The main vegetative organs of higher plants are roots, stems, and leaves, but some plants also have some special vegetative organs that form specialized roots and stems. These organs are usually grown underground and include bulbs, corms, tubers, root tubers, rhizomes, and pseudobulbs. Specialized roots and stems have two functions, one of which is to store nutrients. The organic nutrients produced by the photosynthesis of leaves during the vegetative phase are transferred to the underground part, forming an enlarged fleshy organ that stores a large number of nutrients to survive cold, drought, and other adverse weather, and then use the stored nutrients for germinating and growing in a suitable environment. The other is for vegetative propagation, some plants can use these specialized roots and stems for propagation, and some need to separate and cut these organs manually for propagation.

(V) Production Methods of Seedlings of Perennial Horticultural Plants

1. Layering and Burying

Layering propagation is a propagation method in which branches from the mother plant are buried in soil or wrapped with other moist material to facilitate the rooting of the buried (or wrapped) part of the branch for the formation of a new independent plant. Layering is commonly used to propagate shrubs, fruit trees, and rootstocks. With the dwarf culture of various fruit trees, the propagation of dwarf rootstocks is very important. In order to make the dwarf rootstocks perform consistently, it is not advisable to propagate them by seeds, and the asexual line of dwarf rootstocks needs to be developed, so layering should be used for propagation. The biggest difference between burying and layering is whether the root of the shoot is cut. Burying is to bury very long root-cutting shoots in the soil closer to the surface to promote the shoots' rooting, germination, and growth. The shoots are longer, contain more nutrients, and have no roots, so for the same plant, rooting is more difficult than layering, but the method of burying is labor-saving, and the seedlings are neat, so it is mostly used to propagate plants that are easier to root.

2. Cutting Propagation

Cutting propagation is a propagation method that inserts stems, roots, leaves, and other vegetative organs of plants into the sand, vermiculite, or other substrates *in vitro*. In the absence of the mother, this part of the vegetative organ regenerates the missing other parts and forms a completely new plant under certain conditions. This method of asexual propagation is called cutting propagation, and the seedlings obtained are cutting seedlings.

The types of cutting propagation are branch-cutting, root-cutting, leaf-cutting, etc. Branch-cutting is the most widely used in nursery production practice, so this is often called cutting. Root-cutting and leaf-cutting are rarely used and are generally only used in flower propagation.

The characteristics of the cutting propagation are fast propagation, multi-season and mass nursery, economical use for maternal reproduction material, and generally simple; it can be more consistent in maintaining the genetic characteristics of the mother without the problem of grafting propagation of rootstock affecting the scion; the seedlings grow quickly and the root system is more developed; start flowering and fruiting earlier than the seedling plants. It is commonly used on fruit trees and flowering plants such as grapes, cherries, winter jujubes, Chinese roses, chrysanthemums, kiwis, and pineapples.

3. Grafting Propagation

Grafting is a biological technique that combines two plant parts so that the two parts form a whole and the two plants become the same plant. In grafting, the lower part usually forms the root system, which is called the rootstock; the upper part usually forms the canopy, which is called the scion. When the scion is the shoot, it's called stem grafting, while when the scion is a bud, it's called budding.

The application of grafting technology for horticultural plants has many advantages: First, for the fruit trees, flowers, and garden trees that are not easily rooted by cutting, grafting can realize the purpose of maintaining good characteristics and making the offspring population traits neat and consistent by grafting to achieve asexual propagation. Second, it can achieve high yields in the early stage of fruit trees and forest trees, which is manifested in the early fruiting of fruit trees, and the early seed collection of some forest trees, mainly because the scion used for grafting has passed the juvenile period. Third, the selection of different rootstock types in grafting can make fruit trees achieve dwarf culture, and some dwarf trees grow taller, as well as some ornamental species become pendulous canopy (such as Chinese pagoda tree, *Ulmus pumila* L. cv. Tenue, *Prunus persica f.* Pendula, *Prunus serrulata* cv. Pendula, *Zizuphus jujube* cv. Tortuosa, *Morus alba* cv. Pendula, etc.). Fourth, grafting can improve the cold, drought, flooding, salt and alkali, and pest resistance of plants with the help of rootstock. For example, using *Vitis amurensis* as rootstock can improve its cold resistance. By grafting, melon plants can improve the ability of fusarium wilt resistance and other soil-borne diseases and improve the cold and drought resistance of greenhouse overwintering melon cultivation. At the same time, grafting cultivation is also a means of rapidly propagating asexual lines. Theoretically, each bud can be propagated into a tree as long as there are sufficient rootstocks, thus achieving rapid seedling breeding. In addition, it can also achieve the purpose of changing the bad into the good and improving the cultivars through grafting, and it can save dying trees through bridge grafting technology.

4. Sucker Propagation

In the asexual propagation of plants, sucker propagation is the simplest and easiest method. Many plants have the characteristics of sucker propagation, which enables many sprout tillers (stem tillers) to grow at the base of the stem, forming many small plants connected to the mother that form large bushes together and can be cut separately into a number of small plants by ramet; some arbor species can easily produce root seedlings, such as jujube tree, *Rhus typhina*, *Ailanthus altissima*, *Robinia pseudoacacia*, etc., which can be used to propagate by sucker propagation; strawberry, Indian strawberry, and some Gramineae herbaceous plants can produce stolons, and stoloniferous seedlings can be produced on these stems, which can be used for sucker propagation; some tropical plants such as bananas and pineapples have the characteristic of meristematic plants, which can also be used for sucker propagation.

Review and Reflection Questions

(1) What are the types of horticultural plant seeds?

(2) What are the current levels of seed classification in China?

(3) What are the circular selection methods in the production of original seeds of horticultural plants?

(4) What is the production procedure of good-quality seed for conventional cultivars?

(5) What are the methods of accelerated propagation?

(6) What are the genetic characteristics of asexually-propagating crops?

(7) What are the characteristics of asexual line seed production?

(8) What are the production methods of seedlings of perennial horticultural plants?

Learning Scenario II Hybrid Seed Production

I. Conditions and Categories for Hybrid Seed Production

(I) Conditions for Hybrid Seed Production

In the use of heterosis, regardless of the pollination method of horticultural plants and the type of hybridization, in order to obtain excellent hybrids and give full play to their role, the following conditions must be present.

1. Hybrid Combinations with Strong Advantages

There should be a strong hybrid combination before utilizing heterosis, which is the basis and premise of utilizing heterosis. Otherwise, if the hybrid performance is mediocre or even inferior to conventional cultivars, the hybrid will lose its value. The strength of the heterosis and the quality of the hybrids include yield advantage, resistance advantage, quality advantage, adaptability advantage, and fertility period advantage. Generally speaking, as a cultivar, hybrids should have better stability, adaptability, and overall agronomic traits in addition to ideal yield traits.

2. Hybrid Parents of High Purity

The purity of the parents is directly related to the expression of the advantages of hybrids. Only when the parents have high homozygosity can hybrids be neat and consistent and give full play to their increasing yield performance. On the contrary, when the parental homozygosity is poor, the F_1 hybrids exhibit segregation, which will affect yield. With the promotion of the hybrid, parental purity may worsen year by year, which requires the establishment of a parental reproduction system in the formulation of hybrids at the same time and strict measures and management system to prevent mixing and preserve purity for seeds.

3. Low Propagation Seed Production Cost and the Easy and Simple Process

First, there should be a simple and easy propagation method for parent cultivars. Hybrids need to produce seeds annually, so it is necessary to ensure that sufficient numbers of parents are provided each year to meet seed production needs. Among them, mother propagation is more important. For example, if the male sterile line or self-incompatible line is used for seed production, other methods for propagation are required because the mother cannot be self-fruitful. This particular method of propagating should be simple, easy, low-cost, and effective. Second, there should be simple hybridization techniques. Most horticultural plants are hermaphrodite. The mother emasculation should be made to prevent self-crossing, which is the primary issue in the formulation of hybrids. A simple and easy emasculation method is necessary to produce hybrid seeds on a large scale to meet production needs. If the hybridization techniques to obtain hybrid seeds are very complex, which require a lot of investment in human and material resources, or cannot solve the emasculation problem well, these will not guarantee the quality of seeds or will reduce the cost of seeds, thus affecting the promotion and application of hybrids. The hybrid seed production technology has become easier and more convenient with the research and utilization of male sterile lines and chemical hybridizing agents, which bring great convenience for self-pollinated crops to take advantage of heterosis in production.

4. Higher Seed Yields

Seed yield is directly related to the cost, which will greatly affect the promotion and application of hybrids. Generally, the yield increase benefits of hybrids should be sufficient to compensate for the increased inputs of producing hybrids. Many factors affect the seed yield of hybrids, such as parent growth vigor, whether the anthesis can properly meet the ability of the mother to receive pollen, the dispersal time of the father, the ratio of the parents, and whether the cultivation management is appropriate.

(II) Types of Hybrids

Hybrids can be classified into several types due to the different types and numbers of parents involved in formulating hybrids.

1. Interspecific Hybrids

Interspecific hybrids refer to hybrids that are composed of two parental cultivars. Within the cultivar population of the self-pollinated crops, the genotype is homogeneous, and the phenotype is consistent among individuals. The hybrid is a superior cultivar as long as the combining ability is more desirable. As for the interspecific hybrids of cross-pollinated crops, due to the poor homozygous genotype of the parental cultivars, the heterosis is not too strong, the performance traits between individual plants are not neat, and the yield increase is not large, which is only 5%–10% higher than that of general open-pollinated cultivars (conventional cultivars).

2. Top Cross Hybrids

Top cross hybrids refer to hybrids that are obtained by crossing a cultivar (open-pollinated cultivar) with an inbred line. Generally, the superior cultivars promoted locally are used as the mother, and the other is an inbred line, which is used as the father. Top cross hybrids are more adaptable and have higher seed yields. The yield and neatness of top cross hybrids are better than interspecific hybrids but not as good as an interlineal hybrid.

3. Interlineal Hybrid

Interlineal hybrids refer to hybrids that use different inbred lines as parents for mating. They can be divided into the following four types according to the number of parents and different ways of mating.

(1) Single cross hybrids. It refers to the hybrids composed of two inbred lines, and the mating method can be expressed as A×B. The parental inbred lines of single cross hybrids are highly genotypically homozygous, so the traits of hybrids are neat and consistent, with strong heterosis and large yield increase, and the seed production procedure is relatively simple, which is the main type of heterosis utilization at present. However, it requires a long time to select inbred lines, and the seed yield is low, so the seed cost is high.

(2) Triple cross hybrids. It refers to a hybrid mated by three inbred lines. The mating method is (A×B)×C. That is, A and B are crossed first to form a single cross hybrid, and then this is used as the mother and the inbred line C as the father to form a triple cross hybrid. This type of seed production is more productive. The yield performance of the triple cross hybrid is not as good as that of the single cross hybrid, and it even needs one more isolation area for seed production than the single cross hybrid.

(3) Double cross hybrid. It refers to the hybrids composed of four inbred lines. The mating method is (A×B)×(C×D). That is, two single cross hybrids are mated first, and then two single cross hybrids are mated to make the double cross hybrids. The double cross hybrids have greater heterosis, better adaptability, and higher seed yield, but the yield and neatness are not as good as the single cross hybrids, and the seed production procedure is more tedious.

(4) Synthetic hybrid. Synthetic hybrids are the hybrid offspring population with genetic balance formed by a number of good inbred lines (generally not less than eight) or hybrids between inbred lines being mixed and sown in an isolation area and after sufficient free pollination and multiple mixed selections. The heterosis of synthetic hybrids is less than that of single cross hybrid, triple cross hybrid, and double cross hybrid, but the genetic base of synthetic hybrids is broad. Hence, they are highly adaptable and stable and have a simple seed production procedure. The heterosis of the F_2 generation is not significantly reduced, allowing for the continuously planting of a single seed production over many years without the need for annual seed production as with single cross hybrid.

II. The Technical Essentials of Hybrid Seed Production

(I) Land Selection and Isolation

1. Requirements for Land Selection

In order to match the hybrid seed production system, the first step is to determine the number and area of parental breeding fields and seed production fields for the hybrids to be formulated. For example, two inbred-line breeding fields need to be prepared to plant the inbred lines of the mother and father separately for producing single cross hybrids by artificial emasculation. At the same time, a sizeable single cross seed production field needs to be prepared to plant the single cross hybrids of the mother and father for seed production. Generally, the area of the inbred-line breeding field of the mother and father should be determined according to the seed production task of the following year. Then the seed production field area should be determined according to the number of seeds of the existing parents. In order to ensure the quality and quantity of the produced seeds, the breeding field, and seed production base should be selected from places with flat terrain, fertile and uniform soil, convenient drainage and irrigation, stable yields, and no serious diseases, pests, rats, birds, and other hazards, no quarantine pests and diseases, easy isolation, convenient transportation, higher production levels and production conditions, better labor, and technical conditions.

2. The Setting of the Isolation Area

Both parental breeding and seed production of hybrids must be safely isolated. Besides the use of net rooms for small areas, the following methods should be used flexibly according to the actual local situation.

(1) Spatial isolation. It is required that non-father cultivars are not planted within a certain distance around the parental propagation and hybrid seed production area. As for the isolation distance of different horticultural plants, the self-pollinated plants are generally smaller, and the cross-pollinated plants and the often cross-pollinated plants require larger distances. The crops pollinated by wind require a smaller isolation distance than those pollinated by insects. The isolation distance of the parent breeding area should be slightly larger than that of the seed production area. Different crops have different specific requirements. The isolation should be operated in accordance with the national or provincial seed production technical regulations for corresponding crops.

(2) Natural barrier isolation. Use natural barriers such as mountains, villages, houses, and mature woods to isolate.

(3) Time isolation. The isolation could be achieved by adjusting the sowing period to stagger the anthesis of the seed production field or the parent breeding field with that of the crop field where the same crops are grown. The length of the isolation period is mainly determined by the length of the anthesis of the crop.

(4) Isolation by high straw crops. Plant sorghum, hemp, and other high straw crops within a certain range around the seed production area to separate the seed production area from the surrounding fields. This method is usually used as a complementary measure, requiring the high straw crops to be planted at least 20 days in advance.

Determine the area of the seed production field. The area of the parent breeding field and that of the seed production field must be arranged in proportion to the production of hybrid seeds so that the number of various types of seeds is in harmonious proportion. The area of the seed production field is based on the demand for hybrid seeds in the following year and the yield of hybrid seeds in the seed production field. The calculation formula is as follows:

$$\text{Area of parent breeding field (hm}^2\text{)} = \frac{\text{Area of seed production field next year (hm}^2\text{)} \times \text{Sowing rate of mother of father (kg/hm}^2\text{)} \times \text{Row ratio of parental lines}}{\text{Scheduled yield of parent per unit (kg/hm}^2\text{)} \times \text{Pass rate of seed (\%)}}$$

$$\text{Area of hybrid seed production field (hm}^2\text{)} = \frac{\text{Scheduled sowing area (hm}^2\text{)} \times \text{Scheduled sowing area (kg/hm}^2\text{)}}{\text{Predicted yield of seed production area (kg/hm}^2\text{)} \times \text{Row ratio of mother lines} \times \text{Pass rate of seed (\%)}}$$

When determining the area of the seed production field, in addition to the seed demand of the local fields, the export sales volume should also be considered according to the supply and marketing contract. The production shall be determined by sales, and the area of the seed production field shall be determined by the yield per unit area to avoid an imbalance between supply and demand.

(II) Seed Production Field Specifications and Seed Sowing

During hybridization, the row ratio of the mother and father lines should be reasonably determined. We not only have to ensure that there is enough pollen but also to maximize the number of mother lines to improve the yield of the seed production field. Carry out fine soil preparation and ensure soil moisture before sowing to improve the sowing quality. Specific requirements are as follows.

1. Determine the Row Ratio of Mother Lines and Father Lines

The row ratio is the proportional relation between the mother lines and father lines in the seed production field. The row ratio determines the proportion and fruiting rate of the mothers in the seed production field, which in turn affects the yield of the seed production field. As the mother lines increase, the fruiting rate of the mothers decreases, but the seed yield will increase within a certain range. Therefore, the principle of determining the row ratio is to increase the proportion of the mother lines as much as possible while ensuring an adequate supply of pollen from the fathers. When the water and fertilizer conditions are good, and the father plants are tall and have sufficient pollen, the number of the mother lines can be increased appropriately. When the plant heights of the father and mother differ too much or the sowing time is staggered, the row ratio should be adjusted appropriately to prevent the high straw, early-sown parents from impacting the short straw, late-sown parents. When determining the specific row ratio, it should be flexible according to several factors, such as the plant height, pollen volume, and anthesis of the fathers in the seed production combinations. The parents of melons are generally inbred lines that are monoecious. The mother and father should be seeded separately in the proportion 1 : 10; the solanaceous vegetables can be planted according to the number, growth type, and growth vigor of the parental flowers. For solanaceous vegetables, it is preferable to make the father plants account for 10%–20% of the mother plants. To avoid mixing at the time of harvest, the fathers and the mothers can be planted in separate areas.

2. Improve the Sowing Quality

Sow good seeds in both parent breeding areas and seed production areas, aiming to realize a full stand of seedlings. This will not only facilitate pollination after emasculation but also improve seed yield and seed quality. The father and mother lines must be strictly distinguished from each other during sowing. Cases like sequential lines, wrong lines, parallel lines, and missed lines are not allowed. To facilitate the distinction between the fathers and mothers, other crops can be planted at intervals along the father lines as a marker. To provide pollen when needed, a certain number of the fathers should also be sown at different stages near the seed production area as pollen collection areas.

(III) Anthesis Regulation

Whether the anthesis of the plants in the seed production field is simultaneous determines the seed yield and even the success or failure of seed production. Regulating the anthesis of the fathers is a central part of the seed production process.

1. Sowing Period Regulation

Accurate arrangements of the sowing period of the parents are a fundamental measure to ensure that the anthesis of both the fathers and mothers is synchronized. This measure is irreplaceable, especially when the fertility periods of the parents are greatly different. To ensure that the anthesis is synchronized, it must be operated based on the regulation of the sowing period, supplemented by other regulation methods.

Anthesis synchronization index: The criterion for the anthesis synchronization is that the full-blossom stage of the fathers meets that of the mothers, but the actual diagnostic indexes applied in different crops vary.

(1) Factors affecting anthesis & sowing period regulation. To ensure that the anthesis of the fathers meets with that of the mothers, factors that may affect the sowing period should be considered when determining and regulating the sowing period, such as the biological characteristics of the parents, external environmental conditions, production conditions, management techniques, etc., which should be handled and adjusted well in advance in order to reduce the trouble of regulating the anthesis during production and effectively ensure that the anthesis meets.

① Biological characteristics of the parents: The fertility rules of different parents are different. The seeding parents should be observed systematically first to master the phenological period, flowering time, and the process rules so that the sowing period can be arranged accurately. In terms of the fertility periods of the parents, some are the same and some are different, as in the case of the fertility processes. If the fertility periods and processes of the parents are the same, the anthesis will also be the same. The fathers and the mothers can be sown at the same time even if the anthesis of the mothers is slightly earlier than that of the fathers by

1–2 days. However, if the anthesis of the parents is too different, the sowing period needs to be adjusted. That is to say, the parent with a longer fertility period and later anthesis is sown first. The late sowing index of the other parent is decided according to the length of the fertility period and the number of days that are apart in the anthesis between the two parents.

② External environmental conditions: Consider the influence of external environmental conditions and adjust the sowing period according to the place and time.

③ Production conditions and management techniques: Production conditions and management techniques also have an impact on the anthesis. For example, different parents have different reactions to different soil fertility and fertilization amounts, so appropriate adjustments are required to the sowing period. Another example is that high planting density may slow down the fertility process so the sowing period should be appropriately advanced.

(2) Determination of sowing period difference. The determination of the sowing period of the parents should follow the principle of "Better sow the mother earlier than later". The specific methods are as follows.

① Determine the sowing period difference according to the fertility period of the parents: This is a simple method. If the seed production of the hybrid combinations in a certain area is operated in the same season, the length of the parental fertility period is generally relatively stable. However, it will also vary due to differences in climate, anthesis, soil texture, and cultivation measures.

② Determine the sowing period difference according to the foliar age: The growth and development progress is expressed by the number of leaves on the main stem, which is called foliar age. Under certain conditions, the number of leaves on the main stem of the same cultivars is relatively stable, but when the climate and cultivation conditions change, such as low temperature and low leaf emergence rate, the foliar age also changes. It must be noted that when determining the sowing period difference between the fathers and mothers according to foliar age, it is necessary to know the total number of leaves of the parents in the locality and the growth period of each leaf firstly, and secondly to make accurate observations of the fathers in the field in order to achieve synchronized anthesis.

③ Determine the sowing period difference according to the effective accumulated temperature: The effective accumulated temperature of the same parent at a certain growth stage is relatively stable, while the effective accumulated temperatures of different parents are different for the whole fertility period and each growth stage. Accordingly, the first flowering date of the father under certain sowing conditions should be determined first; then, according to the temperature data, based on the day of the initial heading stage of the father, the sowing period of the mother is from that day to the day when the father's effective accumulated temperatures equal to the value of the ones the mother gets from the sowing date to the heading stage. The accuracy of the methods when determining the sowing period difference between father and mother is in the following order: foliar age > effective accumulated temperature > fertility period.

(3) Auxiliary measures to regulate the sowing period. In order to reduce the trouble of staggered sowing time, in terms of the combinations that need a shorter period to stagger the sowing time of the parents, seed soaking for germination can be carried out on early-sown seeds so that the seeds of the other parent and the early-sown seeds can be sown at the same time; for the inbred lines that have a shorter pollination period, such as the father, in order to extend the pollination period, the seeds can be divided into three parts. Soak and pre-germinate one part of the seeds and sow them at the same time as another part of the dry seeds. Sow the third part of the seeds five to seven days later to ensure that the mother is fully pollinated and fertilized. In order to avoid accidents, a separate pollen collection area of the father can be set beside the seed production field. The father should be sown six to seven days later than that in the isolation areas for urgent needs.

2. Prediction of Anthesis

Although the sowing period of the parents is reasonably determined and adjusted according to various conditions, there may still be abnormal changes in the fertility process of both parents after sowing due to the influence of certain factors, resulting in the unsynchronized anthesis. For this reason, careful observation should be carried out during the fertility period to predict whether the anthesis will meet.

The methods of predicting the anthesis mainly include leaf inspection and flower bud (young ear) microscopy.

3. Careful Management

Field management for parent breeding fields and hybrid seed production fields is much more delicate than that for large fields. In general, it's necessary to meet the water and fertilizer requirements, strengthen the intertillage and weeding, and strengthen the disease and pest control so as to ensure that the inflorescence develops early and well, the flowers grow well, and the grains are full enough.

Under the premise of routine management, agricultural technical measures should also be used to promote and guarantee the synchronized anthesis of the father and mother plants. Frequent inspections shall be carried out after seedling emergence to determine whether the anthesis can meet according to the growth of the parents. If it's predicted that the anthesis can not be met smoothly, remedial measures should be taken actively. For example, methods like early seedling thinning, early final singling, strong seedling preserving, and excessive application of fertilizer and water can be taken to regulate the growth and development of the parents with slow growth. For the parents with high growth, water and fertilizer controlling, deep intertillage and other methods can be used to inhibit growth. There are some specific methods to inhibit the flowering and growth of some horticultural plants, such as pinching, pruning, etc. Chemical sprays can also be used to inhibit or promote growth. In conclusion, all technical measures must be taken in the seed production field to ensure that the parents'anthesis is synchronized.

(IV) Strict Roguing and Emasculation

Roguing is an important measure to ensure seed quality. Generally, the first roguing is carried out based on the appearance, leaf color, leaf type, and growth of the mother inbred lines during the seedling-raising period or the period of planting, seedling thinning, and final singling; when plants like melons and solanaceous vegetables begin to grow flower buds, the second roguing should be carried out; the third roguing should be carried out critically before flowering and pollination. This time, great conscientiousness is required during the operation. Special attention should be paid to the mixed and inferior parents. Careful inspections should be carried out one by one to ensure seed quality. Before harvesting and threshing, the fruit clusters of the mother should be carefully selected. Remove the mixed and inferior fruit clusters.

Emasculation is the process of removing the male organs or killing the pollen viability before anther dehiscence. The purpose of emasculation is to prevent unwanted hybridization or selfing, which is one of the key measures to ensure the purity of hybrids. The emasculation work includes three elements: First, remove the male plants in the dioecious mother area, such as spinach and *Asparagus officinalis*; Second, remove the male flowers of the mothers that are monoecious, such as melons; Third, remove the stamens of the plants that are monoclinous, such as cruciferous and solanaceous vegetables. For monoclinous plants, the method of emasculation can be done manually with small forceps, such as tomato, pepper, eggplant, cabbage, kale, etc. Regardless of which method is taken, there must be a person in charge. Patrol inspections should be strengthened to ensure the quality of the emasculation process. The emasculation can be done both in the morning and afternoon and should be done every day regardless of the weather. The removed stamens should be taken out of the isolation area to prevent the dispersed pollen from affecting the quality of the seed.

(V) Artificially Assisted Pollination

It can be carried out by applying the pollen collected previously to the stigma by a pollination tool, directly by smearing anthers that have dehisced and dispersed pollen on the stigma, or by putting the stigma into the pollinator to make the stigma covered with pollen. The most accurate pollination time is the time when the flower is in full bloom. At this time, the pistil stigma has the highest viability. The sticky nutrient solution secreted by the stigma can make the pollen attach itself to the stigma, promoting pollen germination. However, during the seed production of vegetables, sometimes the pollination is carried out at the same time as emasculation, which not only saves labor and time but also avoids unwanted hybridization. The pollination must be done within the expiry period of pistil fertilization. During the day, the best time for pollination is from 8:00-11:00. Pollination work requires exceptional practical skills. The amount of pollen applied to the stigma should be moderate, and the whole process must be operated gently, carefully, and meticulously. Damage to floral organs should be avoided as much as possible. During the pollination work, unwanted mating should be avoided strictly. After changing cultivars or when the fingers and pollination tools are contaminated by unwanted pollen, disinfection should be done immediately.

Generally, the plants in the seed production field have a long growth period, so it's necessary to strengthen the management of the hybrid mother area, provide good fertilizer and water conditions, remove the flower buds without hybridization in time, strengthen disease and pest control, ensure that the hybrid fruits grow and develop well, and pay attention to prevent wind and birds from damaging the plants.

III. Methods of Hybrid Seed Production

After the selection of good hybrid combinations through a series of experiments, it is necessary to produce F_1 hybrids for applications in production. Its main work includes two aspects: One is the propagation and purity preservation of the parents; the other is to set up isolation areas to grow parent combinations and produce F_1 hybrids. The principles of producing F_1 hybrids are that the crossing rate of hybrids is high (preferably 100%) and the production cost is low as far as possible. In this way, the seeds produced can be competitive. There are many ways to produce F_1 hybrids, which can be broadly summarized as follows.

(I) Artificial Emasculation Seed Production Method

It's also known as the simple seed production method. It can be operated by manually removing the male plants among the mother plants or removing the staminate flowers or stamens on the mother plants, and then leaving the father plants to receive natural pollination or artificially assisted pollination. The seeds produced by the mother plants are the F_1 hybrids. In principle, the artificial emasculation seed production method is applicable to all sexually-propagating crops. However, in practice, this is not the case because whether the F_1 hybrids can be promoted is closely related to the difficulty of producing seeds and the seed price.

For dioecious plants, plant the parents adjacent to the inbred lines. When the female and male plants can be identified, remove the male plants that can produce pollen in the mother lines, leaving only the female plants. Generally, the removal of the male plants should be done every 2–3 days to ensure that there are no pollen producers among the mother plants. However, it is actually difficult to achieve, so the seed purity of F_1 hybrids is reduced. Spinach is a kind of dioecious plant. Generally, the ratio of female to male plants within the cultivars stands at about 1 : 1. The plant sex type can be divided into male, male hermaphrodite, female hermaphrodite, and female. These types can be used to breed the corresponding systems, i.e. dioecious lines, hermaphrodite lines, male and female lines, etc. In heterosis utilization, the female lines or hermaphrodite lines with only pistillate flowers are mainly used as mothers to match with the fathers so as to produce F_1 hybrids that can be used for production. Dioecious plants like *Asparagus officinalis* are generally more productive in male plants, so it is beneficial to select and breed F_1 hybrids of male plants. The female plant of *Asparagus officinalis* has XX chromosomes and the male plant has XY chromosomes. In recent years, X and Y haploids have been screened from a large number of *Asparagus officinalis* seedlings or obtained by the culture of the anthers of *Asparagus officinalis*. The pure female plant (XX) and super-male plant (YY) can be obtained by chromosome doubling, which can be used for hybridization to obtain F_1 hybrids of all-male plants with the same genotype and consistent traits.

For diclinous plants like cucurbitaceous vegetables, in the isolation areas where the parents grow, as long as the male flowers on the mothers are removed before the bloom of the female flowers, the seeds produced are the F_1 hybrids. Each melon can produce hundreds of seeds, and the propagation coefficient is from 100–200 times, which is easy to promote. Cucumbers are generally diclinous and have more male than female flowers, but there are also all-female plants that only have female flowers without male flowers. The line with such stable genetic performance obtained through selection is called the "female line", which is found and used in cucumber, melon, and pumpkin. The female lines of cucumber, in particular, have been widely used both domestically and internationally to produce F_1 hybrids. The female line seed production method uses the female lines as the mothers and inbred lines with good economic traits and high combining ability as the fathers, then plants the fathers and mothers according to a certain ratio (generally 1 : 3). In the isolation area, after natural pollination or artificially assisted pollination, the seeds produced by the mothers are F_1 hybrids.

For monoclinous vegetables, it depends on the number of seeds obtained per cross. For example, in solanaceous vegetables, 100–500 seeds can be obtained per cross. A skilled worker can hybridize 1,000–1,300 tomato flowers or 500 pepper flowers a day. According to the Institute of Horticultural, Zhejiang Academy of Agricultural Sciences, each worker can produce more than 20 grams of tomato seeds, which can be planted on the land of about 1 mu. F_1 hybrids of vegetables like these can be produced using the artificial emasculation seed production method, while those of Apiaceae plants and Allium plants can't be produced by that method.

For those cross-pollinated crops that emasculation is hard to achieve, the simple seed production method can also be used to produce F_1 hybrids. The true hybrid rate is often only 50%–70% when the ratio of planting fathers to mothers is 1 : 1, so the yield increase is limited. If the proportion of fathers is increased further, even fewer seeds will be obtained from the mothers, which makes the production of F_1 hybrids difficult and the hybrids uncompetitive, making it difficult to be promoted.

(II) Seed Production Method Using Marker Traits at the Seedling Stage

Based on the differences in a certain botanical trait between the parents and F_1 hybrids at the seedling stage, hybrid seedlings or parent seedlings (false hybrids) can be identified more accurately at the seedling stage. Such botanical traits that can be easily identified visually are called "marker traits" or "indicator traits". The marker traits should meet two conditions: ① The differences of the botanical trait must be obvious at the seedling stage and must be easy to identify visually. ② The genetic expression of this trait must be stable. The seed production method using marker traits at the seedling stage is to select strains with recessive traits as the mothers (such as melon's highly dissected leaves, watermelon's slightly dissected leaves, tomato's yellow leaves, cabbage's glabrous trait, etc.) and make them hybridize with the fathers with relatively straight dominant traits naturally or artificially without operating the emasculation. Then eliminate those false hybrids that show recessive traits in the hybrid seedlings through seedling thinning. For example, the piliferous and glabrous traits of cabbages are a pair of qualitative traits, with the piliferous trait being dominant and the glabrous trait being recessive. When the F_1 is all piliferous and the piliferous lines are used as mothers, the piliferous plants in F_1 are true hybrids and the glabrous ones are false hybrids that should be eliminated. Therefore, in the case of a low natural crossing rate or pollination without emasculation, the true hybrid rate can reach more than 90% by removing the false hybrids with the seedling thinning method. This method has the advantage that it is easy to breed parents and produce hybrids. The production cost can be reduced by omitting emasculation. It can also help produce a large number of F_1 hybrids in a relatively short period of time. The disadvantage is that seedling thinning and final singling work are very complicated. It requires the knowledge of marker traits at the seedling stage and the skills of seedling thinning and final singling. Some vegetables and specific hybrid combinations have not yet shown typical and obvious marker traits at the seedling stage. Although some traits are obvious, the heritability is complex, so it is not easy to apply.

(III) Seed Production Method Using Chemical Emasculation

Although genetic breeding methods, such as the use of male sterility and self-incompatibility, can solve the problem of seed production and are an economical and effective way, it is sometimes not easy to breed the lines that can be used as hybrid parents, so chemical emasculation is a way to save labor in breeding production. In recent decades, continuous research and exploration have been carried out. With the discovery and synthesis of new agents, the number of chemical emasculation agents is increasing. The main emasculation agents reported so far are as follows: dichloroacetic acid, sodium 2,2-dichloropropionate (dramine), 3-chloropropanoic acid, sodium 2,3-dichloroisobutyrate (FW450), triiodobenzoic acid (TIBA), 2-chloroethyl phosphoric acid (ethephon), maleic hydrazine (MH), dichlorophenoxyacetic acid (2,4-D), naphthalene acetic acid (NAA), chlormequat chloride (chlormequat CCC), etc. It is generally operated with the aqueous spraying method and the spraying period is generally before the flower buds begin to differentiate. In order to achieve a long-lasting effect, repeated spraying is required at appropriate intervals and times a day. When using the chemical emasculation agent for seed production, the agent must have the following characteristics: ① After applying it to the mothers, it will only kill the stamens to make the pollen sterile without affecting the normal growth and development of the pistil; ② It won't cause genetic variation; ③ The treatment must be simple, the agent be cheap, the effect be stable; ④ Harmless to humans and animals.

So far, the practical application in China is limited to the treatment of using ethephon in melon seed production to inhibit the production of male flowers. Sainbhi et al. (1978) had summarized the effects of 15 kinds of chemical emasculation agents on eggplant, green pepper, tomato, and other crops, and concluded that 100–800 mg/kg of MH had good effects on a variety of vegetables and could be used in seed production in practice. However, many of the agents mentioned above have yet to be tested in China.

In addition to the seed production methods mentioned above, male sterile lines and self-incompatible lines can also be used to produce F_1 hybrids.

Review and Reflection Questions

(1) What are the conditions for hybrid seed production in horticultural plants?

(2) What are the technical essentials of hybrid seed production?

(3) What are the contents of the emasculation work in hybrid seed production?

(4) What are the methods of hybrid seed production?

(5) What are the characteristics of the chemical agents required for emasculation in seed production?

(6) What are the main elements of the dominance hypothesis of genetic mechanisms of heterosis?

Module 3 Construction and Management of Seed Production Base of Horticultural Plants

Learning Objectives

(1) Be aware of the significance and tasks of the establishment of a horticultural plant seed production base.

(2) Master the procedures for the construction of a seed production base.

(3) Master the techniques of seed production base management.

(4) Understand the types of seed production bases.

(5) Develop a plan for the construction of a seed production base according to the type and conditions of the horticultural plant seed production base.

Learning Scenario Construction and Management of Seed Production Base of Horticultural Plants

I. Significance and Tasks of Establishing Horticultural Plant Seed Production Bases

(I) Significance of Establishing Horticultural Plant Seed Production Bases

Seed production is systematic work that requires high professionalism and techniques. In seed production, different ecological climate conditions, soil fertility, cultivation management levels, or seed breeding or production techniques of result in differences in seed yield and quality, which in turn affect the yield and quality of the crops in the next season. In order to obtain high-quality seeds of equal quality, it is necessary to establish seed production bases with standardized management, thus realizing the specialization and industrialization of seed production.

The establishment of seed production bases is conducive to improving the level of specialization of seed production, which can promote the improvement of the production level and ensure seed quality; it is beneficial to the unified management and control of seed production, which can ensure the implementation and enforcement of national seed production regulations and help govern the chaos in seed market so as to achieve the legalization of seed management; it is helpful for the organization to ensure uniformity in seed supply so that the seed market prices can be controlled and the seed market can develop in line with the collectivization of seed management, which can improve the competitiveness of China's enterprises of seed production and management in the international market; it is conducive to the realization of the large-scale seed production so that the cost of seed production can be reduced and the efficiency of seed production can be improved; it is conducive to the testing, demonstration and promotion of new cultivars so as to promote the development and use of new cultivars and form a research-production-sales integration model.

In a word, the establishment of the seed production base is not only the main step to gradually realize the specialization of seed production, modernization of seed processing, standardization of seed quality, regionalization of cultivar layout, and marketization of seed management, but also a major measure to realize agricultural and rural economic development, improve traditional agriculture, and increase agricultural economic benefits.

(II) Construction Tasks of Seed Production Base of Horticultural Plants

The quality of seeds is directly related to the farmers' income, agricultural security, and social stability. Since the implementation of the *Seed Law of the People's Republic of China*, the seed market has been gradually thrown open, the number of seed management organizations has increased rapidly, the seed amount on seed market has increased increasingly, and the business forms have varied. The agricultural administrative departments at all levels have made great achievements in regulating the seed market to a certain extent by taking various measures to strengthen the supervision of the seed market, but for a long time, the phenomenon that fake and substandard seeds cause great losses to agricultural production are still common and the problem of seed quality is still not optimistic. In addition to the lack of a strict seed quality control system, the problems of "variety, disorder, mixture" in seed production and "small scale, incompleteness, scattering" in seed management are also important factors. According to the development direction of specialization, commercialization, and socialization, as well as the ecological area and the economic conditions of production and society, using the investment of the nation, locality, or seed enterprise (if possible, appeal to farmers to invest) to establish and improve the national original seed farm and the specialized seed production base that formed based on the crop production situation is important to ensure the seed quality. The main tasks of the seed production base are as follows.

First, propagate new cultivars quickly. Once the new cultivars are approved, they can be promoted for use in field production. The quantity of commercial seeds and parental seeds is small, which often becomes an important condition to restrict the rapid promotion of new cultivars, so rapidly propagating a large number of qualified commercial seeds and parental seeds to meet the urgent demand of field production, turning scientific research achievements into productivity, and bringing the economic benefits of new cultivars into full play have become the main tasks of the seed production base. After successfully selecting and breeding new cultivars, it is necessary to accelerate the propagation of new cultivars in the seed production base to rapidly increase the number of seeds so that the seeds can be put on the market at the fastest speed to meet the needs of agricultural production for a large number of good-quality seeds, which can ensure that the superior varieties can be rapidly promoted according to the plan and play their role in increasing yield as soon as possible.

Second, ensure the characteristics and purity of the superior varieties. During the breeding and production process of superior varieties, mechanical mixing and biological mixing are problems that can't be avoided, which will result in reduced purity and degraded characteristics. In the process of mass propagation and cultivation, mechanical mixing caused by improper sowing, harvesting, transportation, threshing, and storage, biological mixing caused by natural hybridization, and variation caused by the influence of environmental conditions will often result in the reduction of purity and degradation of characteristics. This requires the seed production base to have strict management measures, create reliable isolation conditions, and strictly implement the technical regulations of seed breeding, seed production, mixing prevention, purity preservation, etc., so as to ensure that superior varieties and parental seeds can maintain their purity and characteristics in multiple breeding and production processes.

Third, cultivars should be rationally distributed, upgraded in a planned way, and changed or provided with seeds. In agricultural production, a natural ecological zone should be dominated by one to two superior varieties and supplemented with two to three other cultivars. For such an ecological zone, the seeds for use can be supplied by one seed production base. Relying on the seed production base for seed supply can not only break the boundaries of administrative regions but also make it practicable to make overall arrangements according to the natural ecological zone. Realizing the reasonable layout of cultivars can help overcome the problems of seed variety, disorder, and mixture.

Fourth, reduce the cost of seed production. Specialized seed production operated in the seed production base can not only ensure quality but also bring scale benefits into play, reduce the cost of seed production, lower seed prices, and facilitate the popularization and utilization of superior varieties.

II. Types and Conditions of Seed Production Base

(I) Types of Seed Production Base

The seed production base in China includes two types: the state-owned seed production base and the authorized seed production base.

1. State-owned Seed Production Base

State-owned seed production base includes state-owned farms, state-owned original (good) seed farms, scientific research units, experimental farms of universities and institutions, and teaching farms. This kind of base has a perfect management system, complete facilities and equipment, and a strong technical force, which is suitable for the production of the original seeds, the parents of hybrids, and some precious new cultivars.

2. Authorized Seed Production Base

This is the seed production base established by the seed company and the seed production unit after signing a contract or a letter of agreement on the basis of joint consultation. It carries out specialized production according to the plan of the seed company and will receive technical guidance and inspection from the seed company. The seeds it produces are purchased by the seed company. The authorized seed production base is and will be the main form of seed production base.

(II) Conditions of Seed Production Base

1. Natural Conditions

Natural conditions are the prior consideration when selecting seed production fields, including temperature, light, soil, the frost-free period of the soil, etc. Different horticultural plants require different growth and development conditions, so the seed production field selected must meet the conditions required for the growth and development of horticultural plants to the maximum extent. Only with good natural ecological conditions can the cultivars show their excellent characteristics; the purity of cultivars, parents, and the normal maturity of seed production combinations be ensured; the seed yield be improved; and the production costs be reduced.

The natural ecological condition of the seed production base mainly refers to climate conditions, which include effective accumulated temperature during the fertility period, diurnal temperature difference and frost-free period; sunshine duration and light intensity, annual rainfall and precipitation distribution; soil type and soil fertility; wind and pollinators' activity, etc. Since any new crop cultivar is a good ecological type selected and cultivated by breeders according to the needs of production in a certain region or under specific ecological conditions, this ecological type is adaptive to local ecological conditions. Generally speaking, the location of the breeding cultivar has the most suitable ecological conditions for seed production, and the seed production base of this cultivar can often be built nearby. The ideal area for horticultural plant production should have perfect natural conditions such as sufficient light, moderate temperature and rainfall, and no strong wind. The production base for horticultural plants is usually not built in areas where the temperature in summer is too high and in winter is too low.

2. Equipment Conditions

There should be certain equipment conditions in the seed production of vegetable crops, such as seedling facilities, drainage and irrigation facilities, seed threshing machinery, seed storage, etc.

3. Human Conditions

The producers of the seed production base should attach importance to seed production and have high skills as well as some experience in seed production, or the ability to operate in accordance with the production technology protocols strictly after training. A high level of agricultural production is required in the location of the production base, which means that the farmers should have a high ability to receive new technology and new things. It is conducive to the large-scale technology of seed production, increasing seed yield, reducing seed production cost, and ensuring seed quality. In addition, local leaders should support it, which is very important for safety isolation and coordinating the layout of different cultivars of seed production. The cadres and the masses of the production base should have a certain enthusiasm and be clear about the importance of seed production to facilitate the implementation needed in the process of seed production.

III. Procedures for Establishing a Seed Production Base

The scale and orientation of investment as well as specific requirements for base construction, etc. should be clarified based on a full understanding of the natural ecological conditions, geographical conditions, transportation, agricultural production foundation, the basic quality of farmers, etc. Then, formulate the detailed plan and the implementation scheme for the construction of the seed production base, the economic benefits after the base is constructed, etc. through full investigation and demonstration. The formulated plan should be carried out in cooperation with relevant departments.

1. Full Demonstration

Before establishing the seed production base, a detailed investigation and study should be conducted first of the natural conditions of the base (such as frost-free period, rainfall, isolation conditions, land area and its contiguity, production scale, soil fertility, transportation, etc.) and socio-economic conditions (including land production level, labor, economic situation, the enthusiasm of cadres and masses, etc.). Based on this, determine the scale and the investment orientation of the base, make the design plan of the seed production base, and ask relevant experts to demonstrate it.

The main contents of the design plan are the purpose and significance of the base construction, the planning of the base construction, the implementation scheme of the base construction, and an analysis of the economic benefits expected after the base is established.

2. Planning

On the basis of sufficient demonstration, make a detailed plan for the construction of the seed production base and determine the final scale of the base, the cultivar types, area, and yield levels of the main production crops, as well as formulate the technical protocols for seed production.

The scale of the seed production base can be determined according to the superior variety promotion plan, as well as the number of seeds purchased by the seed company and the base's normal output. The calculation formula is as follows:

$$\text{Sown area}=\frac{\text{Volume of purchase + Base's normal output}}{\text{Average yield}}$$

When the area of the seed production base is planned, more land should be left since the number of seeds in the general seed production plan is about 10% higher than the number of seeds in demand.

In the process of planning and designing the seed production base, the following aspects should also be considered: First, the planning of production communities, the planning of isolation areas, the planning of roads and irrigation and drainage systems, buildings, and sunning grounds, production facilities, etc. should be done in line with the principle of easy management, saving expenses and having less land occupation. Second, establish a mother plantation. The materials of the mother plantation must be taken from the plants with pure cultivars, strong growth, high yield, and no quarantine objects. They should be numbered, registered, and tested on their plantations to prevent confusion. In addition, strengthen management and observe them frequently to eliminate inferior plants. Third, establish breeding areas. Crop rotation must be considered in the planning of this area, because the residue roots of the previous crop, the toxins of decomposition of fallen leaves, soil nematode infestation, the lack of certain nutrient elements in the soil, etc. can make the subsequent crop grow poorly.

3. Organization and Implementation

The main departments involved in the construction of the seed production base are the departments of the planning commission, finance, materials, agriculture, etc., which are responsible for the division of labor and cooperation to be in charge of various work of the base construction and ensuring that the base construction can be completed with high quality and quantity, and be put into use on schedule.

IV. Operation and Management of Seed Production Base

Under the market economy, seed production bases are developing towards large-scale operations, specialization, commercialization, and socialization. Thus, improving the management of the seed production base is conducive to the sustainable development of seed production.

(I) Plan Management

Seed production is commodity production, so the ability to convert all products into commodities depends on both the market demand and the quality of the products.

1. Market-oriented and On-demand Production

As special commodities, the special characteristics of seeds are: On the one hand, horticultural crop seeds are the carrier of life and renewable means of production, so the quality of commodities directly affects the yield of the crops next year and the interests of farmers; on the other hand, horticultural crop seeds have a certain life span and an obvious seasonality, which means the seeds can not be backlogged, otherwise, once the life span expires or the season is delayed, the seeds may lose their utilization value. Therefore, seed production must be implemented as planned.

As for seed producers, in the process of making a seed production plan: First, market awareness must be improved to produce seeds according to the sales; second, quality awareness must be enhanced to do seed production work with a high sense of responsibility and enterprise and strictly control quality standards; Third, efficiency awareness should be enhanced to expand the market as much as possible, reduce costs and form economies of scale.

2. Actively Implement the Contract System, Pre-production, Pre-purchase, and Pre-Supply

In order to build a solid foundation for the on-demand production, protect the legitimate rights and interests of both parties in the deal, coordinate the relationship among production, purchase, and sale as well as improve management and economic efficiency, the pre-order and pre-sale contract system should be actively implemented. The seed company should sign pre-order and pre-sale contracts with the production base and seed-using unit, respectively, to implement pre-propagation, pre-purchase, and pre-supply.

(II) Quality Management

Quality management refers to organizing and producing high-quality seeds that meet the stipulated standards in accordance with the requirements of agricultural production and the quality objectives of seeds expected by users. Quality management is a key part of seed enterprise management. In order to ensure the quality of seeds, on the basis of continuously summarizing practical experience, China has put forward seed testing procedures, seed classification standards, technical operation procedures, and various relevant rules and regulations to improve seed quality, which has played a great role in promoting the quality of seeds.

The method of seed production quality management used by seed companies is mainly implemented through the technical service and supervision of the seed production process.

1. Specialized and Collective Production of Seeds

It will bring the geographical advantages of the base into full play to have specialized and collective production of seeds, contiguous seed breeding and seed production fields, as well as safe isolation; it will bring the talent advantages of the base into full play to have professional technical team, the experience of production practice, as well as high level of production technology; it will bring the technical advantages of the base into full play to make advanced high-yield and pure-keeping measures easily be promoted and applied; it will bring the management advantages of the base into full play to have farmers who are willing to accept technical guidance and seriously comply with the technical operating procedures of seed production as well as a series of rules and regulations to ensure the quality of seed. Therefore, the specialized and collective production of seeds is conducive to improving the yield and quality of seeds and then improving economic efficiency.

2. Establishment of a Quality Management System and Quality Assurance System

To implement quality management all-round, seed enterprises must establish a complete quality management system and quality assurance system in the whole process of production and operation according to the current situation of seed quality management. The quality management system is the systematization of quality management work in the whole process of enterprise operation, while the quality assurance system is the systematization of quality assurance work, such as division of labor, responsibilities, and standards.

(1) Quality management system: The quality management system refers to a system of various rules, methods, and means needed to economically produce seeds

of quality that meet user's requirements. It mainly includes clarifying the quality standards, establishing and improving the organization system, implementing standardized internal management, and strengthening the management of each link.

(2) Quality assurance system: The quality assurance system is to implement organized quality assurance activities in the whole process, from new cultivar planning and production to after-sales service, with the systematic concept and method according to the responsibility of various departments and links and relevant regulations. Its ultimate goal is to make users purchase and use the seeds of the enterprise with confidence and to ensure the quality of seeds from the standpoint of the seed users. As for seed enterprises, in order to produce high-quality seeds stably for a long time, it is necessary to establish and improve the quality assurance system, form a strict organization and management system, establish a quality management organization from the enterprise to the township, village, seed production households, combine administration and technology, and implement various quality management measures; at the same time, it is necessary to develop strict technical standards, management requirements, work systems as well as rewards and punishments, and connect the quality of the work of all links with the quality of the seed to improve the quality of seed.

3. Be Strict with Key Technical Processes

Firstly, isolation work should be strictly done. Secondly, the operation methods must be standardized, and the roguing should be carried out in due course. In addition, we also need to be strict with the quality standard based on the good quality and good price policy.

4. Strengthen the Basic Construction of the Base and Carry Out Strict Seed Testing, Selection, and Processing

Strengthening the basic construction of the base is the basis for achieving seed industrialization. The basic construction includes measures improving production conditions, such as the construction of water conservancy, soil improvement, and improvement of plant protection capacity. It also includes measures to improve the seed quality, such as the construction of warehouses, sunning grounds, and the purchase of seed testing and processing machinery.

Seed testing includes field inspection and indoor inspection. The seed quality and purity mainly depend on field inspection. During the period when the typical characteristics of the cultivars (parental system) are most clearly expressed, check and calculate the authenticity, purity, weeds, and pest infection rates of the cultivars item by item. Indoor inspection mainly tests the purity, cleanliness, germination rate, and moisture of seeds in the process of seed acquisition, transportation, storage, and operation. Seed testing should be carried out in accordance with standard methods and with standard instrumentation to ensure the reliability of test results.

Seed selection and processing include seed drying, threshing, primary selection, selection and grading, coating, and packaging. Past experience has proved that after the selection and processing of seeds, the thousand-seed weight, germination rate, and cleanliness have significantly improved, which is an effective measure to improve the quality of seeds and realize the standardization of seed quality.

(III) Technical Management

The current technical management measures: Seed companies provide the original seeds of the commercial cultivars bred in the authorized bases and on-site guidance on some key technical processes during seed breeding.

These technical processes are as follows: In the sowing period, check whether the breeding farmer's seed breeding field meets the isolation conditions and whether there are other different cultivars of the same crop planted near the seed production field; in the flowering and pollination period, check the flowering of the seed production field and whether the pollen of the bred cultivars is mixed with that of other cultivars; before maturity and harvest, check whether the roguing is done thoroughly and grade the bred seeds; in the threshing, check whether the threshing technology is correct and whether it can cause damage to the seed; in the processing and selection, check whether the processing and selection equipment is running normally, whether it can thoroughly remove the seeds of the mixed cultivars, and whether the sorting machinery will cause seed mixing, etc. If all these processes can be ensured, then the success of seed production is technically guaranteed. In actual production, some processes are often especially emphasized while others are ignored, resulting in unqualified seeds. Therefore, a set of technical support processes for seed production should be developed to ensure that the technical measures in all technical processes of seed production are up to standards. Besides that, the technical support processes, technical support conditions, and the issuance of seed production licenses should be combined to ensure the quality of produced seeds.

Review and Reflection Questions

(1) What is the significance of seed production base construction?

(2) What are the types of seed production bases?

(3) What are the conditions of the seed production base?

(4) What are the contents of the operational management of seed production bases?

(5) What does quality management include?

(6) What are the main contents of the technical management of seed production?

Module 4 Seed Production of Horticultural Plants

Learning Objectives

(1) Understand the seed collection methods for major horticultural plants.

(2) Basically understand the basic theory of horticultural plant seed production.

(3) Understand the flowering and pollination habits of different horticultural plants.

(4) Understand the basic characteristics of seed production, flowering, and fruiting of various horticultural plants.

(5) Understand the importance and role of isolation techniques for seed production of horticultural plants.

(6) Master the requirements and methods of isolation of different types of horticultural plants.

(7) Understand the characteristics and ecological habits associated with the seed production of annual and biennial flowers, bulbous flowers, perennial flowers, woody ornamental plants, and turf grasses.

(8) Master the technical measures concerning the seed production of flowers and ornamental plants.

(9) Be able to correctly perform sexual hybridization of flowers and seed production operations of major flowers.

Learning Scenario I Major Vegetable Plant Seed Production

I. Melon Vegetable Seed Production

(I) Cucumber Seed Production Technology

Cucumber is an annual creeping vine plant in the *Cucumis* genus of the Cucurbitaceae family, coming from the long cultivation of the wild cucumbers distributed from the Himalayan foothills of India to Nepal and Sikkim. Later, it spread all over the world and formed various types and cultivars according to different natural conditions and eating habits in the cultivation areas. The main types of cucumbers cultivated in China are North China cucumber and South China cucumber. In recent years, the cultivation area of Nordic thornless cucumber in China has also been expanding. The edible part of the cucumber is its fruit which smells fresh and is nutritious. People can eat it raw, and cook it, salt it, or sauce it. Each cuisine has its own flavor and is quite popular among people. Cucumbers are commonly cultivated throughout the north and south of China and different cultivars cultivated by different methods are supplied annually.

1. Flowering and Pollination Habits

(1) Flower bud differentiation.

① Flower bud differentiation time: Cucumber flower buds differentiate at the leaf axils at a very early time. In normal cultivation, it takes about 15 days after sowing for the first true leaf to come out. The first flower bud will be differentiated from the third to fourth nodal region. About 20 days after sowing, when the first true leaf is fully expanded, the flower buds of the seventh to the eighth nodal region have been differentiated. When the two true leaves expand 27 days after sowing, the flower buds near the 11th nodal region have differentiated, and during this period, the flower buds at the third or the fourth nodal region start to have sexual differentiation. When the nine true leaves expand 48 days after sowing, the flower buds around the 27th nodal region have differentiated and soon the flower buds around the 16th nodal region start sexual differentiation. The flower bud differentiation of lateral branches takes place very early, but later than that of the main branches. The axillary buds only grow out into lateral branches below the growing point of the seventh nodal region, and flower buds are only differentiated once the axillary buds develop into lateral branches.

② Sexual expression of flowers: Cucumber has three types of flowers, namely unisexual flowers, including pistillate flowers (Fig. 4-1) and staminate flowers (Fig. 4-2), as well as hermaphrodite flowers. Most cultivars used in production are monoecious with different flowers. At the early stage of cucumber flower bud differentiation, the initial forms of the buds are identical, all with androgynous flower primordium. At this time, the sex of the cucumber has not yet been fixed. As time goes by, the pistil will degenerate if the stamen develops, thus a staminate flower will come out; the stamen will degenerate if the pistil develops, and a pistillate flower will come out. In addition to the influence of genetic factors, the sex of the cucumber flower is also affected by the temperature and duration of sunshine during the flower bud differentiation. Usually, the lower temperature and shorter sunshine duration can help the flower grow into a pistillate flower. On top of that, the differentiation of cucumber flowering sex can also be artificially regulated by the hormonal treatment of seedlings.

Fig. 4-1 Pistillate Flowers

Fig. 4-2 Staminate Flowers

(2) Flowering and pollination.

① Floral organs: It has five lobes of corollas and has nectaries. The pistillate flower is born individually or in clusters, with three stamen primordia in the perianth tube, conspicuous ring-shaped nectaries at its base, a short style, a fleshy three-lobed stigma, and a long and inferior ovary, with three to five ventricles. It has several rows of ovules, the number of which ranges from 100 to 500. Staminate flowers are usually clustered in three to five between leaf axils. Three stamens consist of five anthers, two connate, and one solitary. The anthers are densely arranged in an S-shape. In the middle of the pistil, you can see the pistil primordium that has stopped developing.

② The process of flowering and pollination: The pistillate and staminate flowers of cucumber blossom mostly at dawn. The solitary pistillate flowers open for one to two days and don't close at noon. Its fertilization ability is at its best three to four hours after flowering and gradually decreases afterward. Although it can fertilize and bear fruit two days before and after flowering, the seed volume is low. After the staminate flowers bloom, the anthers dehisce, and pollen is densely distributed over the entire surface of the anthers when the temperature reaches 10°C or higher. The pollen grains are not dispersed and are mainly pollinated by insects. Pollen has a short life span and loses its vigor four to five hours after the flower blossoms in high temperatures. However, it is already able to germinate in the afternoon one day before flowering. The cucumber takes four to five hours from pollination to fertilization. It takes about 40 days from flowering to seed maturity. After the after-ripening, seed quality and seed collection volume can significantly improve.

(3) Fruiting characteristics.

① Parthenocarpy: Cucumbers need insects for pollination, and the seeds can only grow after the pistillate flowers are fertilized. The ovaries of some cultivars, especially the cucumber cultivars grown in protected areas, grow and expand as usual without pollination and fertilization. This is called parthenocarpy. Parthenocarpy is a very good thing as it helps flowers fruit in winter without insect pollination. However, during the seed collection, it creates an illusion of tons of fruits. In fact, however, these fruits have no seeds. Therefore, the seed collection fields need to be pollinated.

② Head-back fruit: Cucumbers usually fruit from the bottom up, but some early-maturing cultivars fruit top down. Some pistillate flowers or short lateral branches even show up at the leaf axis during the fruiting. They may blossom and fruit in the first nodal region. Fruits of this kind are called head-back fruits. Head-back fruits (Fig. 4-3) play a big role in improving the capacity of fresh cucumbers on the market. However, due to immaturity or mixing and other reasons, they can't be kept as seed fruits. Instead, they should be harvested in time.

Fig. 4-3 Head-back Fruits

2. Seed Production Technology for Conventional Cultivars

The cultivation methods of cucumbers can be divided into three types: protected cultivation, spring open cultivation, and summer and autumn open cultivation. The production of the original seeds used in protected cultivation should be carried out in protected facilities such as plastic tunnels in early spring. The production of seeds used for production can be carried out in the open field in the spring. The breeding of the seeds used for spring open cultivation can be carried out in the open field in the spring. The breeding of the seeds used for summer and autumn open cultivation can only be carried out in the summer and autumn. The method of quality artificial isolation and mating is adopted for the production of original seeds, while the method of roguing is used for the production of seeds used for production.

(1) Technical points of seed collection in spring open cultivation.

① Isolation of seed collection fields. The spatial isolation distance of the original seeds is more than 1,000 m, and that of the seeds used for production is more than 500 m.

② Selection and roguing. The selection of original seeds should be strictly conducted three times. The first time is before the flowering of the first pistillate flower. Select plants in line with the characteristics of this cultivar, including the nodal region of the first pistillate flower, nodal region interval of pistillate flowers, flower bud shape, leaf shape, disease resistance, etc. The second time is when the seed fruit reaches commercial maturity. Eliminate part of the plant according to the traits of the fruit strip, the number of pistillate flowers, the length of internodes, branching, fruiting, resistance, etc. The third time should be before seed fruit harvesting. Eliminate again the plants and seed fruits that do not meet the characteristics of the cultivar. The method of roguing can be adopted for the production of seeds used for production, of which during the second and third selections, the plants and seed fruits with atypical traits, weak growth, and serious pests and diseases should be eliminated.

③ Artificially assisted pollination. Artificial pollination is a necessary technical measure to improve the number of cucumber seeds. The best artificial pollination time is 7:00–9:00 every day. Interplant pollination is usually required. The individual plant is better not to self-cross. Leave two to three seed fruits per plant.

④ Seed fruit harvesting and seed collection. Seed fruits generally take 35–45 days from flowering to physiological maturity. They can be harvested when the peels become yellow or yellowish-brown. Put the harvested fruits in a cool and ventilated place and perform a week after-ripening. Then, longitudinally dissect the fruits, take out the seeds and pulp, put them in a tank or non-metal container, and ferment them for one to two days to make the sticky material separate from the seeds. When the seeds sink to the bottom of the tank and you don't feel slippery when touching the seeds, you need to pour out the blighted seeds, pulp, and leaven floating on the top, clean the full seeds, put the seeds on the reed mat or gauze net and dry them for one to two days. Do not put them on the cement floor in the sun. When the moisture content reaches 8% or less, the seeds can be stored in bags.

(2) The seed collection method of protected cultivation in spring. The sowing and final planting time of seeds in plastic tunnels should be 7–10 days later than that of vegetable cultivation. Due to the low temperature in early spring, the underdevelopment of sexual organs, few insects, as well as poor pollination and fertilization, artificial pollination must be carried out. Remove the first and second pistillate flowers at the early stage, and conduct pollination starting from the third pistillate flower. Pollinate three to four pistillate flowers and leave one to two seed fruits for production per plant.

3. Seed Production Technology for Stable Cultivar

(1) Cultivation management of seed plants.

① Sowing and seedlings. The goal of seed cultivation is to produce seeds. Therefore, the principle in the sowing period should be that the seed plants are in the optimum environment during pollination and fertilization and the seed fruits can ripen at last. Seed fruits need 30–40 days more to mature than commercial melons in the fertility period, so the seeding time should be appropriately arranged according to different crops. In spring open cultivation, seed breeding should be done early to avoid pests, diseases, and the rainy season at the later stage. If it's autumn cultivation, make sure the seed fruits can fully mature before the frost.

② Final planting and field management. Because the growth vigor of the plants with seed fruits is weaker than the plants of the fruits used for production, the planting density of cucumber's final planting in the seed collection field should be greater than that of the crop field, generally, 4,500–6,000 plants each mu.

Cucumber seed collection fields especially need more organic fertilizer. In addition to pigs and cattle water-logged compost, fine fertilizers like chicken poop and soybean-cake fertilizer are also needed. Besides that, in the fruiting period, phosphatic and potash fertilizers should be added. After the hardening of the seedling ends, add the fertilizer every time after watering and stop watering 10 days before harvesting seed fruits to prevent rotten fruits. After the final planting and reestablishment, set up racks or hang the vines. To prevent excess nutrients, the lateral branches that do not bear the seed fruit should be removed in time. The first pistillate flower of the medium-and late-maturing cultivars appears at the high nodal region, so the first fruit should be kept as a seed fruit. For the early-maturing cultivars, pistillate flowers appear when there are only three to four leaves. It depends on the growth of the plant as to whether to keep the fruit of the first pistillate flower. If the plant is robust, the fruit of the first pistillate flower should be kept; if the plant is weak and the leaves are small, the first pistillate flower should be removed before it blooms to concentrate nutrients on the plant's growth, and the fruit of the second pistillate flower should be kept. If the cucumber has a long suitable growth period, two to three seed melons can be left on each plant, and there is room for adjustment of the seed-saving node, the first pistillate flower can be left unseeded. If the suitable vegetative phase of the cucumber is very short, each plant can only have one seed fruit, and the room for adjusting the nodal region of the seed fruit is small. As a result, the fruit of the first pistillate flower should be kept as early as possible. After two to three seed fruits are set, five to six leaves above the seed fruits should be left for the topping to concentrate the nutrients on the seed fruits and control the plant's continued growth.

(2) Purification and rejuvenation of stable cultivars.

Cucumber is a cross-pollinated crop since its self-pollination rate is only 30%–35%. It needs to be cross-pollinated by insects. Since some insects fly long distances, they will inevitably transfer a small amount of pollen from other cultivars, even with isolation measures in place. In addition, some recessive traits will gradually be separated after multiple generations of reproduction. These are the reasons why cultivars are becoming more and more mixed. Therefore, during the seed breeding of the cucumber, attention should be paid to isolation as well as the purity of the original seeds. The main methods of original seed selection and cultivar purification and rejuvenation are as follows.

① Individual plant mixing selection. The selection is conducted in the following order: original seed → superior variety → crop field. It is a method suitable for fields of the superior variety with a relatively high variety purity and without traits separation. Seed plants shall be selected three times in a large amount during the early bloom, the early maturity, and the full maturity of seed fruits. During the early bloom, seed strains with low nodal regions of pistillate flowers, early flowering, and strong growth vigor shall be selected for early-maturing cultivars; during the early maturity, selection shall be made considering the commercialism, such as colors, warts, shapes, quality, the rate of pistillate flowers, growth vigor, etc.; during the full maturity of seed fruits, further selection shall be made by taking colors, shapes, reticulations, and other aspects. After three times of roguing, the seed fruits selected are mixed and collected, which will be used as the original seeds to reproduce the superior variety in the next year.

② Individual plant selection. The selection is conducted in the following order: individual plant → original seed → superior variety → crop field. In the fields of superior variety with a relatively high variety purity, the excellent individual plants strictly selected will be cultivated into strains as the seeds of a single fruit in the next year. Among different strains, superior strains with traits that meet the requirements shall be further selected and those inferior strains shall be eliminated. After the seeds are kept as the breeder seeds from the superior strains, the breeder seeds will reproduce the original seeds which will then reproduce the superior variety.

③ Purification of self-crossing plants. The selection is conducted in the following order: individual plant self-crossing → separation (excellent plants self-crossing) → plant seed retention → original seed → superior variety → crop field. When the separation of cucumbers happens after self-crossing, the excellent plants after separation will continue to self-cross. After about two to three generations, the traits will be basically stable and the strains will no longer separate. The plants are propagated into the original seeds, which will be propagated into the superior variety.

The purification of self-crossing plants is a very effective purification method. After self-crossing, the viability of cucumber is not significantly weakened. Besides, the more generations there are, the higher the variety purity.

Since cucumber is a cross-pollinated crop, it requires strict isolation and manual pollination during self-pollination and purification. The specific steps are as follows.

A. Selecting plants. When the first pistillate flower blooms while there are no other pistillate flowers, excellent plants with required cultivar traits shall be selected and marked.

B. Binding flowers. To prevent insects from bringing in the pollen of other strains, the large buds that have turned bright yellow of the staminate flowers and the pistillate flowers shall be tied up the day before blooming. In this way, even if the flowers bloom the next day, their buds can not open so that insects cannot burrow into the buds to get the nectar. There are several ways to bind flowers. For example, plastic grafting clips can be used to clip the corollas, or 5A fuse wires can be used to tie up the corollas. The flowers shall be bounden in the middle of the petals. If the binding position is too close to the sepals or at the top of the buds, the flowers will open as usual, which cannot prevent insects.

C. Methods of pollination. First, unbind the pistillate flower in the morning of blooming and pick off the staminate flower. Then, peel off the petals and apply anthers gently on the stigma of the pistillate flower. After that, bind the pollinated corolla of the pistillate flower again to prevent re-pollination by insects. At last, hang a card that indicates the cultivar, plant number, and pollination date. When pollinating another plant, fingers shall be wiped with alcohol to kill the residual pollen on them so as not to carry the pollen to another pistillate flower of other plants.

(3) Isolation, roguing, and assisted pollination.

If the purity of the original seeds is high enough that they do not need further purification, isolation between cultivars and roguing can be used for seed breeding. From the occurrence of the first pistillate flower to the harvesting of the seed fruits, great attention shall be paid to roguing, which means eliminating seed plants that do not meet the traits of this cultivar at any time. If lateral branches are found on the plants, or the nodal region of the first pistillate flower is too different from that of this cultivar, or the fruit shape, warts, color, etc. are different from the traits of this cultivar, the plants shall be pulled out in time. Besides, deformed seed fruits and rotten fruits shall be eliminated. As for isolation, it can be divided into two kinds:

space isolation and net room isolation. In addition, it is also necessary to prevent the introduction of pollen from sporadic cultivated cucumbers planted in front of and behind each house. The net room isolation requires artificially-assisted or insect-assisted pollination.

Cucumbers are parthenocarpic, but there is no seed inside their fruits. Therefore, it is a must to pollinate in order to collect seed fruits. Insects such as bees, butterflies, and flies are all pollinators. If insecticides are sprayed continuously, which kills the pollinators, the collection of seed fruits will be greatly influenced. Therefore, insects shall be protected during the anthesis in the seed collection fields. If conditions permit, bees shall be stocked to increase pollinators. When collecting seeds in a greenhouse or an open field in rainy weather and there are few insects, artificially-assisted pollination is required. The method is similar to the manual pollination of purification of self-crossing plants, except that there is no need to wipe fingers with alcohol when pollinating another plant. Besides, a brush can be used to apply anthers on the stigmas. Repeated daily pollination during the flowering and fruit-setting period can significantly increase seed production.

(4) Harvesting seed fruits and threshing.

① Harvesting seed fruits: About 25 days after the fertilization of the pistillate flowers, the seeds can germinate, but they are not full. Seeds will only be full 40–45 days after fertilization. Therefore, the seed fruits shall be left on the plants to reach full maturity. In order to make the seeds plumper, they shall be kept in the shade for after-ripening for one week.

② Threshing: When cleaning the seeds, the seeds and the pulp shall be taken out together and placed into a tank for fermentation. It is worth noting that a metal container shall not be used as it will make the seeds black. Besides, water shall not be added to the fermented seeds as water will dilute the substances in the pulp that inhibit the germination and the seeds will germinate during the fermentation. The fermentation time depends on the temperature. When the temperature is high, the seeds will be fermented in one or two days. As a result, the seeds will sink after being separated from the pulp, leaving the pulp floating. When the seeds are over-fermented, they will turn gray and lose sheen, and even the germination rate will be affected. When the fermented seeds are rinsed in clean water, the plump seeds will sink and those floating ones are not plump, so they shall be washed away together with the fermented pulp. After being cleaned, the seeds shall be placed on the reed mat or mesh to dry for 1–2 days. Do not expose them to the sun on the cement floor, which will burn the seeds and reduce the germination rate. Qualified seeds are white, without foreign matter, and with moisture content lower than 9%. The germination rate is over 95% and the thousand-seed weight is 25–30 g.

For a small number of seeds or the seeds of individual plants, the seeds can be cleaned directly without fermentation. The method is to put the seeds and pulp into the gauze, which shall be rubbed and washed in a basin filled with water so that the seeds will be separated from the pulp. Then, put the seeds and pulp into the basin; the seeds will sink, while the pulp and non-plump seeds will float. Clear out everything that floats, and clean seeds can be got. By doing this, the seeds are whiter and shinier than the seeds cleaned through fermentation and they have a higher germination rate.

The amount of seeds collected from each cucumber is related to the cultivar and the quality of pollination. Usually, every seed fruit of cucumber contains 100–200 seeds, and the amount can sometimes reach 400–500. The number of seeds has little to do with the size of the seed fruit. As for commonly bred seeds, an amount of 15–30 kg seeds can be collected per mu, and the maximum can reach more than 50 kg.

4. Seed Production Techniques of F_1 Hybrids

The F_1 hybrids of cucumber have strong heterosis and have a significant effect on increasing yield. Most of the new cultivars of cucumber cultivated in recent years are F_1 hybrids. The parental inbred lines for F_1 hybrids shall be produced according to the techniques used to produce the original seeds. If the inbred lines are not high in purity, they shall be purified according to the purification and rejuvenation method of the original seeds. The main methods of seed production of F_1 hybrids are as follows.

(1) Artificial hybridization seed production.

Since cucumbers are monoecious and have large flowers, it is easy to carry out artificial hybridization. Therefore, as long as there is sufficient manpower or the seed production area is small, artificial hybridization seed production is feasible and has been applied in production.

① Parental cultivation. The management techniques and production and cultivation are basically the same in parental sowing and strong seedling cultivation. The differences are as follows: the anthesis of the parental lines shall be paid attention to with the principle that the staminate flower of the father shall open earlier than the pistillate flower of the mother and the father shall generally be sown 5–7 days earlier; during the seedling-raising period, the father shall be appropriately given higher temperature to induce earlier and more occurrence of the staminate flowers of the father, which can meet the needs of the father's pollen during hybrid pollination; when planting, the ratio of the father and the mother shall be 1 : 3–1 : 6. Besides, they can be planted in alternate rows or two places.

② Cross pollination. Before hybrid seed production, the parents shall be selected first. Plants with typical traits of this cultivar, robust growth, and no pests and diseases are required. The mother shall be the second, the third, or subsequent pistillate flowers. Other staminate flowers, staminate buds, pistillate flowers that are open, and root cucumbers shall be removed. In the afternoon before pollination, the pistillate buds of the mother and the staminate buds of the father shall be tied up by grafting clips or other binders. The number of staminate flowers being tied up shall be more than that of the pistillate flowers. On the morning of flowering and pollination, the staminate flower shall be picked off and the corollas shall be removed. Then, open the corollas of the pistillate flower that blooms on the same day, and apply the stamen to the stigma of the pistillate flower. The pollinated pistillate flowers shall be isolated in bunches, and a card shall be hanged or colored lines shall be tied to them as a mark indicating the days of pollination and the authenticity of the seed fruits. Pollination is best done in the morning. Each staminate flower can pollinate three to four pistillate flowers. When the seed fruits are steady, other flowers and fruits shall be removed. Topping can be done after the 20th nodal region. Two to three seed fruit shall be kept on one plant where three to four pistillate flowers are pollinated. During the pollination period and the maturity stage of the seed fruits, it is necessary to constantly check and remove the unpollinated young cucumbers, and pay special attention to the removal of the cucumbers that grow on the lateral branches.

③ Seed fruit harvesting and seed collection. When the pollinated seed fruits start to turn yellow or brown, they are ready to be harvested. When harvesting, the fruits shall be picked one by one according to the mark, and shall not be mixed with the fruits on the father plants. Other requirements are the same as those for the seed production of conventional cultivars.

(2) Seed production by the female line hybrids.

Common cucumbers are monoecious, which means that there are both pistillate flowers and staminate flowers on the plant. However, the cucumber plants of the female line only have pistillate flowers, no staminate flowers. Therefore, when the female line is used as the mother to produce F_1 hybrid seeds, there is no need to carry out emasculation and the purity of the seeds is high. Besides, the female line is dominant to the monoecious cucumbers. Therefore, the F_1 hybrid seeds produced through the female line have strong female traits, such as early maturity, dense flower, concentrated harvesting period, high yields, etc.

① There are two methods of selecting and breeding female lines and preserving cucumbers of female lines. One is field breeding and selection. There are female plants or strong female plants with only 1–2 staminate flowers in cucumbers of many cultivars. Through the purification of multiple-generation self-crossing, cucumbers of female lines can be bred. The other one is trans-breeding, which means that the bred female lines are used to cross with the monoecious cucumber. Through the purification of multiple-generation self-crossing, the female lines with traits that meet the breeder's requirements can be obtained.

The purification and preservation of cucumbers of female lines are based on the principle that the flower buds of cucumbers are not determined in the early stage of flower bud differentiation and they will develop to staminate flowers after being treated with chemicals. When it is not determined whether it is a female plant at the seedling stage during the selection and breeding of female lines, let the plant continue to grow. When it has 5–6 true leaves, remove the growth point of the female plant. Then, place absorbent cotton with 0.2%–0.4% gibberellin on the wounds and let the gibberellin slowly enter the plant to induce the formation of pistillate flowers. After 3–5 days, lateral branches will appear. Leave the strongest one and remove the others. Staminate flowers can be induced at the 5th–6th nodal regions of the branch in about 15 days. The female lines can be preserved through self-crossing or sister-crossing. In this way, the selection of female lines can be carried out after several generations. As for the reproduction of the plants that are already female lines, spray 0.2%–0.4% gibberellin solution on the leaves and growing points when there are two leaves and one bud during the early period of flower bud differentiation, and spray the solution every five days; or spray silver nitrate solution with a concentration of 300–500 mg/L when there are two leaves and one bud, and spray the solution every 3–4 days. A total of 3–4 sprays can induce the growth of staminate flowers. If there are good isolation conditions, insects can be used for natural pollination; if there are not, artificial pollination can be applied. No matter whether the solution is gibberellin or silver nitrate, the induction rate of staminate flowers is generally only about 40%. Besides, there might be the problem of the anthesis not being the same. Therefore, artificially-assisted pollination of sister-crossing should be applied for the reproduction of female lines.

② Seeds can be produced after the excellent hybrid combination of the seed production of female lines is decided. The planting ratio is 3 ∶ 1, that is, three rows of mother cucumbers of female lines and one row of common father cucumbers. Since the cucumbers of female lines bloom early, the father cucumbers shall be sown about seven days in advance to ensure that the parental plants have the same anthesis. The female cucumber itself has a certain proportion of heterozygotes, most of which account for about 20%, while high-purity ones also account for 4%–5%. Generally, for heterozygous plants, when there are 3–4 leaves, staminate buds appear in the leaf axils; while for the pure female plants, when there are 3–4 leaves, neither pistillate buds nor staminate buds will appear in the axils of each leaf. Heterozygous plants can be identified before planting. Heterozygous plants with many staminate flowers shall be eliminated before planting. After planting, there will be strong female plants with a few staminate flowers in the 1st–2nd nodal regions, while there are all pistillate flowers above the 5th–6th nodal region. As for these strong female plants, although they are not pure female plants, there is no need to remove them as long as the staminate buds are picked in time. In the later stage, a small number of staminate flowers may appear above the 10th nodal region because of conditions such as temperature. These staminate flowers should be removed and topping should be carried out in time during the bud stage. When the female line is used as the mother to produce F_1 hybrid seeds, hybrid seeds can be obtained by allowing them to be naturally pollinated during the anthesis, which greatly saves the workload of pollination. On some occasions, in order to increase seed yield, artificial-assisted pollination can also be applied.

(3) Seed production through chemical emasculation and natural pollination.

Large-scale production of F_1 hybrids relies on artificial pollination, which requires a lot of labor and high costs. In order to solve this problem, based on the principle that the flower buds of cucumbers are not determined in the early stage of flower bud differentiation, the method of chemical induction can only allow the mother plant to grow pistillate flowers.

The currently used cucumber emasculation agent is ethephon. When using ethephon for cucumber emasculation, 40% of the ethephon stock solution should first be diluted with water by 1,200–1,500 times, and the concentration should be 250–300 mg/L. For early-maturing cultivars of the mother cucumber, chemical treatment shall be carried out when the two true leaves of the seedlings are just unfolded; for medium- and late-maturing cultivars, chemical treatment shall be carried out when the two true leaves are fully unfolded and there is at least one inner leaf. Spraying shall be carried out in the morning or evening when there is no wind, during which the air humidity is relatively high and the solution can stay on the leaves for a long time for absorption. Pay special attention to the treatment of seedlings from seeds, during which the nozzle shall be lowered. It is strictly forbidden to spray or drift the solution onto the father. After that, spray the solution every 4–5 days, for a total of 3–4 sprays. The early growth of treated seedlings may be inhibited to varying degrees, such as shortened internodes and slower growth, which are all normal. After spraying, the temperature of the seedbed shall be appropriately increased. The temperature shall be kept at 25–30°C during the day and 18–20°C at night. Besides, 0.2% urea or 0.2% potassium dihydrogen phosphate shall be used for foliar topdressing. Seedlings from seeds shall be watered combined with topdressing of 10 kg of urea per mu to promote growth.

The treated seedlings shall be planted in alternate rows according to the parental ratio of 1 : 2 when they grow to the standard of planting, and about 5,000 plants shall be planted per mu. When the concentration of ethephon is appropriate and the treatment is done properly, the flowers that appear on the 10th–15th nodal regions of the mother plant are all pistillate flowers. When the flowers are naturally pollinated by insects, the seeds obtained from the mother plant are the F_1 hybrids. However, when emasculation is carried out through ethephon, due to the purity and concentration of the chemical solution, the cultivars of cucumber, the weather conditions of spraying, etc., there will still be individual staminate flowers on the treated mother plant, especially above the 15th nodal regions. Although there are very few staminate flowers, seed purity will be greatly influenced. Therefore, field inspection must be carried out plant by plant to remove staminate flowers on the mother plant during the bud stage. When the seed fruits are set, topping shall be carried out in time to prevent the occurrence of staminate flowers on the top. In this way, false hybrids can be avoided. The seed fruits' harvesting and threshing of F_1 hybrids are the same as the seed production of stable cultivars.

(II) Pumpkin Seed Production Technology

Pumpkin is native to the tropics and is an annual herb of the Cucurbitaceae family. There are three main types of pumpkins cultivated in China, namely Chinese pumpkins, Indian pumpkins, and American pumpkins. Chinese pumpkins are also known as pumpkins, cushaws, and Fan Gua; Indian pumpkins are also known as winter squash and *Cucurbita maxima* Duch.; American pumpkins are also known as summer squash. In addition, *Cucurbita ficifolia* is also cultivated in Yunnan Province. Pumpkins are very adaptable and can grow on land unsuitable for farming, so they are widely cultivated all over the country. With rich nutrition, pumpkin is a traditional vegetable in China. In particular, the cultivation area of

American pumpkins in protected areas have gradually increased in recent years, and pumpkins have become one of the main species of vegetables grown in protected areas.

1. Flowering and Pollination Habits

(1) Flowering habits.

① Gender expression and positions of flowers. Pumpkin flowers are usually unisexual, and there are two flower types on the plant: pistillate flowers and staminate flowers. Chinese pumpkins generally bear the first pistillate flower at the 7th–15th nodal regions of the main vine, and then every 3–5 nodal regions; the lateral vines usually bear the first pistillate flower at the 4th–5th nodal regions, and then every 3–4 nodal regions. The closer the lateral branch is to the base of the stem, the farther the nodal region where its first pistillate flower is located. The lateral branch that is close to the apex of the main vine often bears the first pistillate flower in the 1st–2nd nodal regions. Usually, a pumpkin can bear about 30 pistillate flowers, and the number of pistillate flowers is much higher than that of staminate flowers. The pistillate flower of summer squash is lower than that of Chinese pumpkin, and the pistillate flower starts from the 4th–5th nodal regions of the dwarf cultivar and the 10th nodal region of the semi-vine cultivar. Indian pumpkins have thick stems and large leaves, with strong growth vigor. Early-maturing cultivars start to produce pistillate flowers at the 5th–7th nodal regions of the main vine, and late-maturing cultivars start to produce the first pistillate flower above the 10th nodal region.

② Flower organ structure. Pumpkin flowers are solitary with large corollas. They are bright yellow and have five-lobed corollas. Their petals are combined to form the shape of a trumpet or a funnel. The pedicels of staminate flowers are slender, with five stamens conjoined into the shape of a columnar, and the pollen grains are large. The pedicels of pistillate flowers are stout and the ovary is inferior, usually with three ventricles. Six rows of seeds are born in the placenta, while some have four ventricles and eight rows of seeds. From the morphology of the ovary, the shape of the fruit in the future can be judged. Chinese pumpkins, like other melon vegetables, generally have staminate flowers first, followed by pistillate flowers. But summer squashes sometimes bloom first with pistillate flowers, which can also be seen in Indian pumpkins, making pollination difficult.

(2) Pollination habits.

① Pollinators. Pumpkin is a cross-pollinated crop. With natural pollination, the fruiting rate of cross-pollination accounts for 65%, and that of self-crossing accounts for 35%. A majority of pumpkin plants fruit through cross-pollination. As its pollen grains are large, heavy, and sticky, the wind can't blow them away. As a result, pollination is generally done by bees and ants. As per the test, the fruiting rate of artificial pollination is 72.6%, while that of natural pollination is only 25.9%. Clearly, artificial pollination is extremely efficient in improving the fruiting rate.

② Interspecific hybridization. Interspecific hybridization of plants belongs to distant hybridization. The seed yield obtained by distant hybridization is not high. However, among pumpkin cultivars, some hybridization may see a high fruiting rate. In particular, the fruiting rate reaches 39% and the reciprocal cross-fruiting rate is 41.8% when the Indian pumpkin is used as the mother and the Chinese pumpkin is used as the father. The interspecific hybridization of pumpkin cultivars has created a new way of breeding and brought about some easily overlooked problems of seed retention and isolation.

③ Suitable time and environmental conditions for pollination. Pumpkin flowers are short-lived. Under favorable conditions, the flowers usually bloom from 5:00–6:00. The corollas begin to shut at noon. The staminate corollas begin to wither in the evening, while the pistillate flowers can last for a slightly longer period of time. The best time for pollination is when the stigma secretes a lot of orangish-red mucus after a pistillate flower blossoms. The fruit setting rate is quite high at this time but low when the pollination is done one day before or after the flowering. The pollen of the staminate flower matures one day before it flowers. The fruiting would turn out to be terrible if the pollen collected two days before the flowering is used for pollination. Therefore, the pollination must be carried out in the morning of the day of flowering, preferably before 8:00. After 10:00, the fruiting rate is very low. The major factor influencing the anther dehiscence and flowering of the pumpkin is temperature. The minimum temperature for its pollen maturation is 8–10°C, and the flowering and fruiting require a temperature at least above 15°C.

(3) Fruiting habits.

The shape, size, color, pattern, the time from fruit setting to ripening, and the number of seeds inside the pumpkin fruit varies greatly among species and cultivars (Fig. 4-4).

Like other melons, not every pistillate flower of the pumpkin can fruit. Although a plant has dozens of potential pistillate flowers suitable for fertilization and a large number of staminate flowers, the fruit setting rate is generally around 50%. Since harvesting young melons can increase the plant's leaf number and area, and thus promote photosynthetic productivity and plant longevity, it is necessary to remove excess fruits during seed collection in a timely manner.

The fruit of the pumpkin matures 60–90 days after flowering. The fruit and the seed grow simultaneously. Usually, the seed is ripe when the fruit matures. Each fruit has 300–400 seeds. Some people observed that a pumpkin needs 4.8–5.9 times the amount of pollen to get a seed. Thus, applying more pollen and artificially assisted pollination can improve the seed setting rate. The seed germination potential of fruits harvested 40 days after flowering and without after-ripening is inferior, while the germination rate and the thousand-seed weight are significantly improved when the fruits are harvested 40–50 days after flowering and with 10–20 days of after-ripening.

Fig. 4-4 Pumpkin Fruit

2. Seed Production of Stable Cultivar

(1) Management of seed collection fields.

① Selection of seed collection fields. The seed collection field preferably has medium soil fertility with a pH value of 5.5–6.8. Too much nitrogen fertilizer may lead to over-nourishment and the falling of flowers and fruit. The distance between different seed collection fields of different cultivars should be 1,000–1,500 m in the case of natural hybridization. At the same time, attention should also be paid to the isolation between different pumpkin species. The distance between Indian pumpkins and Chinese pumpkins must be more than 1,000 m. The odds of hybridization between American pumpkin, Indian pumpkin, or Chinese pumpkin are very small, so there's no need for isolation.

② Field management of seed plants. The cultivation management of the pumpkin seed collection fields is basically the same as those of the crop fields. Seedlings can be transplanted or harvested in the open field. Though it saves lots of work to sow the seeds in the open field in spring, they may get virus disease when they run into hot weather after pollination, affecting the seed yield and intoxicating the seed melons in the worst case. The spring seedling transplanting seed collection method can help the seed fruits avoid the outbreak of virus disease before maturity. It is widely used in original seed production.

The best seedling age for pumpkins is 25 days, which can be used as a basis to project the time of sowing and seedling-raising period. It is easy for pumpkins to grow lateral branches and adventitious roots, so they should be planted and managed according to types, cultivars, flowering and fruiting habits, and different seed collection methods. Each mu of land should have 500–2,000 plants. In the original seed production, topping and pruning are especially important. Chinese pumpkin and Indian pumpkin grow vigorously. The pistillate flowers on the main branch come out comparatively late, but their lateral branches bear fruit early. Therefore, the topping of the main branch could be done early to help the lateral branches grow and fruit soon. After the fruit setting of the lateral branches, leave five to six leaves for topping so that no more branches will come out, which is beneficial to the fruit development. The dwarf cultivar of zucchini mainly fruits on the main vine. This cultivar's lateral branches are not developed, so they don't need topping. In addition, for pumpkins cultivated in a crawling way, besides topping, the branches should be pressed to facilitate the rooting at the internode to stabilize the plant and expand the absorption area. Pest and disease control is the priority during the growth of the seed plants.

(2) Roguing, purification, and rejuvenation.

If the original seeds are of high purity, inspections should be carried out in the seed collection field several times to eliminate the impure and inferior plants. If the purity is low, the original seeds should be purified and rejuvenated in due course. Mark the plants that have features in line with this cultivar in a large crop field after the flowering of the first pistillate flower and before that of the second pistillate flower. Then, artificially isolate the 2nd and 3rd pistillate flowers of the marked plant. It means that one day before the flowering, the buds are tied with thin strings, or the corollas are tied with thin lead wires. Meanwhile, the buds of the unopened staminate flowers should also be tied and have artificial pollination early the next morning. During pollination, open the corolla of the pistillate flower. Remove the corolla of the staminate flower on another plant, and apply the pollen of the stamen to the stigma evenly. After pollination, the corolla of the pistillate flower remains tied and should be marked with a tag at the stalk. After the seed melons mature, they should be strictly selected to make sure that they meet the features of this cultivar. Only this can ensure the purity of the original seeds. Pumpkins harvested are mainly old ones. During seed washing, a quality examination should be carried out.

(3) Artificial pollination and melon retention.

Artificial pollination can significantly improve the fruiting rate and seed yield. Therefore, during the anthesis, artificial pollination should be conducted on the pistillate flowers which are opening that day early in the morning after the staminate flowers have dispersed their pollen. The method is to apply the pollen of the staminate flower evenly to the stigma of the pistillate flowers of another plant. There is no need to bind the flowers before or after pollination. The fruit born by the first pistillate flower on the main vine is small, so it is usually not kept as a seed melon and should be removed during the anthesis. The fruits born by the second or later pistillate flower can be kept as seed melons. Early-maturing cultivars have dense pistillate flowers and small fruits, so three to four seed melons can be kept per plant; two to three seed melons can be kept per plant of medium-maturing cultivars, and one to two per plant of late-maturing big-fruit cultivars. Excess fruit should be removed at the early fruiting stage. At the early and super-ripen stages of the fruit, melons that do not conform to the typicality of this cultivar and inferior melons should be eliminated according to their shapes, skin colors, plant growth conditions, etc.

(4) Harvesting and threshing of seed melons.

① Seed fruit harvesting and after-ripening. Usually, the seed melon physiologically matures about 50 days after pollination. When the fruit stalk appears with longitudinal light yellow shallow grooves and the epidermis is hard with a layer of white powder, the melon can be harvested (Fig. 4-5). Do not bruise the rind when harvesting. After picking the seed melon, do not cut it and take out the seeds immediately. Place it in a shaded and ventilated storage room and conduct an appropriate after-ripening process for 10–20 days depending on the maturity. The after-ripening time should not be too long. Otherwise, the seeds will germinate inside the melon.

Fig. 4-5 Mature Seed Melons

During after-ripening, seeds of rotten melons should be removed right away. Seeds taken from rotten fruits have bad shapes and are not full, so do not mix them with other good seeds. Keep them separate. To quickly find rotten fruits, do not pile a lot of fruits. Spread them out as flat as possible.

② Threshing and drying. Threshing must be carried out in the morning of a sunny day. If it is carried out in the middle of the night or meets the rain, the seeds will be moldy, seriously affecting the appearance of the seeds and reducing their commercial value.

A. Direct threshing. Cut the melon longitudinally with a knife and then break it by hand. Squeeze the seeds out of the pulp and bake them directly after removing the pulp left on them.

B. Water washing threshing. Take out the seeds together with the pulp, filter and squeeze them with a sieve (metal sieve is prohibited) to separate the yellow juice and other debris from the seeds. Then, put them in an appropriate amount of chaff, rub away the spongy material on the seed skin, put the seeds into a non-metallic container, wash them with water, and then fish out the seeds floating up.

C. Drying. Spread the washed seeds on a thin layer of window screen (which should be separated from the ground) and place them in a sunny and ventilated place to dry. Turn them after the surface is dry. Leave them in a ventilated place at night and continue to dry in the sun the next day. When the seeds are dried to the extent that there is a clattering sound when they hit each other and that the moisture drops below 8%, they can be put into cotton bags or woven bags (paper bags or plastic bags are strictly prohibited) and stored in a ventilated and dry place. It must be ensured that accidental damage by rodents, birds, insects, livestock, etc. will not occur.

After drying, move the seeds outside every four to five days on sunny days. Turn to dry them to prevent moisture and humidity.

3. Seed Production Techniques of F_1 Hybrids

F_1 hybrids are mainly used for zucchini and small pumpkins. Due to the large flowers of the pumpkin, artificial pollination is simple and labor-saving. A pollinator can pollinate 200–300 flowers half a day, and one melon can produce 100–400 seeds. The propagation coefficient is high and the production cost is low. Therefore, artificial hybridization is widely used in hybrid seed production. Taking zucchini as an example, the key points of the pumpkin seed collection technology will be briefly introduced below.

(1) Cultivation management of seed production fields.

Zucchini is a short-day plant. Low temperature and appropriate short daylight are conducive to the fruit-bearing of the pistillate flowers. Temperature management should be enhanced in the seedling-raising period to promote the formation and development of pistillate flowers. Generally, it is best to keep eight hours of sunshine per day and keep the temperature at 20–25°C during the day and about 12°C at night. If the temperature is too high, for instance, over 30°C during the day or over 16°C at night, the pistillate flowers will be late in formation and also susceptible to virus diseases. Seeds can germinate at 13°C, but the best temperature is 20–25°C. Temperatures higher or lower than this range will be harmful to seeds' germination.

The best kind of seed production field is sandy loam land with a pH value of 5.5–7.5, moderate fertility, and easy drainage and irrigation. The spatial isolation distance of the seed production fields should be over 1,500 m. When there is not enough space for distancing, shed nets are recommended.

Under the condition that there is no night frost damage, seedlings and field planting should be done as early as possible. Seed production should be finished before the hot rainy season. The seedlings of the parents need to be raised in the facility in mid-March. The fathers are usually sown about one week earlier, mainly to increase the number of staminate flowers at the early stage of pollination. Plant the seedlings in the open field in the first half of May. Before the final planting, make sure that enough farmyard fertilizer has been applied to the seed production field. In addition, apply 25 kg of compound fertilizer per mu, and then harrow the ground to make a 1.2 m wide level border and cover the ground with a mulching film. When planting, punch holes in the film at intervals of 45–50 cm. Each level border should have two rows. Plant one row of fathers every three rows of mothers in order to facilitate the pollination afterward. Plant about 2,000 plants per mu.

(2) Emasculation and pollination.

① Artificial emasculation and natural pollination. If the seed production fields have safe isolation distancing, artificial emasculation and natural pollination can be applied. If allowed, bee pollination can also be used. Growing from three to four centimeters to flowering, it takes the zucchini plants about 15 days. During this period, all the staminate flowers on the mothers should be removed. Afterward, check and remove the newly grown staminate flowers on the mothers every four to five days. All staminate flowers of the mothers must be removed completely. Do not miss a single one of them. This is to ensure that all the seed melons of the mothers are hybridized. The inbred line of the fathers still has pistillate and staminate flowers. They bear melons through interplant pollination in this line, so the inbred line seeds of the fathers can be obtained together with the hybrid seeds. A separate isolation area should be set for breeding the inbred line seeds of the mothers.

② Artificial flower-binding and isolated pollination.

A. Mechanical isolation. If neither space isolation nor net room is allowed, flowers can be bound artificially so as to be isolated. Then, adopt assisted pollination to breed F_1 hybrids. The method is that in the afternoon of the day before flowering, tie the staminate flowers of the fathers and the pistillate flowers of the mothers that are about to open the next day with thin strings or lead wires. Old newspapers can also be used to make paper tube sleeves. Then, clip them on the buds with clamps or paper clips and stick slender bamboo poles next to them as marks.

B. Collect staminate flowers of the fathers. In the evening of the day before pollination, pick the staminate flowers that are to open on the second day (leave the flower stalk as long as possible), bundle them with rubber bands or twine, put them neatly in plastic bags, and bring them home, fill a washbasin with water or sand indoor, and insert these staminate flower stalks in the water or sand. The petals must be put upwards without splashing water on them. Cover them with gauze to prevent insects from polluting the pollen.

C. Artificial pollination. The mothers of Zucchini bloom particularly early. They usually open at 4:00 if the temperature is appropriate. If the anthers of the staminate flowers dehisce, try to carry out pollination at dawn every day. The tools needed for this are staminate flowers, rubber bands, newspaper bags, clamps (or paper clips), and hybridization markers (colored plastic strips). Put them in a basket. When bringing the fathers to the field for pollination, use plastic bags to cover the flowers to prevent out-pollination. The specific steps for pollination are as follows.

a. Remove the paper bags that cover the pistillate flowers and put them in the basket, open the corollas of the pistillate flowers, and then remove the corollas of the staminate flowers by hand, leaving only the stamens. Hold the petals of the pistillate flowers with the left hand and use the right hand to evenly apply the pollen of the stamens to each stigma. In the early stage, there are few staminate flowers. One staminate flower can pollinate three to four pistillate flowers; in the middle stage, more staminate flowers emerge. It is better that one staminate flower pollinates one to two pistillate flowers.

b. After pollination, the pistillate flowers of the mothers should continue to be isolated to prevent insect pollination. Hold the petals with the left hand, gently put the paper bags on them, fold up the lower openings and secure them with clips to prevent the bags from falling off. Tie a hybrid marker string to the node of the upper vine where the pollinated pistillate flower lies. Do not tie the string on the fruit stalk, so as not to cause the fruit to drop.

c. The best time for pollination is from 4:00–8:00. During the pollination, if the pistillate flowers without bags on them have opened, remove them in time to prevent biological mixing from affecting the purity.

D. Post-pollination management. The pollination period usually lasts about a week. Generally, two to three seed melons are kept per plant, depending on the growth of the mother. When two to three seed melons are set, topping can be done at the 30th node.

Check the seed production fields carefully about 20 days after fruit-setting to confirm the hybrid marks. If there are unmarked and inflated fruits, remove them promptly to ensure purity.

(3) Harvesting and threshing of seed melons.

After the collected seed melons are done with after-ripening, cut the melons and take the seeds, the method of which is the same as the seed production of the stable cultivar. Generally, about 40 kg of hybrid seeds can be obtained per mu.

(III) Watermelon Seed Production Technology

Watermelon, the *Citrullus* genus in the Cucurbitaceae family, is an annual trailing herbaceous plant. It is native to Africa and is widely cultivated all over China. It is sweet, juicy, and refreshing, making it an excellent summer fruit. In addition to being an edible fruit, it also has some medicinal value. In the past, it was mostly cultivated in the open fields in the north of China and only supplied in summer. In recent years, with the continuous improvements in facilities, vegetable cultivation technology, and people's living standards, the out-of-season cultivation of watermelon in protected facilities has achieved high economic benefits.

1. Flowering and Pollination Habits

(1) Flowering habits.

① The indication of the flower's sex. Watermelon flowers grow at the axils. The flowers are unisexual, and the plants are monoecious with different flowers. The staminate flowers usually come out first before the pistillate flowers. From the 6th to the 13th node, each node bears one or several staminate flowers. The first pistillate flower of the early-maturing cultivar blossoms at the 6th to the 7th node, while the first pistillate flower of the late-maturing cultivar blossoms after the 10th node. Then, one pistillate flower is borne every seven to nine nodes (for both main and lateral vines). The ratio of pistillate flowers to staminate flowers is about 1 : 10–1 : 20.

② Floral organ structure. A watermelon flower (Fig. 4-6) is bright yellow and has five sepals, five corollas, and a tube-like base. The corollas of a staminate flower are larger and darker, while those of the pistillate flowers are smaller and lighter. A pistillate flower has three anthers, many stamens that separate from each other, and anther cells. When the anther dehisces, the process starts from its back. The pistillate flowers have inferior ovaries, and their sizes and shapes vary with cultivars. The pistil is located at the base of the corolla. The stigma is short and becomes three lobes at maturity. Pistillate flowers of some cultivars and some plants have fully developed stamens and vibrant pollen. These flowers are pistillate hermaphrodites.

Fig. 4-6 Watermelon Flower

(2) Pollination habits.

The stigma of the watermelon pistil and the anther of the stamen both have nectaries. It relies on insect pollination and is a typical cross-pollinated plant. Natural hybridization easily occurs among different cultivars. The ovary of the pistillate flower is already well-developed before flowering. It blooms around 5:00. Three to five minutes after flowering, pollen appears. The best time for pollination is 6:00–8:00. Three hours after pollination, the pollen tube reaches into the style, and after 23 hours, the fertilization process is completed, after which the ovary expands rapidly and bears fruit.

(3) Fruiting habits.

The ovary size of the pistillate flower is related to the nutritional conditions of the watermelon. Usually, at the early stage, the ovary of the pistillate flower is small, while the ovaries of the second to third pistillate flowers on the main vine or lateral vine are larger. When the ovary is large and full, the fruit-setting rate is higher. About 20 days after fruit-setting, the seeds can basically germinate. For the fruit, the time from flowering to seed maturing varies with cultivar maturity traits. Usually, early-maturing cultivars need 28–30 days, and late-maturing cultivars need about 40 days. In summary, seeds mature later than fruits. The fruit of a watermelon (Fig. 4-7) consists of three parts: the rind, the pulp (placenta), and the seeds. It is generally round or oval. The rind has dark green, light green, black, and white colors with stripes or streaks on it. The ripe pulp has red, yellow, and white colors. The seeds are flattened ovoids or ellipsoids in white, light yellow, brown, black, and red colors. The number of seeds in a melon varies, generally 400–500. Some cultivars can have as many as 1,000. The thousand-seed weight of different cultivars varies greatly. The thousand-seed weight of the small-seeded cultivars is about 20 g, medium-seeded cultivars about 50 g, and large-seeded cultivars 80-100 g.

Fig. 4-7 Watermelon Fruits

2. Seed Production Techniques of F_1 Hybrids

(1) Management of seed production fields.

① Soil selection. For watermelon seed production, it is recommended to choose sandy loam of high terrain, with a pH value of 5.5–6.5, loose and fertile soil texture, and easy drainage and irrigation conditions. Plots with melon crops planted in the previous round or in the past two years and plots prone to flooding are not suitable for watermelon seed production. Beware of residual phytotoxicity if herbicide was previously applied to over 200 g of crops per mu. A safe distance of 1,000 m must be maintained among different combinations or cultivars.

② Soil preparation and fertilization. Apply 3,000–5,000 kg of decomposed farmyard manure per mu and then plow the land to make ridges. If the seeds are produced in plastic arched sheds, the plants are mostly cultivated by hanging vines in an upright position. To increase yields, 4,000 plants should be planted per mu. Plow to make wide and narrow ridges. The wide ridge is 0.7 m in width and the narrow ridge is 0.5 m. Make the plant spacing 20–25 cm. Set longitudinal waterways in the middle of the plastic tunnels. The ridges are perpendicular to the direction of the plastic tunnels. The wide and narrow ridges can facilitate manual handling, ventilation, and light penetration. If the seeds are produced in the open field, crawling cultivation can be adopted. 2,000 to 2,200 plants are planted per mu with the ridge 1.1 m in width, and the plant spacing 0.3 m. On top of applying sufficient farmyard manure, fertilizers such as diammonium phosphate and potassium sulfate should be applied during sowing or final planting to promote growth.

(2) Sowing and seedling-raising.

The mothers and fathers of hybrid seeds should be stored and handled separately to prevent mixing. The ratio of the fathers to mothers to be sown is 1 : 4–1 : 5. The fathers should be sown 7–10 days earlier than the mothers. The specific sowing period can be determined by the type of protection measures, the cultivars, and the date of the late frost. When the seeds are to be produced in the open field, sowing can be done in plastic sheds in the first half of April. After sowing and before seedling emergence, the temperature should be maintained at 25–30°C, which is conducive to the growth of the seedlings.

Watermelon parents should be planted separately. The father seedlings are placed in a high-temperature zone (30°C), and the mother seedlings are placed in a medium-temperature zone (25°C). During the seedling stage, temperature management is very important, especially during the flower bud differentiation of the first two to three cotyledons. During the differentiation, if the external temperature is unstable, hermaphroditic flowers will easily emerge in the late period. Hence, after the seedlings emerge, the temperature should be maintained at 11–17°C during the night and 20–28°C during the day. The maximum temperature should not exceed 28°C. Increase the amount of airflow in the late seedling-raising period and before final planting to acclimatize seedlings. Control water during the seedling-raising period, making sure no water is given when there is no wilting. Give less water in the late seedling-raising period. Do not flood the seedlings. The seedling age is about 20 days. The final planting can be carried out when the seedlings have three to four leaves. If the seedling age is too long, the main root will easily penetrate the nutrient bowl, resulting in suberification in the main root when transplanting. As a result, the survival rate of plants is affected.

(3) Final planting and field management.

Furrow the cultivation bed to conduct final planting, and place the plants according to the fixed plant spacing. Spread chemical fertilizer between plants, apply 15–20 kg of diammonium phosphate and 10–15 kg of potassium sulfate per mu, and recover the earth of the ridge after watering the planting. After planting, cover the seedlings with mulch. Use black mulch to prevent weeds.

(4) Watermelon cross-pollination.

① Preparation before pollination. Hybrid seed production should be done in open fields in the northern regions. Mid-June is the best time for pollination. The following preparations should be taken care of before pollination.

A. Roguing in the father field. It should be done plant by plant and thoroughly. Remove the impure plants in time. Since the father plants are only used for pollen supply, they do not need to be pruned and should be left to grow naturally. Remove pistillate flowers and self-crossed fruits in case they affect the growth of staminate flowers.

B. Mother field clearing. The mother field should be checked and cleared plant by plant. Remove the impure plants and inferior plants. When checking, lift up the vines, and carefully remove all the lateral branches, staminate flowers, and self-crossed fruits from the vine. Particularly, the top of the vines and the buried part of the vines must be carefully checked. The lateral branches and flower buds that are removed when clearing the field should be carried out of the field every time to prevent them from rotting and spreading diseases. When producing seeds in the open field, the clearance of adjacent plots must be coordinated and carried out simultaneously to avoid mutual pollution. If the cleaning is thorough, there will be far fewer staminate flowers in the pollination period. After farmers clear the field, technicians must check whether the work is qualified. Only then is pollination allowed to start. If the seeds are produced in a cold shed, the wind vents need to be covered with a screen. Prepare isolation caps and marking ropes and hire workers. Each worker is responsible for pollinating 600–800 mother plants per day.

C. Isolation of pistillate flowers of mothers and handling of hermaphrodite. For watermelon hybrid seed production, the second or third melon is preferred. The first melon should be removed. Half of the first melon can be pinched off as a marker when clearing the field. One day before the pollination, select the pistillate flowers that are to open the second day and that normally develop with fat ovaries and bright and shiny colors. Place paper caps on the flowers for isolation. Make sure the paper caps are the right size. To facilitate finding the marked flowers on the second day, small sticks can be inserted into the ground when placing the paper caps. If hermaphrodites appear on the mothers in the seed production field, they should be removed or emasculated at the bud stage to ensure seed production purity. Generally, when hermaphrodites take 30%–40% of the whole field, it is better to remove them. When they take 60%–70% of the field, they must be emasculated at the bud stage. That means the buds of the pistillate flowers need to be opened one by one to identify their sex when placing the paper caps. If it is a hermaphrodite, remove the male anthers on the side of the pistillate flower's stigma with tweezers. Do not injure the stigma of the pistillate flower during operation; otherwise, the pollination will be affected (deformed melons will easily be produced). After one to two days of pollination, when the operator is gradually skilled, the hermaphrodites can be identified from the appearance of the pistillate flowers' buds without opening the buds. If the pistillate flowers' buds are slightly flat, large, and kind of deformed, they are most likely to be hermaphrodites. The emasculation of the hermaphrodites during the bud stage must be carefully and strictly controlled; otherwise, self-crossing is likely to happen and then affect seed purity.

② Collection of staminate flowers' buds of fathers. Early in the morning (around 5:00) on the day of pollination, select large, bright yellow buds that are not open but will open on that day for pollination. At this time, the staminate flowers' pollen is not yet mature. After one to two hours of after-ripening, it can be used for pollination. The staminate flowers that open after 7:00 or 8:00 have been contaminated by insects and should never be used for pollination; otherwise, it is difficult to ensure purity. Staminate flowers picked the previous evening can also be used for pollination, but the effect is undermined. Therefore, they are not recommended.

③ Cross-pollination. The pistillate flowers usually open from 7:00 to 8:00 in mid-June. The time of opening depends on the temperature of the day. The pistillate flowers are ready for pollination once they open. First, remove the isolation cap and open the pistillate flower's corolla to reveal the stigma while removing the corolla of the father's staminate flower. Second, gently and evenly apply the pollen to the pistillate flower's stigma. Make sure the process is done gently without hurting the stigma. The paper cap will be re-set after pollination, and should be firmly set to prevent pollution when the cap falls off.

If a hermaphrodite is found during pollination and is yellow with pollen on it, this flower must be removed. If an open pistillate flower is found when there is no isolation cap on it or when the paper cap has fallen off, it shows the flower has been contaminated by insects and should be removed. The best pollination time is from 8:00 to 10:00 when pistillate flowers' stigmas are filled with mucus. The effect of pollination will turn out bad if it is done when some oil-like object appears on pistillate flowers' stigmas. The pollination effect is not good when pistillate flowers' stigmas are damaged by external factors.

④ Marking. After the pollination is completed, tie a thin colored string as a marker at the nodal region where the stem of the melon is born while placing the paper cap. The marker can either be placed at the front or the rear of the melon. However, the markers of the whole field should be placed in the same way. The marker must be tied to the vine, not to the petiole; otherwise, the markers cannot be found when the leaves fall off. Change the color of the string every day so that the melons can be harvested uniformly when they mature. Use one string for one melon. Do not tie two melons with one string.

⑤ Key points of field inspection during the pollination period.

A. Check the seed production fields every morning from 3:30 to 4:00 to thoroughly inspect and remove any staminate flower buds left in the fields. At the same time, isolate the pistillate flower about to open with a cap.

B. After the start of pollination, no naturally opened pistillate flower without cap isolation is allowed in the seed production field. If found, it must be removed; otherwise, it is very easy to produce self-crossed fruit.

C. In the field inspection, if a pollinated pistillate flower is found with no isolation cap or its paper cap has fallen off, it should be removed even if the flower is still fresh. That is because biological mixing may occur due to insect contamination when the cap is off. Even if the flowers of the fathers are not enough, no opened staminate flower should be used for pollination.

⑥ Post-pollination management.

A. Water and fertilizer management. When the pollinated melon grows to egg size, apply 15–25 kg potassium sulfate compound fertilizer per mu in a ring shape that is 15 cm from the main root to prevent root burning. At this time, the melon is growing vigorously and must be watered frequently. When it does not rain that much, watering should be done once a week. Be aware of the weather forecast while watering, making sure that there are one to two sunny days after watering. Stop watering 8–10 days before fruit ripening to prevent melon cracking.

B. Pruning and melon retention. Press down the vine at the third node in front of the first melon with a small branch or soil block. When it is confirmed that one melon has been set on each plant, leave five to six leaves before the melon and conduct topping while removing the excess staminate and pistillate flowers on the vine to prevent new self-crossed fruits. A small lateral branch may be left in front of the melon as appropriate.

C. Seed production field clearing. For seed production in open fields, each seed producer should be required to end pollination uniformly to prevent any staminate flowers from opening and polluting neighboring fields afterward. Clear the seed production field again to remove all unmarked, vaguely marked melons, as well as impure plants with different melon types and colors, to eliminate all potential problems that will affect purity. Thoroughly remove the fathers. The father melons are not allowed to be sold as commercial melons to prevent the mechanical mixing of seeds.

(5) Harvesting seeds.

It takes generally 40–45 days for a watermelon to mature from pollination. Before harvesting, pick a few melons to confirm whether they are fully ripe. Collect seeds in stages according to pollination markers. Put the harvested seed melon in a cool, dry, and ventilated place. After five days of after-ripening, the seeds of the melon can be taken out. Put the pulp of the melon into a non-iron container to ferment for 24 hours, after which the seeds can be washed with water. Be careful of the rain and exposure to the sun during the fermentation. When washing, rinse off the blighted seeds and all impurities. The seeds washed on that day should be dried on the same day if possible. Use gauze to dry the seeds and put the seeds in a cool and ventilated place. Stir them frequently and make sure they are not under direct hot sunlight. The dried seeds should not be immersed in the water again; otherwise, they will discolor. When the moisture content of the seeds drops to 8%, they can be bagged and placed in a dry and ventilated place. Seeds of the immature seed melons must be washed after 7–10 days of after-ripening. Seeds produced by diseased and rotten fruits must be harvested separately.

3. Seed Production Technology for Conventional Cultivars

(1) Isolation of seed collection fields. Seed collection fields need to be isolated from the fields of other cultivars at a distance of more than 1,000 m.

(2) Seed melon selection and artificial pollination. When the vine has 8–12 true leaves, the pistillate flower begins to bloom. Leave the second melon on the main vine and the first melon on the lateral vine as the seed melons. Once the seed melons are set, select the melons again. Leave one of the two seed melons on each plant and thin out the other one. Artificially assisted pollination can significantly improve the single melon fruiting rate and the number of seeds. See the section about hybrid seed production for more information. For original seed reproduction, the individual plant should be bagged to make it self-crossed and marked after pollination.

(3) Roguing and selecting. Seedlings must be selected during final planting, and plants also need to be selected before pollination. Impure plants and sick and weak plants that do not match the characteristics of this cultivar should be eliminated. When harvesting seed melons, eliminate the melons that do not match in fruit shape, fruit color, and pulp color or have bad pulp quality.

(4) Seed melon maturity and seed collection. Watermelon generally takes 30 – 40 days from flowering to physiological maturity, so it should be harvested five to six days later than commercial melon. After three to five days of after-ripening of the harvested seed melons, the seeds can be taken out. Cut the melon horizontally,

and put the seeds with the pulp into a tank or non-metal container. Ferment them for half a day, then wash the seeds and dry them in the sunlight. When the moisture content is less than 8%, pack up and store them.

4. Key Technical Points of Seed Production for Hybrid Cultivars

Watermelon seed production normally adopts artificial hybridization. The sowing period of the fathers should be five to six days earlier than that of the mothers, and the final planting ratio of the fathers to mothers should be 1 : 10–1 : 15. In final planting, centralized planting should be carried out. Before cross-pollination, plants that do not meet the standard traits of the fathers should be removed according to leaf color, leaf shape, vine hairs, and especially the shape, color, and stripes of the melons growing out of the fathers. The buds of all open staminate flowers and small staminate flowers should be completely removed. Put paper caps on the second unopened pistillate flower of the main vine and the first unopened pistillate flower of the lateral vine and make temporary marks. Remove the paper cap when pollinating with the staminate flower of the father, replace the cap immediately after pollination, and hang a tag to mark the date of pollination. Pistillate flowers whose caps are blown off by the wind must be removed and cannot be pollinated. One staminate flower can pollinate three to four pistillate flowers. At 6:00–7:00 on a clear day, the corolla of the staminate flower fully expands, the anthers dehisce, and the pollen is dispersed. The fertilization ability is at its best when the pistillate flowers of the mother just bloom. Two hours later, the pollination effect gets worse. After 10:00, the fertilization ability of the mother's stigma weakens when oil-like mucus appears on it, so it should be pollinated before 10:00. When pollinating, gently lift the flower stalk of the pistillate flower to reveal the stigma, and then turn the petals of the selected staminate flower inside out to reveal the stamen. Then, gently apply the pollen to the stigma of the pistillate flower. The watermelon pistillate flower's stigma will split into three lobes. During pollination, pollen should be thoroughly and evenly pollinated on each lobe; otherwise, deformed fruit might occur, resulting in a decline in seed production.

Roguing must be done before harvesting seed melons. Fruits that do not meet the characteristics of this cultivar and have no cross-pollination markers, sick fruits, and deformed fruits are to be eliminated. Seed melons are harvested in batches and stages after reaching their physiological maturity. Place them in a ventilated and cool place, and after three to five days of after-ripening, their seeds can be taken out. After fermentation, cleaning, and drying, when the moisture content reaches the moisture standard stipulated by the state, the seeds can be bagged and stored.

5. Seed Production Method for Seedless Watermelon

Seedless watermelon is a triploid obtained by crossing a tetraploid as the mother with a diploid as the father. Its seed production method and ordinary watermelon seed production method are basically the same, normally using the bag artificial pollination method.

(1) Seed treatment and sowing. Before sowing, the diploid, and tetraploid seeds should be selected, sterilized, soaked and pregerminated. The tetraploid seed testa is thick and hard, and the germ cannot easily break through the testa during pre-germination. So, the seed tip needs to be opened before pre-germination, and the opening size should be about 1/3 to 1/4 of the tetraploid seed.

Tetraploid watermelon seedlings grow slowly, and in order to make their anthesis synchronize with that of the diploid, the sowing period of the tetraploid should be three to five days earlier than that of the diploid.

(2) Make triploid by crossing. To get triploid watermelons, the ratio of fathers to mothers during final planting should be 1 : 3–1 : 5. On the first day of cross-pollination, all staminate flowers and male buds on the tetraploid plant of the mother should be removed so that it can pollinate freely. It should be noted that the male buds of the staminate flowers of the mothers must be removed completely and thoroughly every day. In order to improve seed yield, artificially assisted pollination should be carried out every day, especially during rainy, low-temperature, insect-scarce days. This work will greatly improve the yield of triploid seeds. Wash the seeds directly after collecting without fermentation; otherwise, the germination rate will be lowered. As for other points, check the artificial hybridization seed production section for more details.

II. Solanaceous Vegetable Seed Production

(I) Tomato Seed Production Technology

Tomato is an annual herb of the *Lycopersicon* genus in the Solanaceae family, which originated in Peru, Ecuador, and Bolivia in South America, and was introduced to China in the 17th century. Tomato cultivation developed rapidly throughout China in the 1950s. Because it has good adaptability, high yield, rich nutrition, and a wide range of uses. In addition, it can be eaten fresh, cooked, and used to make sauces and canned food, it has become one of the main vegetables cultivated in facilities and open fields in China.

1. Flowering and Pollination Habits

(1) Flowering habits.

① Floral organ structure. The structure of the tomato's floral organ is shown in Fig. 4-8. There are two different inflorescences of tomatoes depending on the cultivar; common tomatoes have cymes and small tomatoes have racemes. The inflorescence generally has 5–10 flowers. Small fruit cultivars and early-maturing cultivars have more flowers. Tomatoes' flowers are hermaphroditic. A tomato flower has separated green calyxes at the outermost, yellow corollas at the inside, a trumpet-shaped corolla

Fig. 4-8 Tomato's Floral Organ

base, one pistil, more than five stamens, and long tube-like anthers which are closely enclosed outside the pistil. The tube-like anthers are called "anther tubes". They dehisce inwardly at maturity to disperse pollen, which falls on the stigma of the same flower's pistil, thus ensuring self-pollination. However, there are also flowers with longer styles that stick out of the anther tube. These flowers are called long-styled flowers, which have a higher natural crossing rate.

② Flowering process. Tomato flower buds differentiate early, usually 25–30 days after sowing. When the seedling has two true leaves, the first inflorescence differentiation starts; 34–38 days after sowing, the second inflorescence differentiation starts; 43–47 days after sowing, the third inflorescence differentiation starts. To create good conditions for flower bud differentiation, a transplant is needed about 25–30 days after sowing when two true leaves are present. The flowers of tomatoes bloom bottom-up from the base(Fig. 4-9). The interval between two adjacent inflorescences is about seven days. Usually, the base of the second inflorescence opens before the first inflorescence is fully open. Tomatoes tend to bloom in the morning and continue to bloom in the afternoon. The flowers will gradually close on the second day.

A flower goes through several stages before it fully opens from a bud: A. The period after flower bud differentiation and when it can be seen by eyes that the buds are enveloped by sepals is called the bud stage. B. The flower buds gradually

Fig. 4-9 Tomato Flowers

become longer and larger. When the sepals gradually unfold at the top of the flower, gradually pushing the corolla outside, we call it corolla-revealing. C. When the corolla grows for a certain period and the petals spread up to 90°, with the further unfolding of the sepals, we call it open. D. When the corolla spreads up to 180°, we call it full bloom, at which time the corolla is bright yellow.

(2) Pollination and fruiting habits.

① Pollination process. Under natural conditions, when the petals are fully expanded to 180° to bloom, the pistil matures. The pollen is dispersed from the anthers, and the stigma of the pistil also elongates rapidly to contact the pollen. After the fertilization process is done, the fruit is borne. The pistil stigma has the ability to receive pollen one day before flowering, so tomatoes can be pollinated at the bud stage, but the fruiting rate is low and the amount of the seed is small. On the day of flowering, when the stigma secretes a large amount of mucus and the pistil reaches full maturity, the pollen acceptance and fruiting rate are the highest. Therefore, in artificial hybridization, emasculation and pollination can be carried out simultaneously. On the second day of flowering, the stigma is still receptive to pollen, but the fruiting rate is reduced.

The appropriate temperature range for tomato flowering, pollination, and fertilization is 20–30°C during the day and 14–22°C at night. If the day temperature is higher than 35°C and the night temperature is lower than 14°C or higher than 22°C, the pollen will not germinate smoothly, and it will be difficult for pollination and fertilization, resulting in flower and fruit falling. Under room temperature (20–25°C) and dry conditions, pollen remains viable for four to five days.

② Fruits and seeds. Tomato fruits are mainly juicy berries. The shape, size, and color of the fruit vary with the cultivar. The number of seeds within each fruit is proportional to the fruit weight. There are many seeds in the large fruits and few seeds in the small fruits. Generally, 1 kg of seeds can be collected from 200–300 kg of fresh fruits. Most tomato cultivars take 40–60 days from flowering to fruit and seed maturity. But the time varies with the average daily temperature. Below 20°C, it takes 50–60 days; at 20–25°C, it takes 40–50 days. Tomato seeds have vitality before fruit ripening. The seeds do not germinate in the seed fruit because of the germination-inhibiting substances within the fruit. To ensure that the seeds are full, the fruit for seed collection must reach full ripeness. Though after-ripening can be carried out on the green fruits, the seed quality is poor. The average thousand-seed weight is about 3 g.

2. Seed Production Technology for Stable Cultivar

(1) Cultivation management of seed collection fields.

① Selection of seed collection fields. Tomatoes are prone to pests and diseases and are not suitable for continuous cropping. They should be sown alternatively with other vegetables other than Solanaceae plants. Although tomatoes are self-pollinated plants, their natural crossing rate is 2%–4%. Therefore, the original seed field should be isolated from the fields of other cultivars by 300–500 m, and the fields to produce the seeds for production should be isolated by 50–100 m.

② Sowing and seedling-raising. Tomatoes are mostly cultivated in the open field in spring, and most of the seedlings are produced in facilities such as cold frames and greenhouses. In order to improve seed yield, the sowing period and final planting time of ordinary tomatoes should be appropriately delayed, generally by five to seven days compared to those of commercial tomatoes, under the condition that other cultivation measures are the same. In northern areas, seeds can be sown in early and mid-March and transplanted in mid-and-late May.

③ Field management. Water and fertilizer management are the same as those of commercial tomatoes. To promote fruit and seed development, during the fruit growth period, apply 1.5% calcium superphosphate or 0.3% potassium dihydrogen phosphate solution to leaves. The growth-determinate tomato cultivar mostly adopts double-branch pruning, i.e. a lateral branch left under the first inflorescence of the main stem grows together with the main stem. The rest of the lateral branches are removed. The growth-indeterminate tomato cultivar mostly adopts single-branch pruning, and double-branch pruning can be adopted to improve the yield of the seed collection field. Generally, two to four fruits of the fruit cluster should be kept, after which topping should be done to the two leaves left on the upper part of the inflorescence.

(2) Roguing.

The typical traits of the cultivar should be examined throughout the whole growth period of the tomato. During the seedling stage, the seedlings can be selected according to leaf shape, leaf color, and the nodal region of the first flower; during the growth period, the typical plants with healthy growth, no diseases and pests, appropriate inflorescence growing location, same height, and traits in line with the characteristics and features of this cultivar should be selected in the field; during the fruit maturity stage, check whether the fruit size, fruit shape, fruit color, and plant growth type conform to the characteristics and features of the original cultivar. Select healthy plants with high fruit setting rates, consistent fruit sizes, and whose fruit shapes and colors conform to the characteristics of the original cultivar as seed plants. Diseased and inferior plants should be pulled out.

(3) Seed threshing.

After harvesting the fully ripe (all the fruits are colored, but the pulp is not softened and the seeds are fully ripe) seed fruits, cut the fruits horizontally with a stainless steel knife, squeeze the seeds and juice into non-metallic containers, such as porcelain plates, wooden barrels, or plastic buckets, place them in an environment at a temperature of 25–30°C and ferment them for 24–48 hours. During the fermentation, no water should be added to the sap, nor should the seeds be exposed to the sun. They should be stirred two to three times to make them ferment evenly. Otherwise, the seeds will germinate or turn black, reducing the germination rate and ruining the seeds' commercial appearance. When the surface of the pulp grows white mold, the seeds sink rapidly when shaking the container or stirring it, and the pectin and seeds are separated. The seeds should be rinsed with water in a timely manner. If red, green, or black bacterial colonies appear on the surface of the pulp, it indicates excessive fermentation, which can also cause blackening of the seeds or a decrease in the germination rate. At this time, the pulp can be soaked in a 1% dilute hydrochloric acid solution (with a pH value of 0.5–1.0) for 15 minutes. Then, stir it constantly and rinse the seeds with water after the seeds are detached from the pectin substance. The fermented seed juice can be added with an equal amount of water. Then, fully stir it. When stopping stirring it, the seeds quickly sink. Pour off the upper layer of floating liquid, add water to the seed pot, and rinse two to three times with water. If the gelatinous material on the seed surface is not easy to fall off, you can rub it off by hand and then rinse.

After rinsing the seeds clean, put them into a gauze bag to dehydrate, thinly spread them on sieves or bamboo mats, and place them in a ventilated and dry place. Sun them until they are half dry, and then rub and separate them with your hands while turning them several times. Sun them for two to three days until their testae are silver-gray. In the case of rainy weather, they can be dried in a blast drying oven, but the temperature should be controlled below 40°C.

3. Seed Production Technology for Conventional Cultivars

(1) Selection of seed collection fields. Tomatoes do not require strict soil conditions, but good root system development conditions should be created to increase seed amount and seed quality. Therefore, fertile loam soil rich in organic matter and with a pH Value of 6–7 should be selected. Tomatoes cannot be planted continuously. They should rotate with Solanaceous vegetables every three to five years. The isolation distance of the original seeds is 300–500 m, and the production seeds should be isolated by 50–100 m.

(2) Roguing and selection of superior plants. The tomato plants should be carefully selected throughout the whole growth period. Seedlings can be selected according to leaf type, leaf color, and the nodal region where the first flower opens at the seedling stage. Those qualified can have final planting. During the growth period, mixed, inferior, and diseased plants should be removed. Plants that grow poorly, do not conform to the major traits of this cultivar, and have serious diseases or are located at high inflorescence-bearing position must be pulled out. During fruit harvesting, plants with a high fruit setting rate should be selected. On these plants, fruits with no split and small navels, with consistent fruit shape, fruit color, and size, should be selected for seed collection.

(3) Fruit harvesting and seed picking. It takes 45–55 days from flowering to physiological maturity for early-maturing cultivars, and 55–60 days for medium- and late-maturing cultivars. When the fruit reaches physiological maturity or is fully ripe, it should be harvested in time. After one to two days of after-ripening, take out its seeds. Cut across the top of the fruit with a knife or open the fruit by hand to take the seeds, pour the seeds together with the juice into non-metallic containers (smash the fruit with a thresher when there are too many of them), place them in a

place with a natural temperature and ferment them for one to two days. Then stir them two to three times a day. When the surface grows white pellicle and the seeds are not sticky, it indicates that the fermentation is done. The seeds then should be promptly cleaned with water, spin-dried in a washing machine, and sunned. When the moisture content of the seeds is below 8%, they can be bagged for storage.

You can also soak the seeds in a 1% dilute hydrochloric acid solution for 15 minutes, during which the seeds should be constantly stirred. After taking out the seeds, they need to be cleaned by shaking them in the water. Then dehydrate and dry them.

4. F_1 Hybrid Seed Production Technology

Tomato has obvious heterosis. Normally, compared to the yield of its parents, the yield of its F_1 hybrids will increase by 20%–40%. It has strong stress resistance. The hybrid seed production technology of artificially removing pollen is often adopted for its seed production for it is quite simple. Besides the above, its seeds have a high propagation coefficient. Therefore, this production technology has been widely used domestically and abroad.

(1) Final planting management of the parents.

In the north of the Yellow River, with little rainfall, seed production should be done in open fields. In the early spring and plum rain season in the South, hybrid seed production should be done in protected plots as the weather is not in favor of fertilization and fruit-setting. The time of the final planting of tomato parents is about 10 days later than that of the commercial tomatoes. If it is an open field, seeds can be sown in early March and hot beds can be used for raising seedlings with a seedling age of 55–60 days. The appropriate ratio of fathers to mothers is 1 : 4–1 : 5. A total of 3,500 mother plants per mu should be planted. To make the fathers' anthesis synchronize with that of the mothers, the fathers are generally sown 7–10 days earlier than the mothers. If the etiolated seedling is used as the parent, the sowing period should be even earlier, as the etiolated tomatoes grow slowly in the early stage and require high temperatures. In short, etiolated seedlings should be sown about 15 days earlier.

The planting fields of the mothers should be isolated from tomato fields or seed production fields of other cultivars by more than 50 m. Apply sufficient basal fertilizer before final planting. Generally apply 4,000–5,000 kg of high-quality farmyard manure per mu, together with 25–30 kg of diammonium phosphate, plus 50 kg of calcium superphosphate (or 50 kg of lime). Make ridges that are 50 cm in width and plant the tomato seedlings in a wide-narrow row planting pattern. The spacing of the wide rows is 50 cm. Narrow rows are spaced 40 cm apart, with the same spacing between the plants. In the later stage, wide rows can be used as operation lanes for cross-pollination. Dibble the ridges for the final planting. Fill the holes with sufficient planting water. Retain the furrows for irrigation afterward. The fathers can be planted in a plot that is close to home and easy to manage and collect pollen. After the final planting of the mothers and when the plants have readjusted to the environment, "人" shaped racks need to be set up in time; the fathers do not need racks or only need quadruped racks. Adopt double-branch·pruning for the mothers. Tie one branch to the longitudinal rod and another to the horizontal rod. The fathers do not need pruning.

(2) Hybrid seed production.

Based on local climatic conditions, the artificial hybridization of tomatoes should be placed in the season most suitable for flowering, pollination, fertilization, and fruiting. The first inflorescence on the main stem of the mother often produces deformed fruit as the plant is small, does not have enough leaves, and the temperature is low. The fruit is not suitable for hybridization, so usually, the second inflorescence is used for hybridization. Hybrid seed production in the northern region is best carried out from mid-May to mid-June. The specific methods of hybrid seed production are as follows.

① Pollen collection and production.

Pollen should come from the normal flowers in full bloom of the fathers (flowers having opened yet still with bright yellow anthers are also qualified). It is best to collect the pollen on the day of flowering as the pollen has matured by then and has not dispersed. Pollen produced by one staminate flower can be used to pollinate four to five pistillate flowers. The pollen should be collected after the dew has volatilized in case it gets wet. The collection is normally done after 10:00 or at noon on a cloudy day when the pollen is at its highest volume and most vigorous.

After picking the flowers of the fathers, take out the anthers. Fresh anthers have a high moisture content, and the pollen is not easy to disperse. Only after the anthers are dried can they dehisce and disperse pollen. Anthers are dried in the following ways.

A. Natural drying. Scatter all anthers on waxed paper and lay the paper in sieves. The sieve is ventilated top-down, which can help dry the anthers. Under natural conditions, it takes about four hours to dry the anthers in the sunlight. This method is simple and handy and does not require special tools. However, this method is only applicable to windless weather with good sunshine.

B. Lime drying. Fill 2/3 of the lower part of a bucket (bowl, urn) that has a tight lid with quicklime, lay a layer of paper on the quicklime, and then spread the anthers on the paper and seal the top. If the anthers are placed in the evening, the anther stalks can be dry the next morning. Change the quicklime every 10 days or so.

C. Bulb drying. Spread the anthers on a layer of waxed paper, then put the paper in a sieve and hang a light bulb (100–200 W) 30 cm above the anthers. The heat emitted from the bulb can dry the anthers.

D. Oven drying. Ovens can be used for drying. Set the oven temperature to 32°C, and put the anthers in it in the evening. The properly dried anthers will be ready the next morning.

After the anthers are dried, place them on a smooth paper. Smash the anthers with a rolling pin or medicine mill to make it disperse pollen, and then sieve the pollen with a 120–150 mesh pollen sieve. Store the sieved pollen in a vial and put the vial in a desiccator. It can be kept for two to three days at room temperature and 30 days at a low temperature of 4–5°C.

Before pollination, load the pollen into a glass pollinator, which is a hollow glass tube with an open top and a small hole on the side of the bottom. The glass tube is about 6 cm long, with an outer diameter of 5 mm. The small hole is 1.5 mm in diameter and 3 mm from the bottom of the tube. You can make it on your own or buy one from seed companies. The production method is as follows: a. Select a hollow glass tube with a 0.5 cm outer diameter. Quickly burn the top of the glass tube into closing up with the high temperature of an alcohol blowtorch. b. At the same time, blow air into the tube from the other end to make an opening in the place about 2–3 mm from the closed top. c. Continue to heat so that the opening gradually narrows and becomes smooth. Adjust the size of the opening according to the size of the stigma of different crops and cultivars. When the size of the opening is slightly larger than that of the mother's stigma, you can move the tube away from the fire and cool it. d. Cut and break off the glass tube at 6 cm.

When filling the tube, the small hole on the side of the pollen tube needs to be blocked with a finger, and the pollen is filled from the opening at the other end. After filling, use a rubber plug to block the mouth of the tube tightly.

② Emasculation. It is necessary to remove the flowers that are blooming and have already opened, as well as deformed and small flowers from the plants. The appropriate time for emasculation is two days before flowering. Externally, the sepals are dehiscent, the petals are closed, and the corolla is white and not yet yellow. When the corolla is opened, the stamens are yellowish-green. It is important to master the time of emasculation because tomatoes are self-pollinated plants. If the emasculation is too late, the pollen will fall on the stigma, and the flower thus fertilizes itself. Artificial pollination is futile after that, which is one of the reasons for the existence of false hybrids. If the emasculation is performed too early, the buds would be too small to do it, which also reduces the fruit setting rate and seed volume. During emasculation, use the thumb and forefinger of the left hand to gently pinch the bud at the base while using the right hand to open the petals apart to expose the anther tube and strip the anthers with stainless steel-tipped tweezers. Make sure the pistil is not hurt and the emasculation must be done thoroughly. You can also strip the anthers and petals with bare hands.

③ Artificial pollination. Pollinate the flowers in full bloom two days after the emasculation. Pollination should be done when there is no dew in the field. The best time is from 8:00–11:00 every day. If pollination is not completed on a rainy morning, it can be done again from 15:00–17:00. During pollination, pinch the corolla and put the stigma into the pollination hole of the glass pollinator to make the stigma covered with pollen. If there is no pollinator, you can dip a small amount of pollen and apply it gently and evenly to the stigma of the emasculated mother with the eraser tip of a pencil. After pollination, remove two sepals for marking, and the sepals must be pulled off completely from the base; otherwise, they are easily mixed with the naturally dried sepals. Repeating the pollination two to three times can improve the fruit setting rate and increase the number of seeds, which is particularly necessary on rainy days. The optimum temperature for pollination is 20–25°C. When the temperature is lower than 15°C or higher than 30°C, the pollination should be suspended.

④ Plant adjustment after pollination. Plant adjustment should be carried out immediately after the hybridization work. Usually, the first inflorescence is removed instead of being used for cross-pollination. Double-branch pruning is adopted for growth-indeterminate tomatoes. That is, the main branch and a healthy lateral branch under the first inflorescence are retained. Emasculate and pollinate two to three clusters on each branch, keeping five to six clusters in total. Three-branch pruning is adopted for growth-determinate tomatoes. That is, the adjacent branches up and down the first inflorescence are retained. The three branches can keep a total of six to seven clusters. Large fruit cultivars can have two to three fruits per cluster, and small fruit cultivars can have 5–10 per cluster. Remove all the flowers and axillary buds that have not been emasculated and pollinated and leave two leaves above the last hybrid inflorescence for topping so as to ensure that nutrients are concentrated to supply the development of hybrid fruits. This work should be repeated several times.

⑤ Collecting and threshing the seed fruits. They are the same as the seed collection technology of the stable cultivars.

⑥ Several measures to improve the purity of hybrids.

A. In hybrid seed production fields, growth agents like 4-Chlorophenoxyacetic acid are strictly prohibited, or no seeds will be born.

B. To ensure seed purity, it is important to ensure the isolation distance between different cultivars in the seed production field or between the seed production field and the crop field.

C. Emasculate at a proper time to avoid the case that late emasculation leads to self-crossed fruits.

D. Before cross-pollination, all flowers and fruits on the mothers that have not gone through artificial pollination must be removed from the seed production field. After the cross-pollination is over, check the plants and make sure no new buds are on the plants in case of the emergence of false hybrids.

E. When harvesting seed fruits, check whether the fruit is missing two sepals. Any fruit not meeting this requirement must be picked and taken out of the field for deep burial. Any fruit that falls to the ground should be discarded.

F. Mechanical mixing is strictly prohibited during the harvesting of fruit, seed collecting, fermentation, cleaning, drying, bagging, and other links.

Attention should be paid when harvesting more than one cultivar. At each link, labels should be placed, indicating the name of the species and cross combinations, as well as the time and place of collection.

(II) Eggplant Seed Production Technology

Eggplant is an annual vegetable of the *Solanum* genus in the Solanaceae family. Native to India, it has been cultivated in China for a thousand years. The young fruits of eggplant are fit for human consumption. Eggplant is rich in nutrients and can be roasted, fried, deep-fried, steamed, or processed with sauce, pickled with garlic, dried, etc. It is one of the favorite vegetables of Chinese people. It is widely cultivated throughout the country due to its high yield, resistance, and long supply period.

1. Flowering and Pollination Habits

(1) Branching and fruiting habits.

The eggplant is pseudo-dichotomously branched, and the branches bear fruits regularly. For general early-maturing cultivars, when the main stem grows five to seven leaves, the first flower comes out; for medium or late-maturing cultivars, only when the main stem grows 8–14 leaves, does the first flower come out. The lateral branches borne in the leaf axils below the flowers are particularly strong and similar to the main stem, thus forming a "Y" shape. The fruit of the first flower is called "the first eggplant". After the main stem or lateral branch grows two to three leaves, it then branches and flowers. The eggplants borne on the main stem and the lateral branch are called "the pair eggplants". In the same way, the stem and branch continue to branch and flower. From bottom to top, the eggplants borne are called "four funnels" , "eight winds" , and "full of stars". However, in reality, due to nutrient competition, insufficient light, and other reasons, flowers and fruits often fall off. The plant does not bear fruit that regularly after the "four funnels".

(2) Floral organ structure.

Eggplant flowers (Fig. 4-10) are hermaphroditic and usually grow in a solitary pattern, but some cultivars are clustered with two to six flowers. The flowers are purple, lavender, or white. They are large and pendulous, consisting of four parts: calyx, corolla, stamens, and pistil. The calyx has sharp spines, and the number of petals is usually the same as that of sepals (five to eight). It has five to eight stamens, forming an anther tube around the pistil. The anthers are yellow, with two lobes and dehiscent as holes. The pistil is enclosed in the center of the stamens and consists of the ovary, style, and stigma. According to the length of the style, eggplant flowers are divided into three types: long-styled flowers, medium-styled flowers, and short-styled flowers. The first two types of flowers are large and dark, and the pollen dispersed from the anther tube falls naturally on the stigma. They are strong in fertilization ability and are counted as healthy flowers. Short-styled flowers are unhealthy flowers and weak in fertilization ability because the style is shorter than the stamens, making it harder for the pollen dispersed from the anthers to fall on the stigma of the pistil.

Fig. 4-10 Eggplant Flowers

(3) Flowering, Pollination, and Fruiting.

The flower of eggplants mostly blooms in the morning when the stamens mature and the top of the anthers dehisce to disperse pollen. The pollen can fall naturally on the stigma of the pistil, which is called self-pollination. The proper temperature for eggplant flowering is 25–30°C. When below 15°C or above 40°C, fertilization is affected and the flowers are easy to drop. Natural pollination usually happens from 7:00–10:00 on a sunny day. A flower opens for three to five days, and the pollen can germinate from one day before flowering to three days after flowering. From two days before flowering to two days after flowering, the pistil can be pollinated and fertilized. However, if the pistil is pollinated and fertilized on the day of flowering or one day after flowering, the fruiting rate and the number of seeds are the highest.

The fruits (Fig. 4-11) reach commercial maturity 15–20 days after flowering and can be harvested and marketed. However, the seeds develop quite slowly. The seeds' testae are white 25–30 days after flowering. Although the seeds are immature, they already take the form of a seed. The seeds are capable of germinating about 40 days after flowering. At this time, the seeds are slightly yellowish, and the thousand-seed weight is light. The testae have a nice color 50–55 days after flowering.

When the thousand-seed weight reaches a certain value, the germination rate basically meets the requirements. Sixty days after flowering, the thousand-seed weight does not change much, the germination rate and germination potential are the highest, and the seeds are fully mature. After-ripening of fruits can significantly promote seed maturity. Usually, a seed fruit produces 500–1,000 seeds, which weigh about 2–6 g. 100–200 kg of seed fruits can produce 1 kg of seeds. Eggplant seeds are yellowish-brown and shiny, with a thousand-seed weight of 2.8–7.2 g. When the seeds are stored in a dry place at room temperature, they can maintain their germination ability for five years. Most of the newly collected eggplant seeds have a dormancy period, which will be over after two to three months. The dormancy can also be broken when the eggplants are treated with 0.01% gibberellin.

Fig. 4-11 Eggplant Fruits

2. Seed Production Technology for Stable Cultivar

(1) Management of seed collection fields.

① Setup of the seed collection fields. A good seed collection base should be located in the most suitable ecological zone for eggplant growth and development, requiring sufficient light, abundant water, and no continuous rains during the period from pollination to fruit harvest. In the northern region, the seeds are collected when the fruits touch the ground in spring. During the flowering and fruiting period, if the temperature is high, the fruit setting rate and seed volume are high. The seed fruits are ripe before the rainy season. They are less susceptible to rotting, and this method can increase the seed yield. No continuous cropping is allowed for eggplant, nor should it be cropped in succession with other Solanaceae crops. The seed-breeding field must be a new land or have been planted with eggplant at least four years ago. In most areas, the seed collection period of eggplant is in the rainy season, so the seed plant is prone to disease and easy to produce rotten fruit. Hence, the seed field should be on a high and dry terrain with good drainage, ventilation, and light, as well as fertile and soft soil texture. Eggplant is a self-pollinated crop, but it also has a high outcrossing rate. Therefore, the isolation distance of the original seeds in the seed collection fields should be 200 m. They can also be isolated by mesh or paper bags, but the mesh or bags should be removed timely after fruiting in case of poor fruit coloring; the seed production field should be separated from other different cultivars at a distance of more than 50 m.

② Field management of seed plants. The cultivation management of eggplant seed plants is basically the same as that of commercial eggplant production. In Liaoning Province, sowing and seedling-raising are conducted in greenhouses in late January and the final planting is done in the open field in early May. For ventilation, light, and easy management, the planting density should be less than that of commercial eggplant production. Depending on the characteristics of each cultivar, 1,200–2,000 plants should be planted per mu. Generally, for early-maturing cultivars, the row spacing is 66 cm and the plant spacing is 45 cm; for medium-and late-maturing cultivars, the row spacing is 80–85 cm and the plant spacing is 50 cm. Phosphate and potassium fertilizers should be applied to the seed collection field. 5,000 kg of compost, together with 25–30 kg of ternary compound fertilizer, can be applied during soil preparation. After the seed plants are planted and have readjusted to the environment, watering should be properly controlled. Then, the hardening of seedlings should be performed. After the seed fruit is set, it needs to be watered in time. Quick-acting fertilizers need to be added at the fruit expansion stage, 10 kg of urea per mu each time. At the later stage of the development of seed fruits, water should be strictly controlled, so as to reduce rotten fruits and ensure the quality of seeds.

The nodal region where the first flower grows of the early-maturing cultivars is low, and the first eggplant is small. Therefore, this eggplant cannot be used as a seed eggplant and can be removed before the flower buds bloom. The seed fruits of eggplants of medium - or late-maturing cultivars can be kept. The number of the seed fruits retained on each plant should be determined according to the cultivar type: three to five seed fruits for large fruit cultivars (mostly round eggplants), and 6–10 for medium and small fruit cultivars (long eggplant and egg-shaped eggplants). After the seed fruit is set, to guarantee its full development and concentrated nutrients, excess buds should be removed and three to five leaves should be left on the seed fruit to perform the topping. To prevent the plant from lodging and prevent the fruits from rotting after being weighed down to the ground, the plant can be fixed in the middle of the growth period by setting up racks or setting wires or bamboo poles at the edge of the bed. In addition, during the growth period of the seed plants, it is important to address pests and conduct good disease control.

(2) Roguing.

During the growth period of the seed plants, the field inspection can be carried out three times to make sure the impure and inferior plants are pulled out in time. Before flowering, the plant's growth habit, disease resistance, leaf shape, and leaf color should be examined. During the first flower and first young fruit stages, items to be examined are the same as above. Besides these, the density and strength of the spines on the sepals, as well as the shape and color of the young fruit should also be examined. When the fruit reaches commercial maturity during the fruit setting period, fruitfulness, ripeness, disease resistance, fruit shape, size, and the colors of the skin and pulp should be examined. All plants that are poorly developed and sick, and have inconsistent anthesis and deformed fruits should be eliminated in a timely manner.

(3) Harvesting and threshing of seed fruits.

Keep and mark the seed fruits for selection. The seeds of eggplant develop slowly and only mature 50–60 days after flowering when the outer skin turns yellowish-brown. However, there are also cultivars with black outer skins. When the fruits mature, their outer skins remain black. The seed fruits should be placed in a ventilated, dry, and cool place for 7–10 days for after-ripening after being harvested so that the nutrients in the fruits can be fully transferred into the seeds to improve the seed germination rate. It is the rainy season at the later stage of seed retention that makes the fruit susceptible to diseases and rot. Therefore, it is sometimes necessary to harvest the fruit early and at the right time. The after-ripening period can also be properly lengthened to promote the fullness of the seeds and the germination rate.

After going through the after-ripening, the pulp of the fruit becomes soft. Beat the seed fruit with a stick while threshing to separate the seeds from the placenta, cut the fruit across it, put it into the water and take the seeds along the crack, scrub off the mucus on the seeds, clean them, and remove the blighted seeds floating on the water surface. When there are large quantities of fruits, it can be done by smashing the pulp with a modified corn threshing machine. Then rinse them with water and pour out the peel, pulp, and blighted seeds floating on the water's surface. The ones that sink to the bottom are full seeds. Spread out the rinsed seeds on a mat or a piece of sack thinly to dry. Do not place them on a concrete slab floor. Avoid direct sunlight and turn them frequently. After the seeds are sufficiently dried in the sun, pack them for storage.

3. Seed Production Technology for Conventional Cultivars

(1) Selection of the seed collection fields.

Eggplant is a self-pollinated crop, but it has a certain outcrossing rate as well. Therefore, the isolation distance for the original seed reproduction is 300–500 m. Nets and paper bags can also be used to isolate the flowers. The isolation distance between the production cultivars and other cultivars during the reproduction is 50–100 m.

(2) Roguing and selection of superior plants.

The selection of good-quality seeds must be carried out during the reproduction of the original eggplant seeds. For the seeds for production, only roguing needs to be carried out.

(3) Fruit ripening and seed collection.

Eggplant takes 50–60 days from flowering to the physiological maturity of the fruit. The fruit can be harvested when the skin is fully ripe and turns yellowish-brown. After one to two weeks' after-ripening, the seeds can be taken out. When only a small amount of seeds are collected, it can be achieved by beating the pulp soft and cracked with a wooden stick to separate the seeds in every ventricle from the pulp. Then, peel off the seeds after putting the fruit into the water and wash the seeds clean. When there are large quantities of fruits, a modified corn threshing machine can be used to smash the pulp. Then, clean the seeds with water and remove the skin, pulp, and blighted seeds. The ones that sink to the bottom are full seeds. After washing and drying, the seeds can be stored in bags when the moisture content is less than 8%. Generally, a fruit can produce 500–1,000 seeds, weighing 2–6 g, and 100–200 kg seed fruits can produce 1 kg of seeds.

4. F_1 Hybrid Seed Production Technology

Eggplant has great heterosis. Some companies have selected and bred some excellent hybrid combinations that have been applied in production. The cost of producing F_1 hybrids is not high because the hybridization process is easy as the eggplant has large floral organs and a great number of seeds set in the fruit. At present, artificial emasculation and pollination methods are adopted in seed production in China and many other countries.

(1) Management of seed production fields.

① Sowing period and planting ratio of parents. In northern areas, hybrid seed production of eggplant is mostly carried out in the open field. The sowing period of the parents should be determined according to local climatic conditions to ensure that the anthesis of the fathers and mothers is synchronized. The optimum temperature for pollination and fertilization of eggplant is 28–30°C. The anthesis of the mother should be put within the optimum temperature range for pollination and fertilization, from which the suitable sowing period of the mother plant can be deduced. If the father matures earlier than the mother, the father and mother should be sown at the same time; if their maturity time is close, the father should be sown three to five days earlier than the mother; if the father matures later than the mother, the father should be sown 5–15 days earlier than the mother. The ratio of fathers to mothers in the final planting in the field is 1 : 4–1 : 5. To avoid mixing and confusion when harvesting seed fruits, it is best to plant the fathers and mothers separately. Check the purity of the parent plants strictly before final planting and flowering, and pull out miscellaneous, mixed, and diseased plants.

② Field management of seed plants. It is best to use the plastic film mulching method for eggplant seed production in the open field. This method can help roots develop early, facilitate tillering, boost growth, make the plant stronger, improve the fruiting ability, promote seed fruit maturity, and make the seeds full. To facilitate manual handling during seed production, it is preferable to use the wide-narrow row planting pattern, with a 100 cm spacing for wide rows and a 66 cm spacing for narrow rows. The seed production field should be fully fertilized with basal fertilizer and lots of phosphate and potassium fertilizer. To concentrate pollination time, normally, the bud of the first eggplant and all the lateral branches below the bud can be removed, which can promote the early setting of the pair eggplants and the four funnels. Leave two to three leaves after the buds of the four funnels show up and conduct topping on the leaves. After pollination, set up racks in case of lodging.

(2) Cross-pollination technology.

① Field inspection. Before cross-pollination begins, the parents must be checked plant by plant for plant type, plant height, leaf type, leaf color, etc. All plants that do not match the parental traits should be pulled out.

② Pollen collection and preparation. Pollen should be collected before pollination. If only a small number of eggplant hybrid seeds are to be produced, pick the anthers from the open flowers of the fathers on the same day with forceps, pick out the pollen, and put it into use immediately. If a large number of hybrid seeds are to be produced, manually sieve the pollen or use an electric pollen collector. If you want to acquire the pollen with a sieve, you can pick the fathers' flowers that are in full bloom on the same day or whose anthers are still bright yellow, although they have been open for a few days. Take out the anthers, put them into a desiccator to dry, and then sieve the pollen with a sieve (100–150 mesh), collect it, and store it in a small glass bottle. If you use an electric pollen collector, you can align the collector to the anthers of the fathers' flowers that are open on the same day, using the vibrational force of the collector's needle to take out the mature pollen inside the anther cells. Put the pollen taken out in the morning into use in the afternoon. If you take the pollen out in the afternoon, you need to use it the next day. The pollen must be preserved under cool and dry conditions.

③ Emasculation. Select large buds one to two days before flowering, hold the flower stalk lightly in your left hand, use forceps to gently open the petals with your right hand, stick the tips of the forceps into the base of the anthers and gently take out the anthers in two to three times. Do not hurt the stigma and ovary during the operation. It is very important to choose the right-sized buds during emasculation, which is one of the keys to ensuring the purity of hybrid seeds. Attention should be paid to choosing flowers of long stigmas to emasculate, as flowers of short stigmas have a lower fruiting rate. If there are more than two buds on one plant, the strongest bud should be selected for emasculation and the rest should be removed.

④ Cross-pollination. The fruiting rate after pollination varies considerably with the flower ages. The highest fruiting rate is achieved by pollination on the day of mothers, flowering, but the pistils can still receive pollen three days before and after flowering. During the production, while conducting emasculation, you can use your left hand to gently hold the base and stalk of the mother's flower, dip the pollen with a bee stick or cotton swab, and apply it to the stigma of the mother's flower. Then, wait for the day when the petals open (i.e. the flowers open), and then repeat the pollination once. You can remove one sepal at the first pollination and then remove one or two sepals at the second pollination. After pollination, thin strings and tags are tied to the flower stalks as seed fruit marks. Pollination is best carried out in the morning after the dew has dried. If the weather is hot, pollination should be stopped at noon and delayed until the afternoon. If it rains within six to eight hours after pollination, the pollination should be done again after the rain to improve the fruit setting rate. The number of flowers cross-pollinated in each plant should be determined by the cultivar type and seed retention habits. Generally, five to seven flowers per plant are pollinated. Check the plants during the whole process of pollination and remove unpollinated flowers and unmarked fruits in time.

⑤ Harvesting and threshing of seed fruits. After the seed fruits are fully ripe, all the mother plant fields should be checked. Carefully remove impure plants and harvest fully ripe seed fruits with marks. One principle should be adhered to while harvesting, that is, do not pick unmarked, stunted, rotten, and fallen fruits. The threshing of the seed fruits is the same as that in the seed production of the stable cultivars.

(III) Pepper Seed Production Technology

Chili pepper is from the genus *Capsicum*, a member of the Solanaceae family, an annual herb in temperate zones, also a perennial shrub in the tropics, native to Central and South America, introduced to China via the Silk Road and the sea route in the 17th century. It has been cultivated in China for more than 300 years. Chili peppers are widely cultivated in China, with the spicy type mainly cultivated in the South and the sweet type mainly cultivated in the North. Chili peppers have high nutritional value, with higher levels of vitamins compared to other vegetables. Its pungent flavor can enhance appetite and digestion. Chili peppers can be eaten in various ways, such as raw, hot, pickled, or processed into chili powder, chili sauce, chili oil, etc. Dried chili peppers are an important export product.

1. Flowering and Pollination Habits

(1) Floral organ structure.

Pepper flowers are hermaphroditic. Sweet pepper buds are larger and round, and spicy pepper buds are smaller and longer. Most cultivars have white corollas, while a few have light purple corollas, consisting of five to seven petals and a connate base. The flower has one pistil in the center and five to seven stamens neatly arranged around the pistil. Anthers and filaments are light purple. The stigmas and the anthers of the stamens are adjacent. The stamens of general cultivars are equally long as or slightly longer than the stigmas. There are a few sweet pepper cultivars' stigmas longer than the anthers of stamens. This kind of long-stigma cultivar has a high natural outcrossing rate. As soon as the petals of pepper open, the anthers dehisce. The outer of the pollen sac cracks longitudinal, and the pollen spills out. Then, the inner side cracks and the pollen is dispersed. The upper part of the ovary has two to four ventricles containing tens to hundreds of ovules.

(2) Flowering habits.

The first flower's nodal region of the early-maturing pepper is generally in the place from the fourth to the ninth node of the main stem, and that of medium to late-maturing cultivars is in the place from the 10th to the 24th node. As for flowering order, the first flower serves as the center, around which the flowers blossom layer by layer. Usually, three to four days after the first layer of flowers blossoms, the upper layer of flowers opens, and so forth bottom-up (Fig. 4-12). Under normal conditions, the petals change from green to white-green and then fade to white. The buds open when the petals grow to be significantly larger than the sepals. The flowers are mostly open from 6:00–10:00. Generally, a flower takes two to three days from opening to withering. It opens first, and then the anthers dehisce. After that, the pollen is dispersed voluminously. Under high temperatures, anthers tend to dehisce early in the bud.

Fig. 4-12 Pepper Flowers

(3) Pollination habits.

The anthers of peppers' stamens are not the same as the anther tubes of tomatoes and eggplants. Plus, pepper flowers have nectaries that can attract insects, so the outcrossing rate is higher, generally 5%–10%, or even 30% for some cultivars. That is why chili peppers are called "often self-crossed crops".

The pistil can be pollinated and fertilized two days before and after flowering with the pollen two days before and on the day of flowering, but the best result is produced on the day of flowering. During hybrid seed production, it is better to emasculate during the bud stage and pollinate on the same day of flowering so as to improve the fruit setting rate and seed volume of a single fruit. It takes about 26 hours from the start of pollination to the completion of fertilization. The appropriate temperature for pollen germination is 22–26°C. When the temperature is lower than 18°C or higher than 30°C, pollen is hard to germinate. The pollen germination temperature of sweet pepper is low, while that of the spicy pepper is slightly higher. The appropriate relative air humidity is 55%–75%. It is not conducive to pollination when the humidity is below 40% or above 90%.

Pepper takes 50–60 days from pollination to full maturity of the seeds.

Usually, 10 days after fertilization, the cells differentiate intensively, and after another 10 days, the newborn cells expand rapidly, during which the fruit grows rapidly and needs to be fertilized and watered in time.

2. Seed Collection Technology for Stable Cultivar

(1) Cultivation management of seed collection fields.

① Selection of seed collection fields. Pepper is an often self-crossed vegetable with a high natural outcrossing rate. In order to produce a large area of good-quality seeds, it is best to cultivate only one cultivar in an area. Isolate the fields of different cultivars by 500–1,000 m. If a small area of original seeds is to be produced, plastic mesh sheds can be used for isolation, which can achieve a good isolation effect. Besides, the sheds can prevent some pests and diseases, significantly improving the fruit setting rate and seed yield. Peppers cannot be planted continuously with other crops of the Solanaceae family. There should be an interval of three to five years.

② Field management. The cultivation technology of seed pepper is basically the same as that of commercial pepper. If the seed production is performed in the open field, seedling-raising and final planting should be done early. The final planting density is less than that of the commercial peppers. The peppers should better be planted individually at 5,000 seedlings per mu. After the final planting, strengthen water and fertilizer management to promote the tillering of the plant. Ensuring that the ridges are closed before the arrival of the high-temperature season is the key to achieving a good yield. Water and fertilizer management is the same as that in commercial pepper production. During the growth period, the lateral branches on the main stem below the first flower and the first flower or fruit should be promptly removed so as to promote the tillering of the seed plant. Make sure that the upper seed peppers can fully develop to improve the yield and quality of seeds. The "first pair of peppers" and "four peppers" should be kept as the seed fruits because they contain a large number of good seeds. A large-sized sweet pepper plant can keep four to six fruits, while a small-sized pepper plant can keep a dozen to dozens of fruits. Pay attention to cultivating the soil before the rainy season and cleaning the drains to prevent flooding. Pay attention to pest and disease control throughout the growth period.

(2) Roguing.

Pepper is an often self-crossed crop and the seeds are easy to mix and degrade. Therefore, strict selection must be carried out during production. Before flowering, pull out the impure and inferior plants completely in time. Check the typical traits of the cultivar three times during the growth and development period of pepper. At the early stage of fruit-setting, the plants that meet the standards of the original cultivar in terms of plant type, leaf type, leaf color, first fruit-bearing nodal region, young fruit color, and plant openness are mainly selected, and impure and diseased plants are eliminated; at the commercial maturity stage of the fruit, the plants that meet the standards of the original cultivar in terms of plant growth type, disease resistance, fruit size, fruit shape, fruit color, fruit stalk orientation (upright and upward or bent), neatness of fruit at different layers, fruit ventricle number, and fruit setting rate are mainly selected as the seed plants; at the maturity stage when the seed fruit turns red, the plants and fruits that meet the characteristics of the original cultivar in terms of ripeness, disease resistance, fruit size, fruit color, and ventricle number are mainly kept.

(3) Harvesting and threshing of seed fruits.

When the seed pepper is basically red and ripe, the seeds are usually developed and can be harvested separately. After harvesting the fruits of sweet pepper with thicker skin and higher moisture content, place them in a ventilated and cool place for after-ripening for three to five days before threshing to improve the seed germination rate. When threshing, break the fruit into two halves by hand or cut a circle around the white sepal with a knife and lift the fruit stalk to take out the seeds together with the placenta. After impurities, such as the placenta, are removed, place the seeds on a cool mat or nylon gauze in a ventilated place to dry. Seeds should avoid being exposed to the sun on concrete or in a metal dish as it will seriously affect the germination rate and color of the seeds and reduce their quality. After the seeds are cut open, there is no need to wash them with water, because after washing and drying, the color of the seeds is more likely to be whitish-gray. When the seeds are not easy to dry on cloudy days, they are more prone to mold, and deterioration is more likely to happen. The usual method is to dry the seeds together with the placenta, and when the placenta is sufficiently dry, rub it against a wooden board on a plate to separate the seeds from the placenta, then sift out the seeds.

3. Seed Production Technology for Conventional Cultivars

(1) Selection of seed collection fields.

Seed collection fields for spicy (sweet) peppers should be filled with fertile sandy loam soil that is convenient for drainage and irrigation. The pH value of the field should be between 6.5 and 7.0. Avoid continuous cropping with other plants of the Solanaceae family. The interval time should be three to four years. In order to avoid interspecific hybridization, the isolation distance should be more than 400–500 m. For small areas of original seed production, a plastic gauze tunnel can be used to achieve good isolation.

(2) Selection of seed fruits.

The seeds of the fruits of spicy (sweet) pepper that are on the second and third layers have better quality than those on other layers, so the fruits on the second and third layers should be used as seed fruits, and for those cultivars that have strong growth vigor, fruits on the fourth layer are also acceptable. Due to the poor growth of the fruit born from the first flower of each plant, it's not suitable to be selected as seed fruit and should be removed as soon as possible.

(3) Roguing and selection of superior plants.

Spicy (sweet) peppers are often cross-pollinated crops, which means they are prone to natural hybridization and cultivar degradation. Therefore, a strict selection should be made while ensuring the isolation distance. Completely remove impure plants and inferior plants before they bloom; at the early stage of fruit setting, select mainly according to the plant type, leaf type, leaf color, the first fruit-bearing nodal region, young fruit color, plant opening degree, etc., and eliminate impure plants and diseased plants; when the fruits reach their commercial maturity stage, select mainly according to their growth type, disease resistance, fruit size, fruit shape, fruit color, fruit neatness, the number of fruit ventricle and fruit setting rate; when the fruits reach their physiological maturity stage, select mainly according to their maturity, disease resistance, fruit size, fruit color, the number of fruit ventricle, etc. Those plants and fruits that meet the standard will be kept.

(4) Seed collection.

Generally, it takes 50–60 days for the fruits of spicy (sweet) peppers to ripen from flowering. The red color is the sign that the fruits of spicy (sweet) peppers have reached their physiological maturity, indicating that the seeds have developed into maturity and should be collected in batches and stages timely. Diseased and deformed fruits should be removed when collecting. After the collection, place the seeds in a ventilated and cool place for three to five days for after-ripening. Then, break open the fruit by hand or use a knife to cut a circle around the shoulder of the fruit, lightly lift the fruit stalk to remove the placenta that has seeds on it, peel off the seeds, and then place them on a mat or nylon gauze to dry. When the moisture content of the seeds is down to less than 8%, bag them and put them in storage. The number of seeds in a single fruit of spicy (sweet) pepper varies greatly depending on the cultivars, with around 400 seeds in a good one and only around 100 in a bad one. Normally, each hectare of peppers can produce 450–750 kg of seeds.

4. F_1 Hybrid Seed Production Technology

The hybrid of pepper can increase the yield by more than 30%, and it shows early maturity and strong stress resistance. The main methods for producing the F_1 hybrids are the artificial hybridization seed production method and the male sterile line seed production method.

(1) Use artificial hybridization to produce F_1 hybrids.

Pepper's propagation coefficient is high. Generally, 80–200 spicy pepper seeds or more than 200 sweet pepper seeds can be obtained by hybridizing one flower. Besides, the planting of peppers requires a small number of seeds. Only 450–600 g seeds are needed for one hectare, so at present, artificial emasculation of two inbred lines is still the main method in pepper crossbreeding.

① Planting of parents.

A. Cultivation of parents. Plastic arched sheds can be used in the cultivation of the F_1 hybrids of pepper plants as they can isolate the plants from the outside world and the rain. This method can lower the possibility of disease and allow the plant to grow in a better environment, which results in a higher seed yield, but the production cost is higher. For seed production in open fields in spring, seedlings need to be cultivated in the facility in advance.

B. Sowing period of parent plants. To ensure that the anthesis of the father and mother is synchronized during the cross-pollination and to ensure that the father can supply sufficient pollen, the father and the mother plants should be sown separately and their respective sowing periods should be determined according to their maturity. If the parents are similar in maturity, the father plant can be sown about five days earlier than the mother plant. If the mother plant is early-maturing, while the father plant is late-maturing and only has a few flowers, the father plant should be sown 15–20 days earlier than the mother plant to ensure that the anthesis of the parents is synchronized and that there is enough pollen from the father for artificial cross-pollination's need when the mother is in the full-blossom stage.

C. Planting ratio of the fathers and mothers. The ratio of the number of fathers and mothers should be 1 ∶ 4–1 ∶ 6, and they should be planted in stretches. Both the fathers and mothers should be planted on ridges that are 1 m wide, with two rows for each ridge. The plant spacing should be 21–25 cm, and in the final planting, there should be 4,500–5,000 holes per mu. Each hole can be planted with two father plants or one mother plant. If the seeds are produced in plastic tunnels, there is no need for isolation; if the seeds are bred in open fields, they should be isolated from other different cultivars at a distance of more than 500 m.

D. Field management. The field management of the father and mother is the same as that of commercial pepper production. That is, at the end of pollination, pepper enters the full fruiting period, so it is necessary to strengthen water and fertilizer management and increase the application of phosphate and potassium fertilizer. This is to improve the seed yield and quality. At the early stage, the flowers and fruits of the first pepper and the first pair of peppers from the two branches of the mother plant should be removed in time. They cannot be used for cross-pollination. The plastic tunnel where the mother plant is grown should be kept at a suitable temperature and high humidity to form a relatively stable microclimate environment. During the pollination period, water the plants frequently with little water each time and keep the ground moist to improve the pollination effect. Remove excessive flower buds from pollinated plants in time after the pollination and then proceed to perform topping.

② Hybridization period. The hybridization period should be in a suitable season with a temperature between 18–30°C. Pollination is best performed in June for open-field seed production in Northeast China. In addition to the season and temperature, care should also be taken to select sunny days for artificial cross-pollination so that the seed plants have good light for photosynthesis. Weak photosynthesis, high wasting, and low accumulation of dry matter can easily lead to flower and fruit drops.

③ Hybridization techniques.

A. Field inspection. It starts with careful checking of the purity of the fathers and mothers and strict roguing. In addition, the flowers and fruits at the first nodal region of the father and mother plants, the flowers that already bloom and all fruits on the father and mother plants, and the weak and thin branches that are stunted below the first-born pepper of each plant should be removed.

B. Emasculation. Pepper's anthers dehisce and disperse pollen inside the bud when the buds reach their maximum size. Flower buds that have already dispersed pollen cannot be used for seed production. The dispersal of pollen mostly starts when the daily temperature increases and the humidity decreases. In general cases, emasculation can be done from 6:00–10:00 or after 15:00 when the high temperature has passed. Avoid mid-noon when the temperature is the highest in the day. When emasculating, select a large bud on the mother plant one day before it blooms, open the corolla with a pair of forceps and remove all the anthers. The movement must be gentle and the process should be performed without damaging the stigma and ovary. If the anthers are already dehiscent when the bud is opened, remove the bud. If the forceps or fingers are stained with pollen, disinfect them with alcohol in time to prevent the pollen of the mother plant from falling onto the stigma again. When changing the hybridization combination, it is also necessary to sterilize the fingers and forceps with alcohol. Bud pollination can be carried out at the same time as emasculation, or emasculation can be carried out every morning and pollination can be carried out in the morning of the second day. Nowadays, the method of emasculation by bare hands is mostly used in pepper hybrid production.

The specific procedure is to pinch the junction of the pedicel and receptacle, use the right thumb and forefinger to gently pinch the part of the corolla that's near the receptacle, rotate it clockwise, slowly twist off the corolla and anthers (be careful not to damage the style), and then the style and ovary can be revealed. If this method of emasculation is used, pollination must be done right after. In the process of emasculation, once the flowers that have not been emasculated but are in bloom are found, they should be removed immediately to prevent false hybrids.

C. Pollen collection and preparation. Every morning from 6:00–8:00 (The time can be delayed in case of low temperatures or rainy days), remove large white flower buds that are slightly open or about to open or flowers that have just opened and whose anthers have not yet dehisced, from the father plants. Remove the anthers with a pair of forceps, and place them on clean waxed paper to dry. Then, grind and sieve them before putting them in tubes to store, and make them ready to be used for pollination on the next day or the same day. The pollen remaining after pollination on the same day can be stored in an environment with a relative humidity of about 75% and a temperature of 4–5°C for future needs.

D. Artificial cross-pollination. Pollination should be done in windless or breezy weather with temperatures between 20 and 25°C. Pollination is not suitable when the temperature is below 15°C or above 30°C. The pollination time varies every day depending on the temperature and humidity. If the temperature is high and the humidity is low, the flowering time is earlier, so pollination needs to be done earlier; if the temperature is low and the humidity is high, the flowering time is delayed, so pollination needs to be postponed. In general, the period from 7:00–10:00 is appropriate for pollination, with 8:00–9:00 being the best because the stigma secretions increase during this period. Pollination can also be done in the afternoon after 15:00. If the temperature is suitable throughout the day, pollination can be done at any time of the day. Use a special pollination stick or a stick with a rubber head to apply pollen gently to the stigma of the flower that has been emasculated, or use a glass tube pollinator to pollinate. Apply enough pollen. The larger the area of stigma contacting pollen is, the more even the distribution of pollen on the stigma will be, and the higher the seed setting amount of hybrid fruit will be. In general, cross-pollination requests 10–15 people per day for each mu of seed production field (for mother plants) in the full-blossom stage. The pollination needs one month of work before it is basically completed. After every flower is pollinated, it should be marked immediately with a brightly colored thin wire set at the base of the flower stalk or with a colored marker scribbled on the stalk. A ring of thin wire can also be made and placed around the pedicel as a marker. After the pollination is ended each day, the entire mother plant should be cleaned up. If there are flowers or buds that are not hybridized or fruits that are not marked, they should be removed immediately. This work should be done several times to ensure that the hybrid fruits have an adequate nutrient supply, which can improve the seed fullness of the fruits, increase the thousand-seed weight, and avoid false hybrids. If the seeds are produced in a plastic tunnel, a frame is usually required to prevent lodging.

④ Harvesting and threshing of seed fruits. The method of harvesting fruit is different according to different cultivars, planting methods, and seed fruit parts from pollination to maturity of pepper hybrid fruits. In order to ensure seed maturity, it is necessary to harvest fruits with marks only when the seed pepper plant reaches its red stage. Unmarked and marked but deformed fruits should be removed. When harvesting, only harvest the red and ripe batch of peppers and wait for three to four days before harvesting the next batch. At the same time, choose a sunny day to thresh and dry in time.

(2) Use male sterile lines to produce F_1 hybrids.

Using male sterile lines to produce F_1 pepper hybrids can omit the procedure of emasculation during the bud stage, and it does not require marking the pollinated fruits. After the pollination, there is no need to prune the flowers and branches. Therefore, it is a more labor-saving and easier way to produce F_1 hybrids. In addition, it greatly reduces the difficulty of seed production and seed production costs and can improve the purity of hybrids. At present, there are two types of male sterile lines, namely nucleus male sterile dual-purpose line and nucleo-cytoplasmic male sterile line. The main points of the hybrid seed production technology of these two sterile lines are described below.

Key points of hybrid seed production of nucleus male sterile dual-purpose line are as follows.

A. The genetic pattern of nucleus male sterile dual-purpose line in pepper. If a crop has both the properties of a sterile line and a maintainer line, it is called the dual-purpose line in genetic breeding. The sterility of pepper's dual-purpose line is a recessive inheritance controlled by a pair of nuclear genes. The sterile plants are recessive homozygous individuals. F_2 hybrids will show the segregation of fertility with a 3 : 1 ratio of fertile to sterile plants. If a sterile plant in a dual-purpose line undergoes sister hybridization with a fertile plant, the ratio of fertile plant to sterile plant in its offspring is 1 : 1, while if the fertile plant is self-crossed, the ratio of fertile plant to sterile plant in its offspring is 3 : 1, indicating that there is only one gene difference between the fertile heterozygote in the dual-purpose line and the sterile line.

The population of dual-purpose lines shows a 1 : 1 segregation of fertility during the anthesis, in which the male sterile plants are used as sterile lines in parent breeding and crossbreeding; the fertile plants are used as the maintainer line in parent breeding. Use the pollen from fertile plants to pollinate sterile plants in the isolation area, and the seeds collected from the sterile plants are the dual-purpose line.

B. The sowing amount and planting density of hybrid parents. Because pepper's sterile plants and fertile plants of nucleus male sterile dual-purpose line account for 50% each, when using the dual-purpose line as the mothers to produce seeds, the sowing amount and sowing area for mothers need to be one time bigger than those in the seed production by artificial emasculation and pollination. The seed amount required is 80–100 g for each mu. The density of the mother plants should be 8,000–9,000 plants per mu, with the same planting row spacing and the plant spacing reduced by half (12–14 cm). Each hole is planted with one individual.

C. Fertility identification and removal of fertile plants. Before cross-pollination, 50% of the fertile plants in the dual-purpose line should be removed and 50% of the sterile plants should be kept as the mothers for cross-pollination. The fertile plants are usually identified and removed during the anthesis of the first pepper fruit and the first pair of pepper fruits. The main difference between the floral organs of fertile and sterile plants is that when the flowers of sterile plants bloom, the anthers are thin, dry, and not dehiscent or dehiscent without pollen, and the stigma's growth is normal; when the flowers of fertile plants bloom, the anthers are full and fat, and the anthers are dehiscent.

Using the sterile line as the mother and the maintainer line as the father for cross-pollination, the seeds collected from the sterile plant are the newly-bred male sterile line seeds, most of which are used as the mothers in the hybrid seed production field next year, and a few of which are used as the mothers for breeding the next generation of sterile line. The seeds obtained by self-crossing of the maintainer line are the maintainer line seeds. Using the sterile line as the mother and the restorer line that has good combining ability as the father for cross-pollination, the seeds collected from the sterile plant are F_1 hybrids that can be used as seeds for next year's production. Restorer lines can be bred by self-crossing of restorer lines.

D. Breeding of nucleo-cytoplasmic male sterile line. In order to prevent the biological mixing of male sterile lines of pepper in the reproduction process, a strict reproduction system of three lines and two nurseries should be established. All three lines should be isolated by using 60-mesh insect-proof gauze. The breeding nursery should keep a distance of at least 1,000 m from other pepper seed retention fields and crop fields. Corn or other tall crops can also be used for isolation. Plant the cultivated male sterile line and the corresponding maintainer line in a suitable season with a 3 : 1 ratio of sterile line to maintainer line. Plant the sterile line and the maintainer line in two separate fields. Collect half of the maintainer line's pollen for pollination of the sterile line and keep the other half for self-crossing of the maintainer line. Throughout the anthesis, the fertility of sterile line plants should be continuously observed and investigated. Once fertile flowers are found on the plant, the plant and the surrounding four plants should be removed immediately. In the meantime, the growth of the plants in the maintainer line and the sterile line should be continuously observed as well, and abnormal plants should be removed in time. During the fruiting period, the plants that do not meet the requirements in the sterile line and maintainer line should be eliminated according to the typical economic characteristics and botanical characteristics of the sterile line and maintainer line, and then keep the seeds separately. Before harvest, superior plants of the sterile line and the maintainer line should be selected according to requirements: the selected plants of the sterile line and the maintainer line should have the same economic and botanical traits, and the seeds should be kept together. The selected seeds of the sterile line are used as the mothers for breeding the sterile line, and the rest are used as the mothers in the production of F_1 hybrids. The seeds collected from the superior plants of the maintainer line are kept as the maintainer line. In the nursery for the original seeds of the restorer line, bad individuals should be removed according to the botanical traits of the restorer line throughout the fertility period. A thorough roguing should be carried out before the harvest, and the seeds of the restorer line can be obtained after mixing and storing them.

E. Hybrid seed production. Use the male sterile line in the nucleo-cytoplasmic male sterile line to produce F_1 hybrids. Because the sterile plant rate of this male sterile line is 100%, it makes the production of F_1 hybrids easier. The process can be done by removing the impure and inferior plants, the flowers that have bloomed, and seated fruits before pollination. Select the father flower that blooms on the same day to be pollinated and pinch off one or two petals after pollination to mark. As the sterility of the sterile line is unstable due to the influence of environmental conditions, sterile plants may produce fertile pollen under certain environmental conditions. Therefore, close attention should be paid to changes in sterility throughout the pollination period, and sterile plants found to produce pollen should be promptly removed. Or contact the breeding institute to take other measures to prevent the production of self-crossing fruits.

III. Cabbage Vegetable Seed Production

(I) Chinese Cabbage Seed Production Technology

Chinese cabbage, also known as celery cabbage, is of the genus *Brassica* in the Cruciferae family. It originated in China and is a specialty vegetable of China, mainly distributed in North China. Chinese cabbage's main product is its leafy head, which is rich in nutrition. It's one of the favorite vegetables of the Chinese people because of its rich nutrition, high yield, easy cultivation, tolerance to storage and transportation, and long supply period. It is cultivated all over China and also widely in Korea, Japan, and other countries. In recent years, with the promotion of bolting-resist and heat-resist cultivars, the planting area of spring Chinese cabbage and summer Chinese cabbage has gradually increased. The economic outcome has proved to be significant.

1. Flowering and Pollination Habits

(1) Growth characteristics at each stage.

Chinese cabbage is a biennial plant. Under normal circumstances, in the first year, it proceeds vegetative growth, forming vigorous rosette leaves before forming a well-developed leafy head. In the second year, it proceeds reproductive growth, completing bolting, flowering, and fruiting. The conversion of Chinese cabbage from vegetative to reproductive growth requires a certain low-temperature phase, namely the vernalization process. Its floral bud differentiation happens slowly during the head formation stage and dormant stage, during which the embryo of flower stalks and buds are formed. By the time they are planted in the following spring, the mother plant will grow short flower stalks and different sizes of flower buds. The main stem of the flower stalk begins to elongate rapidly when the weather warms up, and the flower buds on it gradually grow and bloom.

Chinese cabbage is a seed vernalization type. The sprouting seeds can pass the vernalization stage after 10–30 days at a low temperature of 2–10°C. During the vernalization, the longer the low-temperature treatment lasts and the older the plant is at the time of treatment, the earlier the flower bud differentiates. After the flower buds are differentiated, long sunshine duration and high temperature stimulate bolting and flowering. But during the vernalization stage, if the low temperature lasts very short, or if the flower buds do not get long enough sunshine duration even if the vernalization is completed, there will be various deformed flowers, or the flowers do not bear fruits. This is the so-called semi-nutritious state, which will greatly affect the seed yield.

(2) Flower stalk's branching and flowering habits.

Chinese cabbage has racemes. Its flower stalk has a strong branching ability. Generally, the branching can occur three to four times, namely the main branch, primary branch, secondary branch, and tertiary branch. Usually, flowers on the main branch bloom first, and then flowers on other branches open in the order of primary, secondary, and tertiary branches (Fig. 4-13). The order of flowering of a single branch is from the base of the inflorescence upwards. Cabbage seed plants have 1,000–2,000 flowers per plant, mainly on the primary and secondary branches. Flowers on the primary and secondary branches can be more than 90% of the flowers on the whole plant. The anthesis of an individual Chinese cabbage is about 30 days. The full-blossom stage comes fast after the first flower blooms. During the full-blossom stage, a large number of flowers on the primary and secondary branches bloom. One individual has 100 or more flowers bloom every day, with a maximum of 200 flowers. The peak of the anthesis of the whole plant happens at the same time as that of the second branching, which lasts only about 10 days, so the pollination work should be started as soon as the full-blossom stage is about to come.

Fig. 4-13 Flowers of Chinese Cabbage

On average, about 200 stem leaves will be born on the seed plant. The quality and existence duration of the stem leaves are directly related to the quality of the fruits born by the seed plant, so care should be taken to protect them and make them grow well.

(3) Pollination habits.

A flower of Chinese cabbage has four sepals, four petals, and six stamens, among which are four long and two short, one pistil, and four nectaries, and it is located at the base of the filament. Chinese cabbage is a cross-pollinated crop with entomophilous flowers. It will bear few fruits or even no fruit by self-crossing, and it shows a decline in the living ability after self-crossing. The pistil can be fertilized and bear fruit from four days before flowering to two to three days after flowering, but fertility is the strongest on the day of flowering. The pistil's pollen has vitality from three days before flowering to one day after flowering, but the strongest vitality is shown on the day of flowering.

(4) Fruiting habits.

The primary branches are directly attached to the main flower stalk, and they are the backbone branches where fruits are born. The flowers attached to primary branches have a high rate of bearing efficient fruits and the amount is large, ranking first among branches at all levels. Fruits born on primary branches have a large number of good-quality seeds, so they are important branches for hybridization and self-crossing, as well as an important part of determining the seed yield for the individual plants; secondary branches are attached to primary branches. The number of these branches is big and they have abundant flowers, accounting for more than half of the total number of flowers on a plant. Although the efficient fruit rate of secondary branches is inferior to that of primary branches due to the large amount of flowers, they are branches that have the highest potential to increase the seed yield for the individual plants. The tertiary branches are at the periphery of the seed plant. They have only a few flowers and their growth is poor, so they are of little significance to seed production.

The number of seeds on one fruit of Chinese cabbage varies greatly depending on the type–up to more than 20 seeds and down to four or five seeds. The seeds mature 30–45 days after flowering, and the seed yield is up to 100 g per plant, generally below 50 g.

2. Seed Production Technology for Stable Cultivar

Chinese cabbage is a cross-pollinated crop with strong self-incompatibility and obvious self-crossing degradation. As a result, there is often a strong heterogeneity within the stable cultivar group. In fact, stable cultivars (including local cultivars and cultivated cultivars) that are currently used in large quantities in the production are a group whose traits are relatively stable and there are large genetic differences between plants. In response to this character, the corresponding technical measures should be determined during the seed-collecting period of the stable cultivar to ensure that the original cultivar's excellent characteristics remain unchanged and even improve from generation to generation without causing vitality degradation or genetic drift.

(1) Seed collection methods.

① Adult plant seed collection method. Adult plant seed collection is also known as large plant seed collection, root-to-seed collection, and large female plant seed collection. Cultivate healthy and strong seed plants in autumn of the first year, carry out strict selection during the maturity stage of leafy heads, and then store or temporarily plant them in winter. In the spring of the second year, plant them in open fields to collect the seeds. Chinese cabbage seeds produced by this method have high genetic purity, and other traits such as disease resistance and consistency are also maintained well. However, the seed yield is not as good as that of the seed-to-seed collection method and the semi-adult plant seed collection method, and it needs to occupy fields for a long time, so the cost is higher.

② Semi-adult plant seed collection method. Semi-adult plant seed collection requires the seeds to be sown 7–10 days later than that in the adult plant seed collection so that the plants cannot completely form the heads before they are selected to be seed plants. Because the plants required in this method are smaller than those in the adult plant seed collection method, final singling in autumn using this method can be 15%–30% more than that in the adult plant seed collection method. The planting density in the next spring can reach 5,000 plants per mu, so the seed yield per unit area is higher than that of the adult plant seed collection method. Moreover, semi-adult plants are more resistant to cold than adult plants, so they are easier to store. They can overwinter in open fields in South China. However, since the selection result of is not as good as that of the adult plant seed collection method, the seed quality cannot be fully guaranteed as in the case of the adult plant seed collection method.

③ The seed-to-seed method is also known as early spring seedling seed collection. Generally, seeds are sown in the cold frame in mid-to-late January and planted in mid-to-late March. The seed plants directly start to bolt, flower, and set seeds without going through the head formation stage. If this method is used, the yield of seeds per unit area is high and the time required to occupy the fields is the shortest, so the cost is low. However, this method makes it impossible to select seeds according to the leafy heads, so the seed quality cannot be guaranteed. Therefore, the seeds can only be produced using high-quality original seeds by the seed-to-seed reproduction method. Seeds collected using the seed-to-seed collection method cannot be used to produce seeds.

(2) Key points for adult plant seed collection method.

① Seed plant cultivation. The autumn sowing period of seed plants for seed collection is slightly different from that for commercial production. In order to make sure that the seed plants can form full leafy heads for selection before they are harvested and are not easy to age to become hard to store, the sowing period should be delayed 5–10 days, which can also reduce the occurrence of disease. The planting density can be increased by 10%–15%, which is 3,000–4,000 plants per mu. In terms of water and fertilizer management, the amount of nitrogen fertilizer should be reduced, and the amount of phosphate and potassium fertilizer should be increased in order to enhance the storability of the seed plants. Watering should be controlled in the mid-and-late stages of the head formation period. Seed plants should be harvested three to five days earlier than those in the crop fields to prevent being frozen. When harvesting, pull the plants out together with their roots, put them on the spot to dry for two to three days, and turn them over once every day. Then, pile them together in the open field with the roots pointing inside. When the weather turns cold, move them into the cellar to store.

② Selection of seed plants. The seed plants need to be selected several times during the cultivating process to improve seed purity and prevent degradation.

A. Selection during the seedling stage. For direct sowing seedlings, abnormal or variant plants can be removed through seedling thinning, while for transplanting seedlings, seedlings can be selected and eliminated before final planting.

B. Selection during the maturity stage. When the seeds are retained in a large area and when the variety purity is high, roguing can be done to eliminate atypical and diseased plants. The selection puts emphasis on the typical traits of the cultivar, such as plant type, leaf color, compacting type of leaf, the presence of prickles, and so on. At the same time, attention must be paid to the selection of seed plants that are healthy, free of pests and diseases, with few outer leaves, and have tight leafy heads. For cultivars that have wrapped heads, select those whose leafy heads have good covers and few exposures of the inner leaves. For cultivars that have vertical heads, select those whose heads are flatter and whose outer leaves are longer than the leafy heads so that they can protect the leafy heads.

C. Selection during the storage period. When the selection is carried out for storage, attention should be paid to eliminating seed plants that are heated, frozen, rotten, and those whose roots are red. At the later stage of storage, especially eliminate seed plants whose outer leaves abscise more, lateral buds sprout earlier, heads crack, and old ones.

D. Selection after the final planting. After the seed plants have already bolted and flowered, plants that do not meet the standard can be further eliminated based on the branching habits of the seed plants, as well as traits of the leaf, stalk, and flower.

③ Pre-planting treatment of seed plants. About 30–50 days before the final planting, cut off the upper part of the leafy head to facilitate the bolting. Those that need temporary planting in the south also need the preparation of the leafy head. Cutting the head too early will cause the heart of the cabbage to frost more easily. It can also cause the seed plant to lose too much water, resulting in poor growth and low seed yield. Cutting the head too late will easily damage the flower stalk and affect the seed collecting amount. There are many ways to cut the head, such as the one-cut method, two-cut method, three-cut method, and ring-cut method, among which the three-cut tower-shaped method is the most advisable. The three-cut tower-shaped method is to cut obliquely three times with an angle of 120° from 7–10 cm above the dwarf stems of Chinese cabbage. The angle between the incision and the longitudinal axis of the cabbage should be about 30°–45°, giving the remaining part a tower shape. Put the heads that are cut into tower shapes in places where they will not be damaged by frost (such as plastic tunnels or cold frames) and can be exposed to sunlight during the day. This is conducive to the healing of the incision and the greening of the leaves, and it can improve the survival rate after the final planting.

④ The final planting of seed plants and the corresponding management.

A. Isolation of seed collection fields. Chinese cabbage is a cross-pollinated plant, so the seed collection fields need to be strictly isolated from the seed collection fields of other large cultivars, or the seed collection fields of rapes, turnips, pakchoi cabbages, and Chinese mustards at a distance of no less than 2,000 m. If there are obstacles, such as woods or mountains, in the isolation area, the isolation distance should be more than 1,000 m. In addition, for crop fields that grow crops like spring cabbage in the isolation area, if signs of bolting are found, harvest them promptly. Moreover, continuous cropping with Cruciferae vegetables should be avoided in the seed collection field. It is best to choose fields that have not been planted with Cruciferous vegetables for two to three years to produce seeds.

B. Final planting. The final planting can be done in early spring when the temperature of the upper 10 cm layer of soil reaches 6–7°C. In Liaoning Province, the final planting is usually done during the period from late March to early April. The final planting should be done as early as possible on the premise that plants will not be harmed by frost damage. This is conducive to rooting and the later growth of the plant. To prevent soft rot, the ridge culture can be used. For the final planting, there should be 3,500–4,500 seed plants for each mu. The planting depth is the best when the incision of the head and the surface of the ridge are on the same level. When planting, the soil must be hilled and stamped down, leaving no gaps. If the soil is not tightly hilled and the main root cannot be close to the soil layer, the new roots will be dried to death due to the "air leak" in the soil.

C. Field management. The seed yield in Chinese cabbage seed collection is mainly contributed by the seed pods on the main stem, primary branches, and secondary branches, so water and fertilizer management should be regulated according to this feature. When the final planting period is early and the soil tilth is good, there is no need to water immediately after the planting in case it will lower the soil temperature. But if the final planting period is late and the soil is dry, water should be applied appropriately. When the main stem elongates about 10 cm, water it with topdressing once until the soil is not arid. During the anthesis, water several times to ensure the plant has sufficient water. During the full-blossom stage, apply nitrogen fertilizer and phosphate fertilizer two to three times. The flowers of Chinese cabbage are entomophilous flowers. The number of pollinators is closely related to the seed yield, so usually, one box of bees for each mu of seed collection field is needed to assist in pollination. After the seed plants bear set pods, it is best to set up a stand to prevent yield loss due to lodging. Pay attention to the prevention and control of pests and diseases during the growth period of plants.

⑤ Harvesting and threshing of seed pods. Chinese Cabbage takes 30–45 days from flowering to seed maturity. Harvesting seeds earlier or later will affect seed yield and quality. As the seed pods are easy to crack naturally after maturity, in order to prevent the seeds from falling when harvesting, or when the seed pods on the main, primary, and secondary branches turn yellow, cut them off from the part above the ground with a reaping hook in the morning before the dew dries. In order to give the seeds in the seed pods that have not yet turned completely yellow a period of after-ripening, the harvested seed plants should be placed on the sunning ground to dry for two to three days before threshing.

(3) Key points for seed-to-seed method.

The sprouting seeds of Chinese cabbage can feel the low temperature, which allows them to pass through vernalization in the seedling stage, and under long sunlight conditions, they can bolt, flower, and produce seeds without having to form heads first. According to the different sowing methods, the seed-to-seed method can be further divided into the following kinds.

① Spring seedling seed-to-seed method. Breed seedlings in cold frames in early spring, then sow the seedlings in mid-to-late January. After the emergence of seedlings, provide a certain period of low temperature to help improve their cold resistance during the vernalization stage. In early April, plant the seedlings in open fields when they have 6–10 true leaves. The anthesis in the seed-to-seed method is later than that in the adult plant seed collection method, and the effective anthesis is shorter, so in order to obtain a higher seed yield, the planting density should be properly higher. Usually, 4,000–5,000 plants per mu is appropriate. The vegetative body of the seed plant used in the seed-to-seed method is small during the final planting, so the vegetative growth of the plant should be stimulated to prevent premature bolting at the early stage. Because the anthesis is late, the later anthesis and the high-temperature season overlap, making it easy to cause the seeds to dry before they grow mature, so attention should be paid to preventing the premature senescence of the plant and trying to extend the effective anthesis. Therefore, its water and fertilizer management differs from that of the adult plant seed collection method. In addition to the seed collection field being fully fertilized, the basal fertilizer should also be added with appropriate quick-acting fertilizer. After the final planting, water the seedlings once and then water the seedlings again after three to five days to help the seedlings re-adapt to the environment. Then, intertillage should be carried out to improve soil permeability and ground temperature and promote the readjustment of seedlings. During the seed-to-seed method for Chinese cabbage seed collection, it is necessary to avoid the hardening of seedlings. When the seed plants get used to the environment, enhance water and fertilizer management to the end, that is, throughout the whole fertility period, make sure the seed plants will not dehydrate and lack fertilizer. It is best to fertilize the plants once after every watering in areas where it is possible, and the main fertilizer should be quick-acting nitrogen fertilizer. The water-stopping period should also be later than that of the adult plant seed collection method. Generally, watering can be stopped during the green pod period in case the seeds get arid. If the watering is stopped during the hot season, a drop in thousand-seed weight will result. Seed collection techniques for the seed-to-seed method are basically similar to those for the adult plant seed collection method.

② Open field overwintering seed-to-seed method. In areas where the average minimum temperature in winter is higher than -l°C, seeds can be sowed directly in open fields before winter or transplanted to the open fields after they are raised into seedlings. Seed collection can be done in the following spring. In order to enable the seedlings to overwinter safely without being harmed by frost damage, seedlings should maintain the size when they have 10 leaves during winter. And if necessary, they can be covered with straw, horse manure, wheat bran, or other appropriate mulches. After a winter of low temperatures, the seedlings will bolt and flower in the following spring. This method makes the anthesis early, the yield high, and the seeds grow mature early. It can facilitate the transportation and use of the seeds in the current year.

③ Spring direct sowing seed-to-seed method. This method means the seeds are sown in spring and harvested in the summer of the same year. In order to meet the low-temperature requirement of the seed plants, make the plant flower earlier, and extend the effective anthesis, early sowing at the right time is the key to this seed collection method. Usually, seeds can be sown in early spring after the topsoil thaw. In the northeast regions of China, seeds can be sown in middle to late March. After the emergence of seedlings, carry out seedling thinning one to two times, leaving 8,000–9,000 seedlings per mu. The field management is basically the same as that of the spring seedling seed-to-seed method.

④ Vernalization direct sowing seed-to-seed method. In the north-central part of Northeast China and some parts of Inner Mongolia, due to the short frost-free period, the sowing period of Chinese cabbage is early. This method is often used to ensure that the seeds collected in the current year are ready for fall sowing. That is, 7–25 days before the best sowing period (before the plow layer of the soil thaws), soak Chinese cabbage seeds in warm water at 45°C for two hours and then put them at a temperature under 25°C to pregerminate. After 20–40 hours when the seeds germinate, put the germinated seeds into gauze bags and then put the bags in a refrigerator at 0–5°C, flatten the gauze bags, and leave them in the refrigerator for 25–30 days. Check the temperature every day and the seed humidity every three days. Each refrigerator can hold 3–10 kg seeds. The seed breeding area can reach one to two hectares. The bolting rate can be more than 95%.

3. Chinese Cabbage Seed Production Technology for Conventional Cultivars

(1) Large plant seed collection method.

It's also known as the adult plant seed collection method, the headed mother plant seed collection method, etc. The seeds are sown in the autumn of the first year to grow into strong seed plants. When the leafy heads are mature, they are strictly selected according to the characteristics of the cultivars, and after overwintering by storing or temporary planting. The seeds are planted in the open field in the spring of the second year for seed collection. The large plant seed collection method can guarantee the seed purity, disease resistance, consistency, head formation ability, and other traits, but the seed yield is low and it occupies the field for a long time, the cellar storage loss is large, and the production cost is high, so it is suitable for the original seed reproduction.

(2) Semi-adult plant seed collection method.

The seeds are sown 7–10 days later than those using the large plant seed collection method to let the plant reach the state of half-heading. The density increases by 15%–30%, and then the seed plants are selected for storage and planted for seed collection in the following spring. When the semi-adult plant seed collection method is adopted, the plant density is higher, the seed yield is higher, and the cold resistance and storage tolerance of the plant are stronger than those using the large plant seed collection method, so the plants can overwinter in the open field in South China. However, the selection outcome is not as good as that of the large plant seed collection method, so the seed quality is not as high as that of the large plant seed collection method.

(3) Seed-to-seed method.

This is the method of sowing and collecting seeds in the same year. This method is further divided into spring seedling seed collection method, spring direct sowing seed collection method, vernalization direct sowing seed collection method, and open field overwintering seed collection method. When the seed-to-seed collection method is adopted, the field occupation time is short and the seed production cost is low, but the selection of leafy heads is not possible, so the quality of seeds cannot be guaranteed. Therefore, it can only be used in the reproduction of the seeds for production.

4. F_1 Hybrid Seed Production Technology

Chinese cabbage has obvious heterosis. The use of superior F_1 hybrids not only has obvious advantages in productivity, disease resistance, and quality but also has obvious advantages over fixed cultivars in purity. This is why Chinese cabbage's F_1 hybrids have been paid more and more attention. As Chinese cabbage is a monoclinous crop with small floral organs, using artificial emasculation cross-pollination to produce the F_1 hybrids is not practical in production. In China, the seed production of Chinese cabbage's F_1 hybrids mainly uses two methods: using the self-incompatible line to produce seeds and using the male sterile line to produce seeds.

(1) Use self-incompatible line to breed F_1 hybrids of Chinese cabbage.

① Reproduction and conservation of parental self-incompatible line. For the reproduction of the parental self-incompatible line of Chinese cabbage, the adult plant seed collection method or seed-to-seed method can be adopted. Seed collection field for parental plants should be isolated from other cabbage cultivars or crops that are easy to hybridize with Chinese cabbage, such as pakchoi cabbages, rapes, flowering Chinese cabbages, and Chinese flat cabbages, at a distance of no less than 2,000 m, or use the plastic tunnel, net room, or bagging and other methods to achieve mechanical isolation. The cultivation management of the seed plants is basically the same as that of the original seed production of the stable cultivars. The key point lies in self-pollination during the bud stage. Usually, when the seed plant enters the anthesis, select the flowers on the middle part of the robust primary and secondary branches, and then apply the pollen from the flowers that open on the same day or the day before on the stigma of the buds of the same line peeled by hand. The most suitable bud age is two to four days before flowering and the best bud size is the size of the ninth bud above the open flower. Do not only strip the large buds to pollinate as this will significantly reduce the fruiting rate. When pollinating during the bud stage, try to damage the flowers as little as possible and do not turn the buds strongly to prevent damage to the flower stalk and thus affect the fruiting.

If multiple materials are planted at the same time in the net room or plastic tunnel, the different materials must be isolated by film or glass. Different materials should be pollinated by separate people. If the pollinator needs to pollinate different materials continuously, the pollinator's hands and tweezers must be strictly disinfected with 75% alcohol, and the work clothes must be changed. When using paper bags to isolate, the paper bag must be set on the flower branch before the flower buds open. Take the pollen inside the paper bag for bud pollination when flowers on the lower part of the spay on the same plant bloom. The isolation paper bag should be reset immediately after pollination. The paper bag should not be removed until all the flowers on the branch have withered. If the seed plants of the self-incompatible line are planted in the isolation net room or spatial isolation area, spray 2%–3% sodium chloride solution on the opened flowers once a day or every other day during the anthesis. Then, pollinate relying on bees or artificial assistance to obtain selfed seeds of the same line. This method is simple and easy and has a low cost, but it is not effective for any self-incompatible line, so it needs to be tested before application.

② Use self-incompatible line for hybrid production. The main method for seed collection for using the incompatible line to produce the F_1 hybrids of Chinese cabbage is the spring seedling seed-to-seed method. That is, sow the seeds of the father and mother plants in early spring at a ratio of 1 : 1 in the cold frame to breed seedlings, and 20–30 g seeds are needed for each mu of seed production field. Separate the seedlings once when they have two to three true leaves, and when the seed plants have 6–10 leaves, the father and mother seed plants should be planted in separate rows in the isolation area. Pay attention to letting the anthesis of the father and mother plants meet; then they can be left behind to pollinate and bear fruits freely. In order to improve seed yield, bees can be used to assist pollination or a chicken feather duster can be used for artificially assisted pollination. The seeds harvested from the self-incompatible line are the F_1 hybrids.

(2) Use nucleus male sterile dual-purpose line to breed F_1 hybrids.

① Reproduction and conservation of nucleus male sterile dual-purpose line. The male sterility of the Chinese cabbage's male sterile materials that have been found so far is mostly controlled by a pair of recessive nuclear genes. The sterility gene is "ms" and the sterile plant's genotype is "msms"; the fertile plant's genotype is "MSins". There are 50% sterile and 50% fertile plants in each generation. Even when fertile plants are used to pollinate sterile plants, 50% of them are still sterile plants. Because this sterile line can stably segregate 50% of sterile plants in each generation to play the role of sterile line, and even after sister-crossing, it can still maintain 50% of sterile plants. Therefore, it also has the characteristics of the maintainer line, so it is called the dual-purpose line.

The adult plant seed collection method or seed-to-seed method can be used to reproduce the nucleus male sterile dual-purpose line of Chinese cabbage. Plant the seed plants in the isolation area, investigate them after the seed plants enter the anthesis, and mark them with a sign on the main stems of the sterile plants. Make sure free pollination takes place in the seed collection field. When the seeds mature, collect the seeds on the sterile plant to gain the original seeds of the dual-purpose line.

② Use nucleus male sterile dual-purpose line to produce hybrids.

A. Final planting of seed plants: The seed-to-seed method is generally applied in the production of Chinese cabbage's F_1 hybrids using the nucleus male sterile dual-purpose line. In early spring, breed seedlings of the father and mother plants at a ratio of 1 : 7. The total amount of seeds used in each mu of seed production field is 25–50 g. Separate the seedlings once when they have two to three leaves and plant them in an isolation area when they have 6–10 leaves. Plant the father and mother plants at a 1 : 3 row ratio. However, the specific planting spacing varies depending on the parental materials. For bigger parental plants, the planting spacing for the fathers should be 47 cm×47 cm and 18 cm×47 cm for the mothers. For smaller parental plants, the planting spacing should be 33 cm×33 cm and 13 cm×33 cm for fathers and mothers, respectively.

B. Removal of mother fertile plants: In the initial flowering stage, the basic characteristics of sterile plants are neat inflorescence, light-colored corolla, small anthers, and no pollen in the anthers when the anthers are exposed. These are the bases to identify sterile plants. There are also sterile types with white anthers. Find the sterile plant that flowers the earliest in the mother lines of the whole field. When the flowers on the first lateral branch are about to bloom, start pulling out fertile plants in the mother lines. After that, the fertile plants in the mother lines need to be pulled out before 9:00 every day continuously until all of them are pulled out, which will take about three to seven days.

C. Cross-pollination and seed collection: The main stems of sterile plants can be topped at the same time when the fertile mother plants are pulled out. This can act as a marker to check the work of removing fertile plants and at the same time, it can delay the anthesis of sterile plants to prevent them from receiving pollen of the same line to form false hybrids when the fertile plants in mother lines have not been completely removed. When the fertile plants in mother lines are completely removed, they can be allowed to hybridize naturally with the father plants and the seeds collected from the mother plants are F_1 hybrids. To ensure that there are no mistakes when collecting seeds, the fathers can be removed about 25 days before the seed collection.

(3) Use the male sterile line controlled by the new nuclear gene to breed F_1 hybrids.

① Reproduction of the male sterile line with 100% sterile plant rate.

A. Genetic mechanism: If the sterile plant of Chinese cabbage's type A dual-purpose line is used as the mother to hybridize with the fertile plant of the type B dual-purpose line, a 100% sterile plant rate can be achieved in some combinations. Feng Hui and some other experts have explained this genetic mechanism by using the Multiple Alleles Hypothesis. That is, there are three multiple alleles at the sites controlling fertility, including MSf, MS, and Ins. MSt is a dominant restoring gene, MS is a dominant sterile gene and ms is a recessive fertile gene. The dominant-recessive relationship between these three genes is MSt>MS>ms. Sterile plants have two genotypes: MSMS and MSms; fertile plants have four genotypes: MStMSt, MStMS, MStms, and msms. Using the 100% male sterile line as the mother in hybridization can eliminate the labor of removing fertile plants and the crossing rate can reach 100%, leading the seed production technology of Chinese cabbage's F_1 hybrid to a new historical period.

B. Use sister-crossing within a dual-purpose line to breed type A dual-purpose line and type B dual-purpose line respectively.

C. Plant the obtained type A dual-purpose line and type B dual-purpose line in segregated rows at a 2 : 1 row ratio. During the initial flowering stage, identify sterile plants row by row and pull out all fertile plants in the rows of the type-A dual-purpose line. In the meantime, top the main branch to promote the development of lateral branches and delay the anthesis. If type A sterile plants are to receive pollen from type B fertile plants, carry out artificial pollination or let them naturally pollinate in isolated areas. Then, the seeds collected from type A sterile plants are the male sterile line with a 100% sterile plant rate and the sterility of the individual plants is also 100% (no fertile chimera nor fertile branches). Type B pollinator can be called temporary maintainer line. Mark the sterile plants of the type B dual-purpose line and seeds collected from the sterile plants can breed type B dual-purpose line.

② Use a 100% male sterile line to produce hybrids: The parental plants are bred by applying the seed-to-seed method. Use the male sterile line as the mother and the inbred line as the father, and plant them in the isolation area at a row ratio of 1 : 3–1 : 4. During the anthesis, let them pollinate naturally or assist the pollination artificially and the seeds collected from the sterile line are F_1 hybrids.

(II) Seed Production Technologies for Common Head Cabbage

Common head cabbage, also known as cabbage, is a biennial plant capable of forming leafy heads in the *Brassica* genus of the Cruciferae family. Its leafy heads are rich in nutrients and can be stir-fried, boiled, made into a salad, or pickled. It is very popular among consumers. Common head cabbage originated from the Mediterranean to the North Sea coast and is a major vegetable in Europe and the Americas. It is now commonly planted around the world. It was brought into China in the 16th century and is now planted throughout the country. It is one of the main vegetables grown in the cooler regions such as Northeast, Northwest, and North China in spring, summer, and autumn. It is also planted on a large scale in South China in autumn, winter, and spring. Sowing can be arranged all year round by selecting suitable cultivars for different regions, and harvest should be performed in stages. In the annual supply of vegetables, common head cabbage occupies an important position.

1. Flowering and Pollination Habits

(1) Growth characteristics at each stage.

Common head cabbage is a biennial crop. Its growth cycle is clearly divided into two periods, including vegetative growth and reproductive growth. The vegetative body (leafy head) is usually formed in the first year. After it passes the vernalization under the low temperature in winter, it sprouts and differentiates, and then in spring, it bolts and flowers.

Common head cabbage is a green body vernalization type plant. Vernalization cannot be carried out on its sprouting seeds under low-temperature conditions. Two conditions must be met to complete the vernalization stages: First, a vegetative body of a certain size; second, to undergo a certain range of low temperatures for a considerable period of time. It is generally believed that seedlings with a stalk thickness above 0.6 cm and a number of true leaves above seven can complete the vernalization stage after going through the low temperature of 0–12°C for 50–90 days. It should be noted that the temperature should not be too low as it will hinder the vernalization. Therefore, if seeds of common head cabbage are sown too early in spring, the inappropriate cultivar is chosen, or the seedlings are too large when overwintering, a great loss is likely to occur because of the "premature bolting" after the completion of vernalization stage. The time required to complete the vernalization stage is related to the plant size and cultivar type. Generally, the larger the plant's vegetative body is, the easier it is for the plant to complete the vernalization stage, and the shorter the time of low temperature is needed to complete the vernalization stage. Most of the oxheart-shaped cultivars and oblate cultivars have stronger winterness. That means, in order to complete the vernalization, on one hand, the seedlings need to be large and the duration of low temperature needs to be long, so premature bolting is not likely to occur in spring cultivation; on the other hand, spherical cultivars have weaker cold resistance, so the seedling size required to complete the vernalization is small and the duration of low temperature needed is shorter.

Long daylight and sufficient sunlight are beneficial for common head cabbage to bolt and flower, but the response to light varies greatly among different cultivars or ecological types. Generally, pointed and oblate cultivars do not have strict light requirements. The seed plants stored in vegetable cellars or buried in the ground in winter can bolt and flower normally after being planted in the spring of the second year. Spherical cultivars require a long sunlight duration. Usually, a certain part of the seed plants stored in the cellar or buried in the ground for overwintering can not bolt and flower.

(2) Branching and flowering habits of seed plants.

① Branching habits. The height of the seed plants of common head cabbage varies depending on different cultivars and cultivation management conditions. For precocial oxheart-shaped cultivars, the seed plant's height is about 1–1.2 m, while for spherical and oblate cultivars, the seed plant's height can reach 1.3–1.8 m. Common head cabbage has a compound raceme inflorescence. Primary branches can grow on the leaf axils on the central main flower stalk; secondary branches can grow on the leaf axils of primary branches. When nutrients are sufficient and management conditions are good, tertiary and quaternary branching can also occur. Different ecological types have different branching habits. Seed plants of spherical cultivars have central main stems that have strong growth vigor and a relatively small number of branches. Oxheart-shaped and oblate cultivars' main stems do not have strong growth vigor like the round spherical cultivars, but their primary, secondary, and even tertiary branches are more well-developed.

② Flowering habits.

A. Number of flowers: A robust seed plant can grow about 800–2,000 flowers, but the number of flowers on the plant also varies depending on the cultivars and cultivation management conditions. Seed plants planted before winter and well managed can have more than 1,000 flowers per plant, while seed plants planted in the open field in the spring of the second year often have only 300–400 flowers if not managed properly. For a plant, flowers on the main stem generally bloom first, then the flowers on primary branches bloom from top to bottom, and the flowers on the secondary, tertiary, and quaternary branches gradually bloom one after another. For one inflorescence, flowers open gradually from bottom to top. Whether the plant has a developed or undeveloped main stem, the primary branches have the biggest number of flowers, followed by the secondary branches and the main flower stalk.

B. Anthesis: The anthesis is greatly different among all cultivars. In general, under the same cultivation management conditions, the anthesis of oxheart-shaped and oblate cultivars is seven to 15 days earlier than that of spherical cultivars. Even for the same cultivar, the anthesis between plants often differs by about seven days. The anthesis of a cultivar is about 30–50 days. For one plant, the anthesis is usually 20–40 days. Plants or cultivars with strong growth vigor have longer anthesis, while plants or cultivars with weak growth vigor have shorter anthesis.

About two to five flowers on an inflorescence open each day (Fig. 4-14). When the temperature is high on sunny days, four to five flowers open every day; when the temperature is low on rainy days, two to three flowers open. The vast majority of flowers open before 11:00, but a few are open in the afternoon. Under natural conditions, each flower usually opens for three days and then withers. In an environment that has higher humidity, such as in a greenhouse or in an isolated paper bag, the flower can open for four to five days.

Fig. 4-14 Flowers of Common Head Cabbage

(3) Pollination habits.

① Floral organ structure. The flowers of common head cabbage are complete flowers, including calyx, corolla, pistil, and stamens, and each flower has four green sepals at the outermost whorl. The corolla consists of four yellow or golden-yellow petals. The four petals are arranged in a cross shape after flowering. Six stamens are attached to the inner side of the petals, two of which are shorter and four are longer. Each pistil has an anther on the top. The anther naturally splits after maturity and disperses pollen. There are four nectaries at the base of stamens. The pistil is in the center of the flower and consists of three parts: the stigma, the style, and the ovary. The ovary has two carpels and can accommodate 25–30 ovules, which are arranged on either side of the central placenta.

② Pollination and fertilization. Common head cabbage is a typical cross-pollinated crop. Under natural conditions, the pollination of the plant is done by insects. Plant two different cultivars together to cross naturally; the crossing rate is generally up to about 70%.

The viability of the stigma and pollen of common head cabbage is generally the strongest on the day of flowering. The stigma can receive pollen for fertilization six days before and two to three days after flowering. The pollen two days before and one day after flowering has certain viability. If the pollen is collected and stored in a desiccator, pollen viability can also be maintained for more than seven days under room temperature conditions and can be maintained for longer under low temperatures below 0°C and dry conditions. Under the conditions of cross-pollination and a temperature of 15–20°C, the pollen tube starts to grow after two to four hours. After six to eight hours, it passes through the style tissue and completes fertilization after 36–48 hours. The optimum temperature for fertilization is 15–20°C: pollen germinates slower when the temperature is lower than 10°C, and it will affect the normal conduct of fertilization if the temperature is higher than 30°C.

(4) Fruiting characteristics.

The fruit of common head cabbage is silique in a cylindrical shape, with seeds arranged on both sides of the septum. Each seed plant normally bears 900–1,500 effective seed pods. On a plant, most of the effective seed pods are on the primary branches and, to a lesser extent, on the secondary branches and the main branch. Each seed pod has about 20 seeds. On a branch, the upper and bottom seed pods have fewer seeds, and the middle and lower seed pods have the most seeds.

The time required for seeds of common head cabbage to grow mature also varies with cultivars and temperature conditions. Seeds of round spheroidal cultivars generally take longer to mature, while oxheart-shaped and oblate cultivars take less time. Seeds mature faster under high-temperature conditions and slower at lower temperatures.

When the fruit ripens, the hull dehisces upward from the base of the fruit stalk, leaving the seeds attached to the placenta, which eventually fall off due to mechanical force. Seeds of common head cabbage are reddish-brown or blackish-brown, with a thousand-seed weight of 3.3–4.5 g. A robust seed plant can generally bear about 50 g of seeds.

2. Seed Production Technology for Stable Cultivar

(1) Seed collection methods.

① Autumn adult plant seed collection method. Sow the cultivar of common head cabbage that needs to be reproduced early in autumn so that the leafy head can basically mature before overwintering. Then, select the good plants to keep. This method can be further divided into the with-head seed collection method and head-cutting seed collection method. The with-head seed collection method is to let the selected plant overwinter with the complete leafy head and then cut the top of the leafy head into a cross shape in the following year so that they can bolt and flower. There are two ways in the head-cutting seed collection method: one is to cut off only the outer part of the leafy head and the outer leaves and let the seed plant overwinter with the central part of the leafy head; the other is to cut off the leafy head completely before or after overwintering and leave the old roots behind, then collect the seeds after the old roots grow lateral buds, bolt, and flower. When the autumn adult plant seed collection method is used, it is possible to select the seed plants strictly according to the plant traits. Therefore, the purity of the collected seeds is high. This method is often used to reproduce the original seeds of autumn common head cabbage cultivars. When using this method, it's impossible to identify the winterness and head formation traits of seed plants under spring cultivation conditions, so spring common head cabbage cultivars should not be continuously collected by this method.

② Autumn semi-adult plant seed collection method. In autumn, sow the common head cabbage late so that it can form an incompact half-enclosed leafy head to overwinter and be ready for seed collection in the following spring. This method requires less field occupation time and lower cost. When this method is used, the seed plant develops well and the seed yield is high. However, it is impossible to select seeds strictly. Therefore, it can only be used to reproduce general-quality seeds for production.

③ Spring old root axillary bud cuttage method. Select superior plants in the spring common head cabbage field, cut off the leafy head and leave the old root and rosette. When the axillary buds grow four to six leaves, cut them off along with part of the old stem tissue for cuttage. In order to improve the survival rate, a shed should be set up to keep off rain and sun after cuttage, and water it frequently to keep it moist. After the new plant roots, the frequency of watering can be reduced. The new plants will form leafy heads in autumn and seed collection can be carried out in the following spring after overwintering. When spring common head cabbage is reproduced in this way, the good characteristics can be maintained because the seed plants have been strictly selected in spring. However, this method of seed collection is laborious and costly. For spring common head cabbage seed collection, the autumn adult plant seed collection method and the spring old root axillary bud cuttage method can be used alternately.

(2) Key points for adult (semi-adult) plant seed collection in autumn.

① Cultivation of seed plants. The cultivation management of seed plants in the first year is basically the same as that of commercial vegetable production. However, the sowing period can be a little later. If using the autumn adult plant seed collection method, the medium- and late-maturing cultivars in northern areas can be sown in late June and the early-maturing cultivars in late July. If using the semi-adult plant seed collection method, the sowing period should be about 15 days later than that using the adult plant seed collection method. The main method for common head cabbage is the seedling transplanting method, which requires 20–25 g of seeds per mu. As the seedling-raising period is in the hot and rainy season, it is necessary to choose a plot with fertile soil and high topography to make high ridges for raising seedlings and build a shade shelter on the plot before the seedlings emerge to prevent heavy rain and exposure to the sun. Transplant the seedlings into larger fields when they grow two to three true leaves. If using the adult plant seed collection method, the planting spacing should be 43 cm×50 cm for the medium- and late-maturing cultivar and 33 cm×40 cm for the early-maturing cultivar. If using the semi-adult plant seed collection method, the planting spacing is 36 cm×43 cm for the medium- and late-maturing cultivar and 27 cm×33 cm for the early-maturing cultivar. From the seedling stage to the time before harvest, in addition to general water and fertilizer management, special attention should be paid to the use of pharmaceuticals to prevent and control the damage of cabbage worms and aphids.

② Selection of seed plants for seed retention. The mixed selection method is usually adopted to select seed plants in the crop fields, and the selection rate is not more than 25%. In specialized seed production fields, roguing can be done by individual plant selection to ensure the purity of the seeds. The selection of seed plants is usually carried out at three stages: seedling stage, leafy head formation stage, and bolting and flowering stage. In order to maintain the characteristics of the original cultivar of common head cabbage, at least 50 seed plants should be included in the mixed pollination.

A. Seedling stage. Select seedlings that are disease-free and robust, with their leaf shape, leaf color, leaf margin, leaf wax powder, petiole, and other traits that match the characteristics of this cultivar as seed plants. The leaf shape of the common head cabbage has some correlation with its head shape. Usually, oblate cultivars have round or oval leaves, while pointed cultivars have longer ones.

B. Leafy head formation stage. Before harvesting leafy heads, select plants with normal growth, no diseases, few outer leaves, large and round leafy heads, and those whose main traits of outer leaves and leaf bulbs are in line with the characteristics of this cultivar for seed retention. In specialized seed production fields, strict roguing must be carried out to eliminate all atypical deformed and diseased plants. If it is the original seed production, roguing should be performed once during the rosette stage.

C. Bolting and flowering stage. Eliminate plants that do not match the characteristics of the cultivar according to the height and branching habits, the color of flower stalks and stem leaves, and other traits of the seed plant during the anthesis.

③ Overwintering of seed plants. In southern China and the southern part of North China, seed plants can overwinter in open fields or overwinter after being slightly covered. In the central regions of North China, as well as the northwest and northeast parts of China, the selected seed plants need to be stored for overwintering. The methods of storage include temporary planting in the cold frame, burial in the dead cellar, and storage in the live cellar.

A. Temporary planting in the cold frame. Dig up the selected seed plants with their roots, remove the old and diseased leaves, hoard them in the cold frame, and then water them thoroughly once. Cover them with straw curtains or cattail mats at night when the temperature is below 0°C and uncover them during the day to keep the temperature in the cold frame at about 0°C.

B. Burial in the dead cellar. Select a higher ground that has better drainage to dig a ditch. The ditch depth is generally 80–90 cm and the width is 1 cm. After the seed plants are harvested, dry them for three to four days and store them in the ditch before they get frozen. As the temperature drops, cover the seed plants gradually with soil or other mulch. The thickness of the mulch varies with the temperature, but the general principle is to keep the seed plants from frost damage. At the same time, do not make the mulch too thick, or the plants could rot due to high temperature.

C. Storage in the live cellar. Remove the old and diseased leaves on the harvested seed plants, dry the plants for three to four days, and then store them in the cellar for cabbage before they get frozen. In order to make full use of the space of the cellar, shelves should be made out of plank or bamboo on both sides of the cellar to hold the seed plants of common head cabbage. Keep the temperature in the cellar around 0°C and the relative humidity between 80% and 90%.

④ Treatment of seed plants. If the seeds are collected by the adult plant seed collection method, the plants must be treated to facilitate bolting and flowering when the seed plants mature in the first year. There are three ways to treat seed plants.

A. Stele retaining method. Cut off the outer part of the leafy head and outer leaves, leaving only the stele, and then transplant the plant with roots. The specific procedure is to vertically cut the heart into a square column with a side length of about 6 cm or to diagonally cut a circle from the base of the leafy head upward and then peel off the outer part of the leafy head. This method facilitates the growth of main stems and lateral branches and promotes the seed yield, which makes it the most effective. However, the leafy heads cannot be sold as commodities.

B. Head-mowing method. Cut off the leafy head from the base, with the cut surface slightly oblique to avoid hydrops and rotting of the pith part. After cutting off the head, cut off the outer leaves when the buds on the inner side of the outer leaves grow to 3–6 cm, and then transplant the plant with roots in the seed collection field. This method allows the leafy head to be used as a commodity, but the main stem is cut, so it mainly relies on the lateral branches to bear seeds, resulting in the low seed yield of individual plants.

C. With-head seed retention method. After overwintering in the open field or in the cellar, cut a cross on the top of the leafy head before spring to a depth of about 1/3 of the head height (on the premise of not hurting the growth cone) in order to help the bolting. This method produces a slightly higher seed yield per plant than the head-mowing method, but the leafy head cannot be used.

⑤ Management of seed collection fields.

A. Isolation conditions. Common head cabbage is a typical cross-pollinated plant and is easy to hybridize with other cabbage crops. In order to ensure seed purity, the seed collection field of common head cabbage should at least be strictly isolated from seed collection fields of other common head cabbage cultivars, or cauliflower, bulb kale, mustard, broccoli hold kale, collard greens, etc. The isolation distance of the original seed production field should be more than 1,600 m, while for the field to produce seeds for production, the isolation distance should be more than 1,000 m.

B. Final planting. At places where the plants cannot overwinter in the open field, the final planting should be done in the spring of the second year. Normally, plant about 4,000 plants per mu. After the final planting, the mulch can be used to help the seed plants adjust to the new environment as soon as possible and increase the seed yield. After two times of watering to help the plant re-adapt to the environment, intertillage should be done in a timely manner to increase the ground temperature and promote the development of the root system of the seed plants.

C. Cultivation management of the seed plants. After the seed plant bolts, the lower old and yellow leaves can be removed. After flowering, remove the weak lateral branches and flowers at the end of each branch to concentrate nutrients to supply the siliques. To prevent flower branches from breaking and lodging, brace a rope or set up a support around the plant. Throughout the growth period of the seed plants, special attention should be paid to prevent damage from aphids, chard moths, and cabbage worms. After the first flowering date, apply further fertilizer and water. Dress 20 kg of ammonium sulfate and 10 kg of calcium superphosphate per mu. When the seed plants enter the full-blossom stage, water every five to seven days and apply fertilizer properly, with about 15 kg of ammonium sulfate and 10 kg of calcium superphosphate per mu. After entering the podding stage, although watering can be reduced appropriately, timely irrigation is still necessary in case of hot and dry weather. Stop watering when the seed pods are about to mature to prevent the seed plants from sending forth the second stubble of flower branches. 10 mg/L aqueous solution of boric acid can be added during the anthesis for disease prevention to promote fertilization and improve the seed yield.

⑥ Harvesting of seeds. Harvesting can begin when 1/3 of the seed pods start to turn yellow. Because the mature seed pods are easy to crack, the harvest cannot be too late. The harvest should be done before 9:00 to avoid losses caused by the cracking of the seed pods. The harvested seed plants can be placed in the sunning ground for three to five days for after-ripening, but pay attention to keeping turning the seed pods and shelter them from rain, so as not to cause seeds to rot. Threshed seeds should be dried in time, but do not put them under the sun on the concrete ground or plastic mulch for too long. In the drying, threshing, cleaning, and bagging processes, there should be dedicated staff to be in charge of preventing mechanical mixing. Normally, around 50 kg of seeds per mu can be obtained.

3. F_1 Hybrid Seed Production Technology

The heterosis of common head cabbage is obvious, but like Chinese cabbage, it has small floral organs and few seeds on a single flower, and the usage of seeds is large during production. Therefore, artificial emasculation and cross-pollination are not feasible when producing the F_1 hybrids. This is why many countries use the self-incompatible line of common head cabbage to produce the F_1 hybrids. Self-incompatibility is a characteristic of hermaphroditic plants with normal male and female sexual organs, which can set seeds normally after they are pollinated between plants of different genotypes. However, they cannot set seeds or have a very low seed setting rate when self-pollinating during anthesis. The inbred line that has the characteristics of self-incompatibility and can be inherited stably is called the self-incompatible line. When using the self-incompatible line to produce F_1 hybrids, the parental plants can both be the self-incompatible line, or the mother can be a self-incompatible line and the father an inbred line. The seed production process consists of two main parts: parental maintenance and F_1 hybrid production.

(1) Maintenance and reproduction of parental plants.

The technology of the reproduction of parental plants' self-incompatible lines (or inbred lines) is basically the same as that of the original seed production. The main difference is that the reproduction of self-incompatible lines requires pollination during the bud stage to obtain seeds.

① Management of seed plants. As the parents of the F_1 hybrids are selected through multiple generations of self-crossing, their stress resistance is generally poor, so it is best to use the semi-adult plant method to retain seeds, which helps to reduce seedling diseases and facilitate storage and overwintering. In order to ensure the purity of parental seed plants, the seeds should be selected strictly according to the botanical characteristics of incompatible lines during the seedling, rosette, and heart-wrapping stages. Choose superior seed plants with pure typical traits and without diseases to reproduce parents and eliminate impure and diseased plants. Plants that have extraordinarily strong growth vigor, dark leaves, and extra-large heads are generally hybrids and should also be eliminated. The method of seed plants overwintering is the same as that used in the production of the stable cultivar.

Most of the original seeds of self-incompatible lines of common head cabbage are currently reproduced under protected conditions such as the glass solar greenhouse or cold frames, which not only makes it easy to isolate the seed production fields from other cabbage cultivars, but also makes the seed plants flower earlier, and the problem of pollination labor is easy to solve. In order to facilitate pollination, a path for pollination should be left out during the final planting; the plant spacing should be about 30 cm and the line spacing should be 30–40 cm. The first flowering date of self-incompatible lines of the round spherical type is relatively late. If the temperature management is inappropriate during reproduction in the solar greenhouse, the plants are likely to be incapable of flowering and bearing seeds. In recent years, the seed plants have reproduced under the yarn cover in the cold frame. The seed plants are planted in the cold frame in late October of the first year or in the mid-or late February of the following year. Cover the seed plants with a yarn cover before they flower and do not let the flower branches make contact with the yarn cover after they flower so as not to cause impurity of seeds because of insect pollination. In order to ensure the purity of the original seeds, the seed plants are selected once during the bolting and flowering stage according to the flowering characteristics of the plant.

② Bud pollination.

A. Flower bud selection. Flower buds being too big or too small will both result in poor fruiting. According to the time of flowering, the best time for pollination is two to four days before flowering, but there are some differences between different strains. According to the position of the buds on the flower branch, the 5th to 20th buds above the open flowers are the best for pollination and have the best fruiting result. For plants with weak growth vigor, the 5th to 15th buds have the best fruiting result.

B. Pollen selection. Pollen for pollination should be fresh pollen from flowers that open on the same day or the day before. In order to avoid excessive degradation of viability due to too many generations of self-crossing, mixed pollen from plants within the line can be taken for pollination.

C. Artificial pollination. First, peel off the top of the bud with forceps or bud stripper to expose the stigma, then take the pollen of the same line to pollinate the stigma of the bud. Pollination requires great attention to avoid injury to the flower stalk and stigma. The whole process needs to be carried out by dedicated staff to prevent mixing. When pollinating from one incompatible line to another incompatible line, hands and forceps must be sterilized with alcohol. The solar greenhouse or yarn cover where the seeds are kept must prevent bees or other insects from flying into it. When using the bagging isolation method to reproduce, make sure to bag the flower branches with sulfuric acid paper bags before the flower buds open. After the flowers at the lower part of the flower branches open, take the mixed pollen in the paper bags of the same strain for bud pollination, then bag the flower branches immediately after the pollination and mark them.

Using the method of bud pollination to reproduce the original seeds of common head cabbage of self-incompatible line requires more labor and is costly. In order to overcome this shortcoming, in recent years, some institutions have tested out that spraying 5% saltwater during anthesis can help overcome the self-incompatibility and improve the fruiting rate of self-crossing. The seeds reproduced from the parental plants are best to be used for generations to prevent mixing and degradation.

The harvest of the seeds of parental plants is the same as that of stable cultivars.

(2) Production of F_1 hybrids.

As for the seed plants for producing the F_1 hybrids of common head cabbage, some procedures such as sowing, seed selection, storage, and field management are basically the same as those used for the reproduction of the original seeds. The following issues must be noted.

① Selection of seed breeding methods. At present, the main seed production methods are as follows: open large field seed production, cold frame seed production, modified film cold frame seed production, open field & cold frame (cold frame path) alternating seed production, etc. For combinations in which the anthesis of the father and mother plants is the same, all of the above methods can be used. If the seed collection area is large, the open field seed production method is preferred. For those companies that do not have excellent seed collection technology, when the father and mother plants have a large difference in anthesis or they are in areas where it is not possible to overwinter in the open field, it is better to use the cold frame or modified cold frame for seed production, so as to adjust the anthesis. When the seed production area is large and there is a lack of equipment such as straw curtains and films, open field & cold frame (cold frame path) alternating seed production method can be used.

② Management of seed production fields. The self-incompatible line of common head cabbage is more likely to receive pollen from other cabbage cultivars after flowering, so its seed production field should be separated from the seed collection field of cauliflower, turnip cabbage, cabbage mustard, and other common head cabbage cultivars by more than 1,500 m.

If both parental plants are of self-incompatible lines, the father and mother plants can be planted at a ratio of 1 ∶ 1 in alternate rows. For combinations where the father and mother have large differences in growth vigor, plant the parental plants in alternate two rows to facilitate bee pollination. The row spacing is usually 50 cm and the plant spacing is usually 30–40 cm. If the father is of the inbred line, the row ratio of the father to mother plants should be 1 ∶ 3.

③ Adjusting the parental plant anthesis. If the anthesis of the parental plants does not overlap during the production of the F_1 hybrids, it could not only affect the seed yield of the F_1 hybrids but also seriously affect the quality of hybrids due to the low crossing rate. In order to solve this problem, the following measures can be taken.

A. Use semi-adult plants to produce seeds or open the head in advance. For spherical-type parents whose anthesis comes late, the anthesis of their semi-adult plants comes three to five days earlier than adult plants. For pointed-type and oblate-type parents whose anthesis comes early, the first flowering date and full-blossom stage of their semi-adult plants come later than adult plants. So sowing the parents appropriately later and using the semi-adult plants to produce seeds is conducive to the safe storage of seed plants over the winter and can also help the anthesis of the father and mother plants meet. If the spherical-type parents have formed heads before winter, a feasible practice is to cut open the leafy heads in advance before winter so that the leaves can see the sun and grow green after the anthesis to help them overwinter. This is also conducive to early flowering in the next spring.

B. Final planting before winter. In North China, planting the seed plants in the cold frame or modified cold frame from late October to early November could not only make the seed plants grow vigorously and increase seed yield, but also make the types that flower late start flowering much earlier than those planted in the following spring. However, due to the strong growth vigor, the anthesis of each plant is longer, which delays the later stage of anthesis, and this is conducive to the meeting of late anthesis of parental plants.

C. Use different microclimates of the windbreak and cold frame to regulate the anthesis. The different locations of the windbreak and cold frame have different temperatures and light, so the parental plants that bolt late can be planted on the north side of the cold frame near the windbreak before winter so that it grows and develops under higher temperatures and better illumination condition, encouraging it to flower earlier. Similarly, the seed plants that bolt and flower early can be planted on the south side of the cold frame so that their growth and development are inhibited and the anthesis is delayed.

D. Adjust the anthesis by pruning branches. If the case that the anthesis does not meet has already occurred during the seed production, the anthesis can be adjusted by pruning branches. The degree of pruning depends on how much the anthesis meets. If the difference in anthesis is not big, cut the main stems of the parental plants that flower early. If there is a difference of 7–10 days, the main stems and the apical inflorescence of the primary branches of the parental plants that flower early should be removed, which can delay the anthesis by about seven days. At the same time, nitrogen fertilizer should be applied heavily to promote the development of the secondary and tertiary branches. It can also be achieved by removing the buds of the main stems and the apical buds of the primary branches on the parental plants that flower late to promote the remaining buds to flower earlier. When the later stage of anthesis is not consistent, remove the end of the flower branches on the parental plants that have long anthesis after all flowers on the parental plants that have short anthesis open in order to promote the anthesis to meet and improve the fullness of the seeds.

④ Water and fertilizer management of the seed production field. The water and fertilizer management of the seed production field of common head cabbage's F_1 hybrids is basically the same as that of the regular cultivars, but several aspects should be noted as follows.

A. Roguing. The seed plants should be strictly selected before and after the final planting and during the bolting and flowering stage to eliminate hybrids that do not match the traits of the strain and diseased and inferior plants so as to ensure the quality of the F_1 hybrids.

B. Ensuring bee source. In order to hybridize the two parents, insects such as bees are needed for pollination. The practice has proved that it is best to provide one box of bees for every mu of the F_1 hybrid seed production field of common head cabbage for pollination. This can greatly improve the seed yield and the crossing rate of the F_1 hybrids.

C. Building frameworks to prevent lodging. Lodging not only affects insect pollination but also causes seed pods to rot, thus affecting seed yield and quality in the later period. Therefore, frameworks made of bamboo or plank should be built before the first flowering date to prevent the seed plants from lodging.

⑤ Harvesting of F_1 hybrids. If both the father and mother are of the self-incompatible line and the traits of the progeny from direct crossing and reciprocal crossing are the same, the F_1 hybrids on both the father and mother can be harvested together. If the father is of the inbred line, only seeds on the mother plants can be harvested. To improve the evenness of the F_1 hybrids, seeds from the father and mother can be harvested separately. Especially for hybrid combinations where the anthesis of the parental plants does not coincide, seeds on the parental plants that flower earlier can be harvested first before harvesting seeds on the parental plant that flower later.

IV. Seed Production of Other Vegetables

(I) Seed Production Technology of French Bean

French bean, also known as green bean, kidney bean, common bean, etc., is an annual vegetable in the *Phaseolus* genus of the Leguminosae family. It is native to Central and South America and was introduced to Europe and Asia during the 16th to 17th centuries. It is now planted in both North and South China. Its cultivation area is second only to soybean's cultivation area among all legume crops. French beans have a good flavor and high nutritional value, and their young pods contain 6% protein. Similar to animal protein, French bean protein is rich in lysine and arginine and contains vitamin C, carotene, fiber, sugar, etc. French bean is tolerant of storage and transportation and is suitable for dehydrating and quick freezing. It is a very common crop in China. Due to the long history of cultivation, it is widely consumed by people of all ethnic groups in China and is in great demand. With the improvement of living standards, the market demand for a balanced supply of French beans on an annual basis has become stronger and stronger. Therefore, in recent years, the use of solar greenhouses and plastic tunnels to cultivate French beans has developed rapidly. French bean is now the third-class protected production species after cucumber and tomato.

1. Flowering and Pollination Habits

(1) Branching and flowering habits.

The flowering and podding habits of French beans are closely related to their cultivars. Dwarf French beans are self-capping, with short main branches, few inflorescences per plant, and short anthesis. Generally, flower bud differentiation begins when the compound leaves are unfolded, and then each node differentiates flower buds. After the main branch has unfolded four to five compound leaves, the apical inflorescence is formed and the branch's growth stops. The lower nodes can produce lateral branches. After the lateral branches grow a few nodes, their growing points also form flower buds, and then the terminal buds stop growing. The apical inflorescence has five to six or even more flowers, and the following inflorescences have a reduced number of flowers. The flowering order of dwarf French beans is irregular. Some plants' top flowers open first, some plants' lower flowers open first, and some plants' flowers on the main and lateral branches open at the same time. Even the same inflorescence has an irregular flowering order. The inflorescence of the trailing French bean is axillary and its inflorescence forms successively as the stem grows, so the total number of inflorescences is larger and the anthesis is longer. The main vine usually grows slowly at the beginning of the growth of the trailing French bean and starts to bolt from the 3rd to 4th leaf node. Usually, the main vine has strong growth vigor and the axillary buds at the base do not easily sprout into lateral branches. The flower bud differentiation starts when the trailing French bean unfolds two compound leaves. Due to the strong vegetative growth, the flower buds of the basal nodes often cannot fully develop flower and bear pods. Only after it grows to the 4th to 5th node will it bear pods more easily. Lateral branches bear flowers at lower nodal regions than the main vine and usually have inflorescences at the 1st to 2nd nodes. The flowering order of the main vine and lateral branches of trailing French beans is more regular, usually opening one after another from the bottom up.

(2) Floral organ structure.

Each French bean plant has a raceme, with each inflorescence having a few to 10 or more flowers. The single butterfly-shaped flowers come in white, yellow, purple, and red. The flower consists of five petals, with the largest one being the flag petal, the two on either side being its wing petals, and the lower two combined being the keel petals. The spirally curled keel petal encases the pistil and stamens within. There are 10 stamens, nine of which are combined in a tube-like orientation, and the remaining solitary stamen forms the diadelphous stamen. The single pistil has a brush-like fluffy stigma, which can stick to the pollen (Fig. 4-15).

Fig. 4-15 Flowers of French Bean

(3) Flowering, Pollination, and Podding.

The French bean is a strict self-pollinated crop with a natural outcrossing rate of 0.2%–10%.

Their flowering time begins from 2:00–3:00 and ends at around 10:00, with their peak blooming period at 5:00–7:00. The stigma is fertilized three days before flowering, but the fertilization capacity is the strongest on the day before the flowering phase and can be maintained until the second day of flowering. The anthers mostly dehisce and disperse pollen a day before flowering, so self-pollination is performed before it is in bloom, but it is still possible for the stigma to receive pollen after flowering. A large number of flowers bloom on the plants. Each trailing French bean produces 80–200 flowers, and the dwarf bean produces 30–80. Although the budding number is high, usually only 30%–40% of pods are produced. From the environmental perspective, the suitable temperature for flowering and podding of French beans is 20–25°C, and when the temperature is higher than 30°C or lower than 15°C, the germination rate of pollen decreases, and the pollination effect is poor. In addition, the water resistance of its pollen is very weak. As a result, important factors such as temperature fluctuation and high rainfall during the anthesis affect seed yield and pod rate.

Each French bean pod produces 5–12 seeds (Fig. 4-16). They can germinate 20–25 days after flowering, and the seeds are fully mature in 35 days. With after-ripening, the seed germination rate of harvested fruit pods can be improved. The seeds vary in color, pattern, shape, and size due to different cultivars, with the thousand-seed weight between 200 g and 400 g and a shelflife period of two to three years.

Fig. 4-16 French Bean Pod

2. Seed Production Technology for Stable Cultivar

The cultivation management techniques of breeding French beans and producing commercial beans are largely the same, but breeding beans is for the purpose of producing seeds. To maintain the excellent traits of cultivars, there are specific requirements for the isolation of seed collection fields and the selection of seed plants.

(1) Cultivation management of seed collection fields.

① Selection of seed collection fields. The French bean is a self-pollinated crop that has a certain hybridization nature. The rate of cross-pollination is related to temperature change and bean cultivars. In general, the natural crossing rate is below 4%. This is why the different cultivars should be kept at a certain distance during breeding. In the original seed field, the isolation distance should be more than 100 m; in the field producing superior variety, more than 50 m; and in the seed production field, 10–20 m. High straw crops such as corn, planted between different bean cultivars, can play the role of isolation.

The plots with a deep soil layer, high terrain, and good drainage and ventilation should be selected as the bean seed retention fields, and the soil should preferably be sandy loam or clay. The French beans should not be planted continuously. An interval of two to three years is required.

② Soil preparation and sowing. In the seed collection field, deep tillage should be done in autumn while harrowing should be done in spring, with soil well prepared and basal fertilizer sufficiently applied. The fertilizer must be fully decomposed so as to prevent maggot infestation. At least 2,000–3,000 kg of fertilizer per mu and another 20 kg of calcium superphosphate should be applied. Sowing should be carried out as early as possible on the premise that there is no frost damage after the seedling emergence so that the wet season can be avoided when the French beans are mature. Sowing can be done in late April in Liaoning Province. Seeds that are large, full-sized, uniform, and have a glossy shine are used for sowing. In order to kill bacteria, the seeds can be disinfected. Soak the seeds in 1% formalin solution or 0.1% potassium permanganate solution for 20 minutes, and then clean them with clean water and dry them. As for French beans in the northern regions of China, dry seeds are generally sown directly in large fields. In early spring, the ground temperature is low and it is slow for the seeds to break through the soil, which can cause seeds to rot. To avoid this, seedlings can also be transplanted.

French bean seeds are mostly sown by ditching or digging holes, and the holes are about 4–5 cm deep. The planting density of the seed collection field should be slightly smaller than that of the crop field. Generally, dwarf French beans are planted 50 cm apart per row and 30 cm per hole; while the trailing beans are spaced 60–70 cm apart in rows and 25–30 cm apart in holes, and a wide-narrow row planting pattern can be adopted for management convenience. Four to five seeds are sown per hole, leaving three seedlings per hole after the seedlings have emerged. Before the French beans grow unearthed, do not let the soil harden. Therefore, if the soil is found to be deficient in moisture, it should be watered before planting to keep the soil moist. No watering shall be carried out after sowing until they grow unearthed.

③ Field management. After the watering is done when the seedlings of the directly sown French beans are basically grown out, or when the seedlings of cultivated French beans have readjusted to the environment after the final planting. If the weather is not very dry before the pods are born, watering is not generally carried out, and intertilling and hardening of seedlings should be carried out. When the French bean bears pods and the plant enters its vigorous growth period, fertilizer and water management should be strengthened. Soil surface shall be kept moist, and fertilizer with the water shall be applied once or twice. Watering should be controlled to promote early seed maturity during the late podding stage. For trailing French beans, the support shall be erected in time before the vine spreads, and intertillage shall be carried out once before that.

(2) Roguing.

In order to maintain the excellent traits of bean cultivars, roguing should be done constantly to breed original seeds in the seed breeding field. In addition to paying attention to seed selection before sowing and eliminating seeds with irregular shapes, bad colors, and mixed seeds of other cultivars, three times of selections shall be carried out in the breeding process. The first selection should be carried out in the seedling stage, and combined with the final planting. It is necessary for the seedling breeders to eliminate the hybrid plants that do not conform to the traits of this cultivar or the plants that are not correct in color and are sick and weak according to the embryonal axis color. The second selection should be made during anthesis. In addition to removing diseased, malformed, and poorly growing plants, remove the trailing plants and semi-trailing plants in the dwarf cultivars. In particular, plants of other cultivars should be eliminated according to the color of flowers, leaving plants whose growth habits, stems, leaves, and flower characteristics are in line with the characteristics of this cultivar for seed collection. The third selection shall be carried out when the tender pod reaches the commercial maturity. In this selection, plants in line with the characteristics of this cultivar should be left according to the shape, color, and other traits of fruit pods, and plants that are not of this cultivar and other undesirable plants should be eliminated. Mixed, mutant, and diseased plants can also be eliminated according to the traits of mature pods and beans before collecting seeds. One or two selections can be made in superior variety fields, and one selection can be made during anthesis in seed production fields.

(3) Harvesting and threshing of pods.

After flowering, the French bean seeds take approximately 30 days to mature. Harvesting generally begins when the seedlings turn from green to yellow and gradually dry out. Thus, it is best to harvest in batches every time the seeds turn mature. Otherwise, the seeds tend to germinate and rot in the seed pods if harvested too late. Dwarf French beans can be harvested at a time when about half of the seed pods of the plant have dried out, which reduces seed loss and ensures a high germination rate of seeds produced by later flowers through after-ripening. The best time to do this is in the morning when the dew has not yet dried. Harvesting dwarf beans can be done by uprooting the bean seedlings, bundling several plants into a small bundle, and laying them upside down on the ground for two to three days so that they can dry and go through after-ripening. The pods on the lower part of the trailing beans that mature first should be picked first, and when the pods on less than two-thirds of the plant turn yellow, they should be pulled up and dried for a few days before being threshed. Reserving the middle to lower parts of the trailing beans for the seed pods is preferred, as the pods in these parts are in a suitable environment for growth and have full grains and high seed quality. Young pods borne at a later stage should be removed as early as possible to concentrate nutrients in the seed pods.

Generally, for trailing French beans, 100–200 kg of seeds per mu can be collected, and for dwarf beans, 50–100 kg per mu can be collected. The threshed seeds can be stored in a dry, pest-free, and rodent-free area once their moisture content drops below 12%.

3. Purity Maintenance and Degradation Prevention of French Beans

Although during the sowing period, the focus has been made on seed selection and roguing in the seed collection field, degradation could still occur after prolonged reproduction. Therefore, a comparative selection between strains should be carried out after four to five years of reproduction to maintain variety purity. The methods for the selection are listed below.

① Individual selection method. Select several superior individual plants from the original field and thresh them separately. Sow the seeds of each strain separately in the following year. Compare the agronomic traits of each strain. After eliminating the unfavorable strains, mix the retained strains and breed original seeds with them.

② Mixed selection method. In a large original seed field, select several superior individual plants, mix them, and thresh them for seed retention. After identification in the following year, those that meet the criteria can be extended for breeding as the original seeds.

③ Plant hole selection method. For the French beans in the original seed field, final planting and thinning should be done in each hole so that the numbers of the plants in each hole are the same, and then use the hole as a unit, compare the plants in each hole and eliminate some bad ones, and use the plants in the left holes to reproduce the original seeds.

④ Pod selection method. Select some superior seed pods randomly from the original seed field and the number of pods selected should be slightly more, with roughly an equivalent number of holes in the original seed field. One pod should be sowed per hole in the following year. After the seedlings emerge, follow the above method to conduct final planting and thinning. Compare and eliminate seedings, using the hole as the unit, and then use the seeds from the remaining plants to reproduce the original seeds.

The first two methods can be used on dwarf French beans. The latter two methods can be used on plants of trailing French beans when the plants are intertwined and difficult to divide. No matter which method is used, it is ideal to do this at least once every four to five years.

(II) Onion Seed Production Technology

Onion is a biennial vegetable of the *Allium* genus in the Liliaceae family. Onions, native to Central Asia, have long been widely cultivated in Europe and the Americas, and have been cultivated in China for more than 100 years. Onions have a high nutritional value; in addition to vitamins, minerals, proteins, and sugars, they also contain oily and volatile organic sulfide-containing compounds, which are pungent. Onions can enhance appetite and have an effect on disease prevention and sterilization. With strong adaptability, onions are cultivated using simple techniques. In the growing period, onions are less likely to be affected by diseases and pests. Hardy and resilient to transport, onions also have a long storage period. In addition to being sold as fresh food, onions are also a major commodity vegetable exported from China.

1. Flowering and Pollination Habits

(1) Growth characteristics at each stage.

Onion is a biennial vegetable, and it takes two to three years to go through the initial sowing, the onion head (bulb) harvesting, then bolting and flowering, and the seed collection. The whole growth cycle of onion, from seed sowing to bulb harvesting, is the vegetative growth period; after the bulb is formed, it enters the natural dormancy period; from bulb bud differentiation to seed maturity, it is known as the reproductive growth period. When the onion bulb diameter is more than 0.8 cm, under the low temperature of 2–5°C, after 60–70 days through the vernalization stage, the vegetative growth is transformed into reproductive growth, the growth cone begins flower bud differentiation, and then it goes through bolting, flowering, and fruiting. In the northern parts of China, anthesis usually begins in late May to early June, with seeds maturing by the end of July.

(2) Inflorescence and floral organ structure.

Each onion plant can produce around three to six flower stalks, sometimes reaching a maximum number of as many as 20 stems. The flower stalk can reach a height of 120–150 cm, is firm and hollow in structure, and has a fusiform expansion at the base. The globose umbels (that is, buds) are borne on the top of the flower stalks and are covered with leathery bracts. When the flowers grow up, the bracts are dehiscent, and the buds open in random order from the outside to the inside, with an average of 600–800 flowers per inflorescence (Fig. 4-17). Single inflorescence flowering continues for 12–18 days, and the anthesis of the plant with more inflorescences continues for about 30 days. Its full-blossom stage starts three to seven days after the initial bloom date and lasts for about 10 days. Onion florets have long pedicels, and each floret has six white perianths with green vascular bundles in the middle. Each stamen is inserted to the position opposite the perianth. There are six stamens. A needle-shaped smooth stigma grows in the middle.

Fig. 4-17 Onion Flowers

(3) Pollination and fruiting habits.

Onion is a cross-pollinated crop whose stamens mature first, and only after the pollen is shredded could the stigma have the ability to receive pollen. When it just blooms, its style is about 0.1 cm long and will grow to its maximum length (0.5 cm) when the pollen is completely dispersed. The effective pollination period lasts up to seven days after the anthesis, but it is most effective during the first two days of anthesis. The pollen has a short life span and is only effective on the same day. Pollen quickly loses its germinability at 100% relative humidity. The stamen pollen is dispersed between 9:00 and 10:00, but for cloudy or colder climates, the dispersal time is delayed until the afternoon. Rain, fog, as well as hot and dry winds during flowering are not conducive to pollination. Onion is a cross-pollinated crop with a low cross-pollination rate and has a considerable proportion of self-pollination rate. It can be pollinated by its entomophilous flowers, honeybees, scoliid, wild bees of various families, and many kinds of flies. The ovary is in the upper part. The fruit is a three-lobed capsule with three ventricles. Each ventricle has two ovules, and each floret produces six seeds. In the late growing stage, the plant withers, the capsule dries up, and the seeds mature. From flowering to seed maturity, it takes about 70 days. The standard thousand-seed weight of onion is 3.8–4.2 g, and seeds with a thousand-seed weight of 2.8 g or less are chaff seeds.

2. Seed Production Technology for Stable Cultivar

(1) Seed collection methods.

① Summer sowing and three-year planting of bulbs seed collection method. Sow the seeds in mid-May of the first year in level borders at a rate of 500 g/mu. When the plant grows four to six leaves, it becomes dormant in high temperatures and under the long summer sun. The plant doesn't appear to grow new leaves. The old leaves turn yellow in early and mid-July and small bulbs with a diameter of 1.5–3.0 cm are formed in late July. Plants that have formed small bulbs are harvested at the end of July and then dried until the leaves wither so that the nutrients are retracted into the bulbs. The small bulbs are then screened for storage. To prevent vernalization during storage, attention must be paid to temperature regulation. In late March of the second year, the small bulbs are planted after a screening process and the onion cultivation commences. Small bulbs can be properly densely planted, usually planted after earthing up. They are planted at a spacing of 12 cm between rows and 10 cm between plants. In mid-to-late July, onions are harvested and then stored in warehouses by August. In the third year of spring, onions are further screened and final planting should be done for those that meet the requirements in mid-to-late March. The seed plants bolt in May and the seeds can be harvested from late July to early August. If this method is used, multiple field selection and storage period selection can be carried out, and better selection can be carried out according to variety purity, storability, bolting performance, disease resistance, cold resistance, etc. Moreover, the quantity of the seed for collection is high and the seed quality is good. This method can be used for the purification and rejuvenation of superior varieties. However, the disadvantage is that the field occupation time is long. It takes 26–28 months from planting to harvesting the seeds, and the cost of seed production is high.

② Autumn seedling and three-year planting seed collection method. In the autumn of the first year, they are sown and grow into autumn seedlings of appropriate size, which are overwintered in the open field or stored and overwintered after germination. In the second year, they are planted in spring, and after they grow into full and plump onions, they are stored after roguing. In the south of North China, the final planting of onions can be done from September to November of the same year, while in the northeastern regions of China, onions can be kept in storage until the mid-to-late March of the third year, and after selection, the plants are planted. They bolt in May and are harvested from late July to early August. In this method, several selections will be made, so that it can be used to produce original seeds. It takes 21–23 months from sowing to seed harvest.

③ Spring seedling and two-year planting seed collection method. In the early spring, sowing and seedling-raising are carried out in the greenhouse. After the external soil thaws, they are planted in the open field. In the autumn of that year, onions used as seeds are harvested and stored for overwintering. In the spring of the second year, they will be planted for seed collection. It takes 16–19 months from sowing to collecting the seeds.

④ Two-year semi-adult plant seed collection method. Sowing is done before early August of the first year. Make the ridge 66–100 cm wide. One row of seeds shall be sowed on the ridge platform, with seedlings about 3.3 cm apart, ensuring they are not too packed. Large seedlings with six to eight leaves are harvested before winter. The seedling age is seven to eight leaves or more, and the stem diameter is 1–2 cm, so it is called a semi-adult plant, and vernalization is completed in the process of winter storage. In the south of North China, plants can overwinter in the open field during winter, while in Northeast China, it is better to dig out large seedlings and then store them for overwintering after selection. In mid-to-late March of the second year, final planting is carried out. They bolt in May and the seeds are harvested in late July to early August. Because the plants are directly put to bolt, bloom, and seed before they have grown into onions, they are also known as the kid seeds. This method of seed collection without the selective process is suitable for rapid reproduction and can reduce costs. Though degradation is easy to occur to the seeds over successive generations via this method, they can be used to reproduce seeds for production based on high-quality original seeds. It takes 11–13 months from sowing to seed harvest.

⑤ Continuous seed collection method. It is also known as the meristematic bulb collection method. Several small bulbs formed at the base of the seed picking plant after the bolting and seed setting period are harvested, dried, stored, and used as onions for seed collection in the next year, or they overwinter in the open field and seed collection continues in the next year. There is only one year between the last seed collection and the next, and the field occupation time is four to nine months. The two batches of seeds obtained by this method are not two generations of parents and children, but only two batches of seeds harvested by stages of the same sexual generation. Therefore, the purity is not affected. Using this method, planting is carried out in the first year and seed collection is carried out in the second year, which reduces the cost, shortens the seed collection cycle, and ensures the quality of seeds. However, it is necessary to adopt high ridge cultivation and strengthen field management to prevent the lateral bulb from rotting.

(2) Key techniques of autumn seedling and three-year seed collection method.

① Cultivation of onion for seed collection.

A. Sowing and breeding seedlings. Most areas of China adopt the autumn seedling method to collect seeds. Final planting should be done to seedlings before winter and then they overwinter in the open field or are stored until the spring of the following year when the final planting can be done. With the onion being a green body vernalization type crop, when the seedlings reach a certain size and are under low temperature, they will quickly pass vernalization, resulting in a pre-bolting phase. For this reason, proper sowing is an important measure to prevent premature bolting. As for the autumn sowing period, such as in southern Liaoning, it is advisable to sow in mid-to-late August, and the more in the South, the later the sowing; the more in the North, the earlier the sowing. Sowing too late causes the seedlings to be too small, which are then susceptible to freeze when overwintering. Due to the slow unearthing and small root quantity of these onion seedlings, it is ideal to select loosely packed, fertile soil with strong water conservation as seedling beds. Prepare the ground before sowing by plowing, raking, and applying sufficient fertilizer to make a level border. Scatter the seeds and cover them with soil after sowing. Generally, the amount of seed used per mu is 3.5–5 kg, which can be transplanted to the seed production field of about 10 mu. Water once every two to three days after sowing to keep the soil moist. After the seedlings emerge, watering should be properly controlled. Water every five to six days. The hardening of seedlings starts when the seedling is about 10 cm tall. If the basal fertilizer is sufficient, no further fertilizer needs to be applied during the seedling stage. However, if the seedlings are looking weak, apply diluted manure water or 10 kg of urea per mu. Wee two to three times in the seedling stage and prevent onion flies with dimethoate and trichlorfon or other insecticides. When overwintering, if the seedlings reach three to four leaves and are approximately 18–24 cm in height, with the pseudostem thickness being less than 0.7–0.9 cm, the premature bolting rate in the next spring can be greatly reduced.

B. Overwintering of autumn seedlings. In southern China, the final planting can be carried out before winter. Generally in mid-to-late November, seedlings are planted. After the seedlings re-adapt to the environment, let them overwinter in the open field. In southern Liaoning Province, seedlings can be dug up from the seedbed between early and mid-November, densely planted in shallow trenches that are 20 cm deep temporarily, and taken out for final planting in mid-to-late March of the next year.

C. Final planting. Before the final planting, plow the land, apply 5,000 kg of decomposed organic mixed fertilizer per mu, and add 40 kg of calcium superphosphate and an appropriate amount of potassium fertilizer to make level borders. Strictly select and grade seedlings before they are planted, and eliminate diseased seedlings, dwarf seedlings, seedlings with excessive growth, weak seedlings, tillering seedlings, and base curved seedlings. These chosen seedlings are graded according to size, and then planted in separate beds and managed accordingly. The distance between plants should be (13–16) cm×(10–13) cm, with about 30,000 plants per mu of land. Onions are suitable for shallow planting, with an optimum depth of 2–3 cm. It is best when the stems can be buried after final planting, after the plants are covered with soil, and when the plants do not fall after watering.

D. Field management. If onions are planted in autumn, the temperature is low during the period from final planting to overwintering, so irrigation should be properly controlled, and intertillage and soil moisture retention should be mainly carried out. Fill the soil with sufficient frozen water when it starts to freeze. Sometimes, it can be covered with horse manure and fertilizer to protect the roots from the cold. After the weather warms up, prompt watering should be applied to reproduce the return of growth and diluted manure water is suggested. After the final planting of spring-planted onions, water should be provided to facilitate readjusting, followed by intertillage and the hardening of seedlings. Whether planted in autumn or spring, due to the lower temperature in early spring, it is necessary to water less often and the amount of water each time should be small. From the readjusting stage of seedling to bulb enlargement, light tilting should be done two to three times to improve ground temperature and promote root development. When entering the peak leaf growing period, it is critical to strengthen its water supply and apply fertilizer. Apply 1,000–2,000 kg of farmyard manure, 10–15 kg of ammonium sulfate, or 5–7.5 kg of potassium sulfate per mu. When the overground functional leaves are basically formed, the growth slows down, and the bulbs are obviously expanded. When it is about 3 cm in diameter, timely fertilizer is applied with 15–20 kg of diammonium phosphate and 10 kg of potassium sulfate per mu. Watering should be frequently done throughout the bulb expansion period and can be stopped seven to eight days before harvest.

② Harvest, selection, and storage of onions.

A. Harvest and selection of onions for seed collection. Onions are harvested in the northern region in mid-July. In general, when the three leaves on the onion stem begin to turn yellow, the pseudostem gradually loses water and becomes soft, the bulbs stop expanding, and the outer scales are leathery; it is the perfect time for harvest. Harvesting too late or exposure to the rain can easily lead to pseudostem rot and rupture of the outer skin, making them intolerant to storage. Selecting a sunny day and proper onions is the key to onion seed collection. The plants should be selected in the field. During the vigorous growth period, hang labels on the plants that are compact and symmetrically shaped, with normal leaf color, thin and short leaf sheaths, and rapidly expanding fruits. Make a second selection of bulbs from labeled plants at harvest time. The plants with moderate size, no separation of bulbs, correct shape, and pure color when the pseudostem is lodging, which is in line with the typical characteristics of this cultivar; the plants whose outer scales are not cracked during harvest; the plants that are not damaged or harmed by diseases and pests, as well as the plants without bulb germination and lateral buds in the pseudostem will be selected for harvesting and storage.

B. Treatment and storage of seed plants after harvesting. After harvesting, the onion plants enter a natural dormancy period. The length of the dormancy period varies with the cultivars, the degree of dormancy, and the temperature of the outside world, generally 60–70 days. The harvested plants can be dried for four to five days (avoiding direct sunlight). Braid the pseudostems when they become soft, and then dry them for a few days before stacking and storage. Alternatively, hang the braided onions in rows in a ventilated, cold, dry, empty room or pergola to avoid rain and sun exposure. In the Yellow River basin and most areas of the North China Plain, the seed bulbs are stored over summer, and after the natural dormancy, the final planting is carried out between September and October. During storage, onions can be selected several times to eliminate the onions that germinate, develop diseases, and rot during storage. In Liaoning Province, the onion can be stored in the warehouse for overwintering, so that the onion enters the forced dormancy state, and the final planting is carried out after the land thaws in the following spring. To ensure a state of normal dormancy, onions must be stored in a dry, well-ventilated, cool place, and preferably on a shelf. Do not pile up too thick. Close doors and windows during the cold season to properly keep warm and insulate against the cold. The optimal storage temperature is 1–6°C. The vernalization of onions stored by freezing for a long time is slow. However, if the storage temperature is higher than 10°C, the nutrient consumption in the onion is too large. Before final planting, eliminate onions that have sprouted, rotted, or suffered from cold or heat in storage.

③ Management of onion seed collection fields.

A. Selection of seed collection fields. Onions prefer cooler climates, but the cool season suitable for growth in most northern regions has too short of a season to meet their growth needs. The high-temperature damage after mid-June is evident. It affects pollination and makes the plant weak and prone to premature senescence, which is reflected in the late anthesis or seed filling stage. Only the bolting flower stalk remains and the leaves all wither. Rainy conditions, dry and hot winds, or hail downfall during the anthesis can lead to crop failure. In rainy or foggy conditions during the seed filling stage, onion purple spots and downy mildew occur. Spraying agents on the flower bud is not preferable. The control effect is not good, and the seed production will be affected. Therefore, when onion seeds are collected on a large scale, particular attention should be paid to the precipitation of the planting area. Only the area with a rainfall of less than 150 mm during anthesis is suitable for onion seed collection. A clayey-fertilized soil with strong water conservation properties should be selected for the seed collection field. Proper irrigation and drainage conditions should be in place as drought will affect the bulb's fullness, harvest yield level, and germination potential of the seeds. Isolation should be carried out at 1,000 m surrounding the field. Plants of other cultivars should not be planted together with the onion in the seed collection field, so as to avoid flower hybridization and disease or pest contamination and to ensure the purity of the collected seeds.

B. Pregermination of onion plants. The storage temperature should be raised to 15–18 ° C at least 15 days before the spring planting of the onion plants. The onions should be pregerminated when they sprout with green leaves emerging towards the head but not grow into bulky buds.

C. Final planting of seed plants. The seed production field can be applied with sufficient fertilizer before winter, with 4,000–5,000 kg organic fertilizer, 25–35 kg of calcium superphosphate, and an appropriate amount of potassium fertilizer per mu. In early spring after the soil thaws, remove the snow in time, plow, apply basal fertilizer, carry out rotary tillage, and harrow. Keep wide rows 50–60 cm apart and narrow rows 30 cm apart. Dig trenches about 8 cm deep. Plant wide plants 20 cm apart and small plants 15 cm apart. Take the soil between the large rows to cover the narrow rows. The rows should be covered with plastic mulch or covered with small arch sheds for a short time without irrigation.

④ Bolting and fruiting management.

A. Intertillage and weeding. After the removal of mulch or small arch sheds in middle to late April, intertillage and soil cultivation are carried out to remove weeds and prevent too many weeds from affecting plant growth. Alternatively, it can be sprayed with 100 g 50 % prometryne diluted with 60–100 kg of water per mu for chemical weed control.

B. Irrigation and fertilizer management. Irrigation should be properly controlled prior to the seed bolting period, and intertillage and moisture retention should be carried out to prevent the bolting flower stalk from being excessively thin and weak. Irrigation and top dressing should be done once in early April, and an amount of 15 kg of diammonium phosphate should be applied per mu. In early to mid-May, the plants bolt one after another. When they stop bolting, a ternary compound fertilizer of 15–20 kg per mu should be applied with water to promote the differentiation of flower buds. During the flower bulb formation period, when the involucre breaks and the flower bud starts to bloom, apply 15 kg of diammonium phosphate and 10 kg of potassium sulfate per mu, water every six to seven days to keep the soil moist. After flowering, spray a 0.3% solution of potassium dihydrogen phosphate once every 7–10 days in combination with pest control. This can meet the needs for phosphorus and potassium fertilizer in the filling stage of seeds, make the grains full, and prevent premature senescence of seed plants.

C. Plant adjustment and assisted pollination. Since each ball of seed can produce 1–20 flower bolts, those that bolt late often flower in the rainy season, resulting in nutrient dispersion. So, some of the bolts need to be removed. Leave only three to four flowering bolts per plant. Artificially assisted pollination is performed during anthesis, usually by tapping on the inflorescences with a piece of styrofoam or by stroking all the flower bulbs with gloves on hands. It could also be done with a one-meter-long bamboo tied to a strip of rag to sweep the ball. Assisted pollination should be carried out between 8:00 and 9:00 once the dew has dried and should be performed once a day. Because the flower bolt on the onion is thin and the bulb is heavy, it is easy for the bulb to lodge as the plant grows. Make a support frame with bamboo poles in the field, or surround the plants with ropes and bamboo poles to reduce lodging losses.

⑤ Seed collection, threshing, and storage. The onion seeds must be harvested in batches in multiple stages because the bolting period and anthesis vary greatly between different plants in the onion seed collection field, and the seed maturity stage is also inconsistent. The best time to collect seeds is usually about 20 days after the full-blossom stage when a few capsules turn yellow and crack at the top of the bulb and the seeds have not scattered. During harvest, the bulb is to be cut 30 cm below the stem, but if the lower capsule has not yet matured, after-ripening is needed. After the flower bulbs are collected, they should be protected from rain and spread out under the sun to dry. After the capsules are fully dried, they are threshed by repeated rubbing and pounding. After threshing, they still need to be spread out and dried, but exposure to the sun is restricted. Seeds need to be cleaned after threshing so as to remove the sand, fruit stems, peels, and chaff seeds. Then, they are put onto the market and can be sold in less than half a month. Generally, 50–80 kg of seeds can be harvested per mu.

Onion seeds cannot be stored for a long time. Usually, seeds older than two years cannot be used. Dry seeds can be stored for one year in the constant temperature storehouse and for two years in the refrigerated storehouse for surplus and deficiency adjustment.

3. Seed Production Techniques of F_1 Hybrids

Onion is one of the earliest vegetable for which F_1 hybrid seeds are used in production. Its advantage is obvious, and the yield increase effect is 20%–50%. The male sterile lines must be used as the mothers in hybrid seed production because of onions' small floral organs and the small number of seeds per fruit. However, this has not yet been successfully implemented in China. According to the experience of foreign seed companies, the hybrid planting techniques are introduced as follows.

(1) Breeding of parents.

① Reproduction of the male sterile lines. Male sterile lines are reproduced using a three-year single-generation seed collection method. The sterility lines and maintainer lines that have undergone strict roguing are planted in the seed production field in early September or in the spring of the following year at a row ratio of (4–8) : (1–2) or 7 : 1 with the usual row spacing. The seed production field should be located in an isolated area with no other flowering cultivars within a distance of 1,000-2,000 m. Attention should be paid to the timely and thorough removal of fertile plants in sterile plants at the early bloom stage, and then natural pollination or artificial assisted pollination should be carried out. After artificial pollination, the seeds harvested from the sterile plants are the seeds of the sterile lines. These seeds are mainly used for seed production, with a small amount for the next year's sterile line reproduction. The seeds harvested from the maintainer plants are the seeds of the maintainer lines. When harvesting, the harvest of different lines should be separated to prevent mechanical mixing. After the seeds are dried, the moisture content of the seeds should not be above 12%, preferably below 8%, to facilitate storage and maintain a good germination rate.

② Breeding of the father inbred line (restorer line). A father inbred line needs to undergo multiple generations of artificial control of self-crossing, and F_1 individuals should be strictly selected and planted in a single plant system using the same method of artificial control of self-crossing. After three to four generations, several excellent inbred lines with good economic traits can be obtained, and these inbred lines need to undergo the combining ability test with sterile lines to get the father with the highest combining ability as the restorer line.

Plant the seed bulbs of the father inbred lines in isolated areas with no other flowering plants within 1,000–2,000 m, make it undergo natural pollination or artificial assisted pollination, and then collect seeds. It is necessary to use the three-year generation method or the two-year generation method to breed the seeds.

(2) Hybrid seed production.

Since the parents have been selected and purified for many generations, the production of F_1 hybrid seeds can adopt the semi-adult two-year seed collection method. If this method is adopted, the seed collection cycle will be short and the production cost will be low, but because the plants bolt and bloom directly without going through the bulb hypertrophy stage, there will be a complete lack of bulb selection. For this reason, this method is not as good as the three-year heading large plant seed collection method in terms of persevering good seed characteristics.

① Sowing in advance, cultivating large seedlings, and ensuring proper bolting. Proper early sowing, strengthening management, and making the diameter of seedling pseudostem above 1 cm before overwintering are good ways to ensure the bolting. The sowing period for the semi-adult plant seed collection method in the northern region is more suitable at the end of July, which basically allows more than 90% of the seedlings to reach a height that is greater than 1 cm with more than 70 days of low temperature. In early November, seedlings (remove seedlings lower than 1 cm) are planted temporarily in the open field for overwintering and subject to natural low-temperature treatment.

② Management of seed plants and assisted pollination. In early March of the second year, the male sterile line of onion cultivars (line A) and the father inbred line (line C) are planted at a row ratio of 4 : 1 or 6 : 1, with an isolation radius of more than 1,000 m. Since semi-adult plants do not have large bulbs, they are not nutritionally nourished and have fewer flower buds. Therefore, the number of seedlings per mu can be increased to 18,000–20,000. At the same time, compared to the seed collection of adult plants, intertillage, water, and fertilizer management should be strengthened to ensure the smooth bolting, flowering, and fruiting of plants. Before flowering, the plants that are impure, diseased, or inferior should be completely removed, along with fertile plants in sterile lines. (Normally, the filaments of sterile plants are short, and their anthers are shriveled and not dehiscent. The color of the anthers is grayish-brown, and the anthers are transparent and green in the early stage. Anthers have no pollen, and some lack stamens.) During the field inspection, inflorescence can be hand-touched for further identification. Natural pollination is generally performed, but artificially assisted pollination can also be carried out to ensure the crossing rate and seed purity. To maintain the same anthesis of male sterile lines and father inbred lines, the inflorescence with inconsistent anthesis should be removed, which can also ensure an increase in the crossing rate. In this way, the seed collected from the sterile line is an F_1 hybrid, and the seed collected from the father inbred line is still used as the father.

The semi-adult plant seed collection method to breed F_1 hybrids should not be used for consecutive years. When the conditions are available, hybrid seed production can be carried out by using the three-year or two-year large plant head formation method, which can improve the quality of F_1 hybrid production so as to improve the yield and quality of commercial onions.

(III) Production Techniques for Radish Seed

1. Seed Production for Conventional Cultivars

The radish seed collection method is also divided into the adult plant seed collection method, semi-adult plant seed collection method, and seed-to-seed method.

(1) Key points for adult plant seed collection method.

① Sowing and cultivation of seed plants. The sowing period of radish coincides with the hot and rainy season, so it is advisable to select loam or sandy loam soil on high terrain that has a good drainage and irrigation system, with a deep, loose, and fertile soil layer and a pH value of 5.8–8.6. Avoid continuous planting of the same crop in the same field. Radishes should rotate with Cruciferous crops in the same field every three to five years. Apply 45,000–75,000 kg of decomposed organic fertilizer, 450 kg of diammonium phosphate, and 225 kg of potassium sulfate per hectare. Generally, ridge culture is carried out, with a row spacing of approximately (40–50) cm×(25–30) cm.

The sowing period for radish seed collection can be the same as that of vegetable cultivation or three to five days later than that, and the principle is to ensure that the fleshy roots are fully mature before harvest. In Liaoning Province, the sowing period starts from early to mid-July; in Hebei Province, it starts from late July to early August. The first seedling thinning should be carried out when the cotyledons are fully unfolded and the true leaves are exposed, leaving one plant every 3 cm if they're planted in rows. The second seedling thinning should be carried out when there are two to three true leaves, and the seedlings should be set at a spacing of 13–16 cm. The final singling should be done when there are five to six true leaves. Selection shall be carried out at the time of seedling thinning and final singling. Meanwhile, roguing is necessary to ensure the purity of seed plants.

② Selection and storage of seed plants. Before harvesting the fleshy roots, remove the diseased and insect-damaged plants, as well as plants with bad leaf color, uneven leaf shape, and too many or too few leaves. Fleshy roots that meet the characteristic features of this cultivar, with few lateral roots, smooth surface, pure color, upright shape, and no harm from diseases and insects shall be selected. After harvesting, the petioles of 1–2 cm should be left as seed plants for seed collection.

For fleshy roots, the burying method is often used during winter. Dry the fleshy root in the field and bury it in a ditch prepared in advance around the start of winter. In the northern areas, they are placed in the cellar. During the initial stage, there's no need to bury them in sand piles. However, as the temperature in the cellar drops, the storage method can be changed, in which layers of sand and carrots are stacked alternately. Then, cover the layer with wet, fine sand. Generally, five to six layers are appropriate. The most suitable temperature for winter storage of radish is 1°C.

③ Planting and field management of plants. Land preparation, fertilization, and spacing between plants are the same as before. Buried radish is taken out before final planting and seed plants with blackened, bran hearts and plants that are rotten or bolt too early shall be removed. In south-central Hebei Province, final planting is done in early March, and in north-central Liaoning Province, it is done in early and mid-April. For plants with long fleshy roots, adopt a horizontal planting position or cut off a quarter to a third of the tail of the long roots before planting. Radishes that are red in color and round in shape are planted vertically, with 3–4 cm of mulch covering them. Each pit is covered with a layer of horse manure. If the soil is dry, water it when planting. When the seedlings have readjusted to the environment, and the new leaves and buds have emerged from the soil surface, water may be applied to the seedlings. Then, intertillage, soil loosening, and hilling are carried out. After the plant bolts and grows buds, ample water should be supplied. While watering, apply 150 kg of urea and 225 kg of potassium sulfate per hectare. After the full-blossom stage, reduce the watering frequency when it enters the podding period. Remove flowers and buds from the top of the inflorescence. To prevent the seed plants from lodging, supporting structures should be inserted before it blooms.

④ Seeds harvesting. Pods can be harvested when they turn yellow. It is mostly harvested in early and mid-June in south-central Hebei Province and in early and mid-July in north-central Liaoning Province. It is difficult to thresh the radish seeds, so it can be achieved by shedding the testae first, using a rice mill to thresh them next, and then drying them. Seeds are stored when the moisture content reaches below 8%. The amount of seeds collected for radish is about 750–1,200 kg per hectare.

(2) Technical key points of the semi-adult plant seed collection method. At present, the method is mostly used for the seed reproduction of conventional cultivars. Compared to the adult plant seed collection method, its characteristics are as follows.

① Its sowing period in autumn is 15–30 days later than that of the adult plant seed collection method.

② The planting density of the semi-adult plant seed collection method is relatively larger than that of the adult plant seed collection method, with a planting spacing of (20–25) cm×(40–50) cm, allowing an increase of 15%–30% in plant number.

③ The frequency of water and fertilization management needs to be reduced in autumn as well as in spring.

④ If the semi-adult plant seed collection method is adopted, the seed plant disease is less, the fleshy root is storable, and the seed collection amount is higher.

2. Seed Production Technology for Hybrid Cultivars

The parent-producing hybrids include inbred lines, self-incompatible lines, and male sterile lines. Male sterile lines are mainly used for seed production in China.

(1) Reproduction and conservation of male sterile lines. The male sterile lines of radish belong to the type of cytoplasmic-nucleic sterility and are bred by natural population selection. In order to breed sterile lines, the maintainer line is needed. The adult plant seed collection method can be used. In the autumn of the first

year, the sterile line and the maintainer line should be planted separately. After the fleshy roots are formed, remove the inferior ones, mark and store the good ones respectively; in the next spring, plant the sterile line and maintainer line at a ratio of 4 : 1–5 : 1. In order to increase the amount of pollen in the maintainer line, reduce planting spacing and increase the number of plants. The isolation distance around the seed collection field should be more than 2,000 m. When flowering, pollinate the sterile line with pollen from the maintainer line so that seeds harvested from the sterile line remain sterile. If the maintainer line is not needed, remove all the plants in the maintainer line after the full-blossom stage and only harvest the seeds of the sterile line. Set up another isolation area for breeding the maintainer line.

If the parents are self-incompatible lines or inbred lines, the plants can be grown in a yarn shed and the artificial mixed pollination method can be used to reproduce the original seed at the bud stage or anthesis.

(2) Hybrid seed production technology. The adult plant seed collection method or seed-to-seed method can be used to breed F_1 hybrid seeds with the male sterile line. In North China, the method of spring seedling is generally adopted, while in the South, it can be directly sown in the open field.

① Sowing and seedling-raising. If the method of spring seedling is adopted, seeds can be sown in cold beds or cold frames. It is necessary to cover the ground with a film 15–20 days before planting to warm it up. In south-central Hebei Province, seeds are usually sown from early to mid-January, while in Liaoning Province, from early to mid-February. Nutrient bowls or nutrient soil blocks can be used for seedlings breeding, and the sowing amount of the male sterile line and the father line can be arranged at a ratio of 3 : 1. Raise the temperature before the emergence of the seedlings, and maintain the temperature at 20–25°C during the day and 8–10°C at night. After the seedlings emerge, the temperature should be lowered to 20°C during the day and 5–8°C at night. Make sure to let in some wind. When the seedling has five or six true leaves, lower the temperature and let in more wind to develop its resistance to cold temperatures.

② Final planting and field management. Seed production fields for radish should be separated from other easy-to-hybridize cultivars at a distance of more than 2,000 m. The ground should be prepared and fertilized in time after the soil has thawed. In south-central Hebei Province, final planting is generally done in early and mid-March; In Liaoning Province, final planting is done in early and mid-April. The row spacing is (40–50) cm×(25–30) cm. The final planting ratio of the male sterile line to the father line is 4 : 1. Seedling selection should be carried out before final planting. Diseased, impure, and inferior seedlings need to be eliminated. After final planting, water promptly. After the seedlings re-adapt to the environment, loosen the soil in time, improve the ground temperature, and strengthen the hilling. Plants should be fertilized and watered after they bolt, and the plants of the father line should be removed after the full-blossom stage.

③ Anthesis and plant adjustment. During the hybrid seed production, attention should be paid to the synchronization of the antheses of the father and mother plants. If the antheses of father and mother are not synchronized, it is better to adjust the sowing period. Otherwise, it is necessary to remove the inflorescence of the main stem of the early flowering parent and delay its anthesis to coincide with that of the other parent.

Radish takes 40–50 days from flowering to seed maturity. In order to promote the early maturity of seeds and provide new seeds for the production of that year in time, plastic film mulching cultivation can be used in northern China. In the anthesis, it is best to release bees for pollination to increase seed yield.

(IV) Production Techniques for Carrot Seed

In the seed reproduction of conventional cultivars of carrots, the root-to-seed method and seed-to-seed method are mainly used. In the original seed production, it is necessary to use the root-to-seed method. The seed-to-seed method is commonly used to produce the seeds used for hybrid production.

1. Key Points for Large Plant Seed Collection Method

(1) Cultivation of seed plants.

① Ground preparation and fertilization. Carrot plants have well-developed roots that can go deep into the soil, so it is better to choose deep, fertile, loose, well-drained loam or sandy loam with a pH value of 5–8. Do not plant carrots in succession in one field. Apply 60,000–75,000 kg of decomposed organic fertilizer before cultivation, along with 450 kg of diammonium phosphate and 225 kg of potassium sulfate per hectare.

② Sowing. The sowing period begins from mid-July to early August in North China and from June to July in Liaoning Province. The sowing period in the South should be progressively delayed. Broadcast sowing in low, level borders and line sowing on high ridges are both acceptable. When line sowing is adopted, keep the distance between ridges at 55 cm, and when broadcast sowing is adopted, dig ditches 15–20 cm long and 3–5 cm deep. The sowing amount is 22.5–30 kg per hectare. In order to ensure the neat emergence of seedlings, it is necessary to rub the surface of the seeds to remove the prickly hairs and eliminate the immature seeds by wind, grain, and water selection before sowing. This can significantly improve the germination rate of the seeds.

③ Field managemen. In order to ensure the survival of all the seeds sown, the sowing density is often large, so thinning should be performed early and is generally done twice. The first is done when the seedling has one or two true leaves with a height of 3 cm, and the second is done when there are four or five true leaves. Remove densely packed small seedlings and diseased, weak, and impure seedlings. Distance between seedlings should be kept at 12–14 cm for small-sized cultivars and 14–16 cm for large-sized cultivars. Remove weeds in time during thinning. It is better if herbicides are applied before sowing.

The seedling stage of carrot plants is in the rainy season, and seedlings are afraid of waterlogging, so after rain, it is necessary to carry out timely drainage. Fertilizer is applied about two to three times during the growth period, and 300 kg of human manure or urea is recommended per hectare.

(2) Harvesting and storage of seed plants seed.

Plants need to be harvested before frost, in mid-October in north-central Liaoning Province and in early November in central North China. Selection is done during harvest. Plants with regular leaf color, few leaves, smooth fleshy roots, small root apex, bright color, no cracks, and plants that are non-lodging, free of pests and diseases, and have characteristics of the cultivar should be reserved as seed plants. Remove the leaves from selected plants, leaving only petioles that are 1–2 cm long. At the beginning of harvest, when the weather is warm, ditches should be dug in the field for temporary planting and covered with a straw curtain or a thin layer of soil. In order to avoid heating, a bundle of sorghum stalks for ventilation and heat dissipation is buried vertically every 1.5–2 m in the pile or shallow ditch. When the temperature drops to 4–5°C, store them in the cellar. The sand layer method is mainly used in the cellar, in which clean layers of sand and carrots are stacked alternately up to about 1 m high. The storage temperature in winter is 1–3°C.

In southern China, the ditch storage method is generally used. The ditch is 80 cm deep and 100 cm wide, and the length is unlimited. Put 35 cm thick carrots in it. When the weather becomes cold, the ditch is covered with soil about 7 cm thick at the beginning, and then the soil should be covered at any time as the temperature drops. The total thickness of the soil is about 50 cm. In the following spring, mulch is gradually removed as the temperature rises.

(3) Final planting and management of seed plants.

① Selection of seed collection fields. The plots with fertile and loose soil, good structure, higher ground, and convenient drainage and irrigation should be selected as the seed collection fields, and the same crop should not be continuously planted on the same field. For every hectare, apply 75,000 kg of basal fertilizer, 450 kg of diammonium phosphate, and 225 kg of potassium sulfate. For the original seeds, the isolation distance is more than 2,000 m, for the seeds for production, more than 1,000 m.

② Field planting. When the soil temperature rises to 8–10°C after thawing, the final planting can be carried out. The final planting can be carried out in mid-March in central North China and in April in north-central Liaoning Province. The plant spacing varies from region to region. Generally, 45,000–47,500 plants are planted per hectare, with a spacing of (40–50) cm×(25–30) cm. When planting, the growth point should be level with the ground. The top should be covered with soil that is 3–4 cm deep or horse manure, and stamped down to prevent freezing and rodent damage.

③ Field management. After final planting, if the soil is dry, water it to promote fast root growth and readjustment of seedlings. Maintain the field water capacity at 70%–80% throughout the fertility period. There are many weeds on a carrot field. After final planting, intertillage, and weeding should be carried out in combination with watering until the ridge closure. During the growth period and anthesis, 300 kg of ternary compound fertilizer per hectare should be applied once each. Attention should be paid to the prevention and treatment of aphids and diseases after bolting.

(4) Pruning.

In order to promote large inflorescence, improve yield, and prevent field lodging in shade position, it is necessary to prune in time. Generally, keep the inflorescence of the main branch and three to four strong inflorescence of primary lateral branches, and remove all the other lateral branches, or only keep the primary lateral branches, and remove the inflorescence of the main stem and the lateral branches above the secondary level. Pinch the branches off with fingers or cut them off with shears. Do not tug on the roots.

(5) Harvesting and threshing of seeds.

In north-central Liaoning Province from early to mid-July, and in the Yangtze River basin in late June, the seeds are gradually mature. When the inflorescence turns from green to yellow, the outer margin rolls inward and the lower stem turns yellow, the floral discs can be cut off. Be sure to harvest in batches. After harvest, two dozen branches are grouped into a bundle and placed in a well-ventilated space for after-ripening for a week or so. Mechanical threshing or manual rubbing can be used to sieve out inert matter, and then dry the seeds. When the seed moisture content drops below 8%, the seeds can be bagged for storage. Generally, 600–900 kg of depilated seeds can be harvested per hectare, with a thousand-seed weight of 1.4–1.6 g.

2. Key Points for Seed-to-Seed Method

The parents should be sown a month later than that using the root-to-seed method. Increase plant density and maintain an approximate row spacing of (25–30) cm×(1.5–2) cm. The proportion of parents planted in spring should be at a

1 : 3 ratio. When controlling pests and diseases, do not spray pesticides during the full-blossom stage, as this can also kill insects and thus affect pollination; set up a support frame. Carrot plants are tall, so in order to prevent mutual lodging and mixing between plants, it is necessary to set up strong support to maintain their independent growth; roguing is also needed. In the process of carrot seed production, the father is the male fertile plant, and the mother is the male sterile plant. It is necessary to strictly check the floral disc of the mother in the anthesis, and the plants that disperse pollen should be completely removed to avoid mixing with the seeds of foreign plants. If the father is monoclonal, it can be removed after pollination.

Review and Reflection Questions

(1) What is the pollination process of cucumbers?

(2) What are the technical points of seed collection in spring open cultivation for cucumber seed production?

(3) What are the methods of selecting and breeding female lines and preserving cucumbers of female lines?

(4) What are the pollination habits of pumpkins?

(5) What are the methods of artificial flower-binding of pumpkin for isolated pollination?

(6) What are the key points of field inspection during the watermelon pollination period?

(7) What are the techniques for producing seeds of conventional watermelon cultivars?

(8) What is the seed production method for seedless watermelon?

(9) What are the seed production techniques of melon hybrid cultivars?

(10) What are the seed production techniques for conventional tomato cultivars?

(11) What are the measures to improve the purity of tomato hybrids?

(12) What is the hybridization technique for producing pepper hybrid seeds?

(13) What are the techniques for performing the seed-to-seed method on cabbage?

Learning Scenario II Ornamental Plant Seed Production

I. Annual and Biennial Flower Seed Production

(I) Ecological Habits of Annual and Biennial Flowers

1. Annual Flowers

Annual flowers are also known as spring sowing flowers, spring flowers, and less cold-resistant flowers. They are native to the tropics and subtropics and cannot withstand low temperatures below 0°C. The seeds are sown in spring and grow to produce flowers, bear fruit, and die. In the vernalization stage, higher temperatures are required, and seeds can pass through the vernalization stage after 5–15 days at 5–12°C. They are short-day plants and bloom under 8–12 hours of sunshine in autumn. Annual flowers are heliophilous and shallow-rooted.

2. Biennial Flowers

Biennial flowers are also known as autumn-sowing flowers, winter flowers, and cold-resistant flowers (they are sown in autumn and bloom in spring). They are generally native to temperate or cold regions and are able to withstand low temperatures below 0°C. The most abundant are the biennial flowers that are native to the Mediterranean coast, such as snapdragon, pansy, violet, etc.

The vernalization stage of winter flowers requires low temperatures. Seedlings pass the vernalization stage in 30–70 days at 0–10°C. The process takes place most quickly below 0°C, so biennial flowers will not bloom properly if they are planted in the warm spring. If biennial flowers are treated with low temperature (vernalization) before spring sowing, they will bloom in the current year, but the plants will be short and the pedicels will be too short to be cut flowers. Biennial flowers grow robustly under short-day conditions in autumn and winter and bloom in long-day conditions in the following year. A photosensitive period of 14–16 hours is generally required.

(II) Mixing and Degradation of Flower Cultivars and Prevention Measures

1. Causes of Degradation of Flower Cultivars

(1) Mechanical and biological mixing.

The mixing and degradation of self-pollinated flowers are mainly due to unreasonable crop rotation, field management, seed collection, sunning, storage, packaging, transportation, and other processes, which lead to mixed seeds or seedlings of different cultivars, reducing the purity of the cultivar. Biological mixing occurs in all kinds of flowers, but it is most common in cross-pollinated flowers. This is caused by natural outcrossing between cultivars or between species due to inadequate isolation. For example, the plants of dwarf *Zinnia* cultivars are extremely short and almost flat on the ground, which are good materials for arranging flower beds, flower borders, and flower tables. However, when they are biologically mixed with tall *Zinnia* cultivars, the plants will show uneven height and disorder of plant shape.

(2) Genetic deterioration.

All kinds of flowers have low-frequency mutations under natural conditions, but most of these mutations are inferior, resulting in the mixing and degradation of cultivars. For example, the cockscomb shows atavism after mutation, losing its large corolla and changing its inflorescence back to the feather cockscomb inflorescence of the original cultivars.

(3) Unsuitable cultivation environment.

Most of the excellent cultivars of flowers are obtained under artificial conditions by selecting and breeding wild species directly or indirectly, and the related genetic genes of their wild traits often still exist, but they cannot be shown under cultivation conditions because they are recessive. When the cultivation technology is improper or the external environmental conditions cannot meet the requirement of good seed quality of that cultivar, the good cultivated traits will gradually be replaced by the latent wild traits and degradation will occur.

(4) Inappropriate reproduction methods.

In the case of asexual reproduction, though all parts of the plant are genetically identical, plants grown from different parts of the reproduction material do not have exactly the same effect in production. For example, in terms of chrysanthemums, plants reproduced from foot buds without old roots are stronger than those grown from cuttings using ramet seedlings with old roots or old branches. Dichromatic foliage plants often lose their original typical traits when they are reproduced using cutting parts with no representative traits. In sexual propagation, seed selection from different parts will also have a great impact on seed quality. For example, the capsule weight of snapdragon and petunia decreases progressively from the lower part of the inflorescence to the top. The seeds produced by the radial florets of coreopsis are large and heavy, while the central disk seeds are light and small. If the seeds of the upper part of the inflorescence or the central disk are used for reproduction, their offspring will grow poorly and the seedlings will be weak.

(5) Virus infection.

Many flowers, especially bulbous and greenhouse flowers, will decline in vitality, growth vigor, and ornamental value after being infected by the virus.

2. Preventative Measures for Flower Degradation

(1) Prevent mechanical and biological mixing.

In sowing, transplanting, seed collection, and other seed production links, strictly follow the technical operation procedures to prevent the mixing of cultivars due to human error. In the seed production process, it is necessary to pay attention to roguing work and to prevent biological mixing through space or time isolation. The distance isolation for flowers varies according to different cultivars and pollination habits. The flowers with the minimum isolation distance of 400 m include coreopsis, calendula, garden nasturtium, marigold, etc.; the flowers with the minimum isolation distance of 350 m include hollyhock, *Cheiranthus cheiri* L., *Dianthus* L., etc.; the flowers with the minimum isolation distance of 200 m include petunia, snapdragon, zinnia, etc.; the flowers with the minimum isolation distance of 50 m include *Salvia splendens* Ker-Gawler, *Clarkia amoena*, *Scutellaria barbata*, *Callistephus chinensis*, *Lathyrus odoratus* L., etc.; the flowers with a minimum isolation distance of 30 m include pansy, delphinium, etc.

(2) Purification and rejuvenation.

At each of the flower growth and development stages, according to the typical traits of the cultivar through observation and comparison, excellent plants are selected to be reserved for seed collection.

(3) Choose a suitable cultivation environment.

Choose suitable sites for cultivation and change planting seasons if needed to reduce the impact of unsuitable growing conditions on the cultivar.

(4) Strengthen field management.

Create a suitable environment by preventing diseases and pests, performing weeding and fertilization, carrying out soil disinfection, avoiding continuous cropping, and such measures. Sometimes, topping and pruning can be done to increase the yield of solid bulbs, bulbs, and tubers, increase sprouting tillers, and continuously improve the quality of seeds or seed bulbs.

(5) Select appropriately.

Select seeds or vegetative organs with high typicality for reproduction.

(6) Use tissue culture technology for detoxification.

Many viruses cannot be transmitted through the shoot apical meristem of plants, so the biotechnology of shoot tip tissue culture of plants can be used to produce large quantities of virus-free test-tube seedlings, solving the virus-causing degradation. At present, commercial production has been carried out on flowers such as lilies, carnations, chrysanthemums, dahlias, petunias, *Freesia refracta* Klatt, narcissus, irises, *Sinningia speciosa*, *Clivia miniata*, and Orchids.

(III) Production Technology of Original Seeds of Annual and Biennial Flowers

The original seeds are the basic material for reproducing superior flower varieties. Therefore, the requirements for their purity, typicality, vitality, and quality are particularly strict. The main method of producing original seeds is to select the best through purification (i.e. through purification and rejuvenation).

For the original seed production of annual and biennial herbaceous flowers, the method of the plant (strain)-to-row selection and purification is widely used, which is simple and easy to operate and has a good effect. The general procedure includes: selecting superior individual plants, and then conducting comparison and identification between plant rows; selecting superior plant rows, and then conducting the comparison test between the strains; mixing the selected strains conforming to the typicality of the original cultivar to produce the original seeds in the original seed nursery.

For often cross-pollinated flowers, the three-level purification method is used to produce the original seed, that is, plant selection in the first year, plant row comparison and identification in the second year, and strain comparison test in the third year; for self-pollinated flowers, the original seed can be produced by the procedure of two-level purification method without going through the strain comparison test, that is, plant selection in the first year, plant row comparison test in the second year.

When producing the original seed by plant-to-row selection and purification, it is necessary to bag the typical superior individual plants of cross-pollinated flowers selected from the selection nursery for self-crossing. In the plant row nursery, it is necessary to select the typical superior plants from the selected typical superior plant rows and make them perform selfing. After harvest, they are selected, mixed, threshed, and multiplied to produce the original seeds of inbred lines.

In the production of original seeds, attention must be paid to the following links: select the basic materials needed for the production of the original seeds; select superior individual plants according to typical good traits and other main traits; the soil fertility of each field should be uniform so as to carry out accurate comparison and identification; pay attention to safety isolation, carry out roguing and adopt other methods to prevent impurity and keep purity, so as to prevent new biological mixing; adopt measures to accelerate reproduction, so that the original seeds can be rapidly popularized and applied.

(IV) Basic Procedures for Good-Quality Seed Production of Annual and Biennial Flowers

1. Select Superior Mother Plants for Seed Collection

The mother plant for seed collection should be robust, well-developed, free from diseases and pests, and have good characteristics of this cultivar or species. For herbaceous flowers, special seed collection plots should be set up in the nursery, and they should be properly cultivated and carefully managed to obtain good seeds. For cross-pollinated flowers, depending on the characteristics of the species and cultivar, it is sometimes necessary to arrange suitable pollinators around them with artificial pollination or to set up isolation areas to avoid biological mixing.

2. Seed Collection at the Right Time

(1) The processes of seed ripening: The ripening of seeds involves two processes.

① Physiological maturity. When the seed has the ability to germinate, it means that it has reached physiological maturity. But at this time, the nutrients inside the seed are still in a soluble state, the accumulation of dry matter remains insufficient, and the moisture content is high; the testa is soft and unable to protect its embryo, which is why it crumples when dried. Such seeds are not good for storage.

② Morphological maturity. At this time, the seed moisture content is reduced and the accumulation of internal nutrients is about to terminate; the physiological activities are reduced, the testa is firm and dense, and its resistance to adverse external conditions has improved. In the external form (shape, color, size, luster, etc.), the seeds show the inherent characteristics of maturity. Such seeds are of good quality and easy to store.

In general, the seeds first reach physiological maturity and then morphological maturity, but there are also cases where the order reverses; that is, although the seed objectively shows mature characteristics, the embryo is still very small and not fully developed. It is necessary to make it through a certain period of storage and after-ripening, so that the seed embryo continues to grow, such as the seeds of *Colchicum autumnale* L., *Panax ginseng* C. A. Mey., and *Panax quiquefolium* L.

In addition to the genetic characteristics of the plant, the specific time of seed collection is also affected by its growing environment. Generally, seeds mature earlier in barren land than in fertile land, earlier in dry years than in wet years, and earlier in years with high temperatures than in years with low temperatures. The climate changes from year to year, which can make the seed mature earlier or later, so it is impossible to determine the exact time of seed collection for a flower.

(2) Marks of mature seeds.

① Color: During growth, the seed color changes from light to dark, from green and white to brown, dark brown, or variegated. Some berries change color from green to bright red, blue, purple, or black.

② Taste: Generally, mature fruits are less acidic and astringent, more fragrant, and sweeter.

③ Firmness: The berry becomes soft and juicy when it matures. Dry fruits such as capsule fruits, legume fruits, and silique fruits crack.

Because the specific morphological characteristics of various flowers are different when they mature, the cultivator should accumulate experience constantly. The maturity can also be tested by cutting the seed to see whether the kernel is full, hard, or empty.

Flowers (seeds) with infinite inflorescences, where the upper part is blooming, and the lower part matures, shall be harvested in stages and repeatedly for flowers such as snapdragon, cockscomb, *Cleome spinosa*, etc.

For those species whose fruits mature quickly and are prone to cracking, seeds should be harvested when the fruit loses its green color and becomes light yellow, but collected when the seeds are not dry enough, such as the likes of impatiens, pansy, and sorrel.

3. Seed Collection Methods

(1) Picking from the plant: The fruits are picked as they ripen, and then gathered for processing.

(2) Picking from the ground: For mature and easy-to-fall species, spread plastic film on the ground around the plant and shake the plant by hand to knock down the fruit or seed.

(3) Trimming inflorescences and plants: Generally, the mother plant grows robustly. The seeds from the plants that first flower and those near the top of the trunk are preferred.

After seed collection, the number, name (Chinese name, Scientific name), date of collection, place of origin, environmental conditions, and collector shall be recorded.

When drying the seeds, do not spread them on metal plates or cement ground; otherwise, it is very easy to scald the seed embryo under strong sunlight. The best time for seed collection is in the morning of a sunny day.

4. Seed Life

(1) Seeds with long shelf life can last around four to six years, such as cockscomb, marigold, *Cheiranthus cheiri*, *Cucurbita pepo* var. *ovifera*, *Gypsophila, Impatiens, Cypress vine*, etc.

(2) Common seeds can last two to three years, such as Corn poppy, Cornflower, *Gomphrena globosa* L., *Zinnia elegans* Jacq., *Helichrysum bracteatum*, petunia, snapdragon, morning glory, *Dianthus chinensis* L., etc.

(3) Seeds with short shelf life can last around one to two years, such as *Consolida ajacis* (L.) Schur, *Phlox drummondii* Hook., *Kochia scoparia*, *Gerbera jamesonii* Bolus, *Lantana camara* L., and *Lathyrus odoratus* L.

(4) Extraordinarily long-lived seeds, such as lotus seeds, can live 1,000–2,000 years long.

II. Production of Bulbous Flower (Seed Bulb)

(I) Types and Habits

1. Types

(1) They can be divided into five types based on the shape of underground roots.

① Bulb type. The underground stems shorten and shrink into a disc, oblong, spherical, or conical shape, consisting of scales that are fleshy. Taking daffodil as an example, the part below the scales is called the stem disc, and the root grows below the stem disc. The stem disc is the real stem; that is, the axillary buds can be produced between the scales and the leaves, which can be alternately old and new. According to different arrangements (or different attachments) of scales, it can be divided into the following types.

A. Tunicated bulbs (layered bulbs, firm bulbs): The scales are arranged in several layers with some coriaceous scales on the outside. This kind of scale does not lose water easily, and it includes daffodils, hyacinths, tulips, onions, *Lycoris radiata*, and squill.

B. Bulbs without skins (flaky bulbs, loose bulbs): The scales are flaky and do not cover the whole bulb, such as lilies. According to bulbs' life span, bulbs can be divided into two kinds.

a. Annual bulbs: For example, tulip bulbs die in the year when they are planted, and large bulbs will be formed from axillary buds.

b. Perennial bulbs: For example, daffodils and hyacinths usually develop axillary buds at the base of scales and grow true leaves, whose lower scale leaves gradually increase and become hypertrophic and then separate from the old bulb.

② Corm type. The corm is formed by modification and shortening of the underground stem (the enlarged part is equivalent to the stem disc of a bulb). It's spherically shaped, with an interior that is of parenchyma tissue. Only a few layers of membranous outer skin are marked with ring-like patterns, and that's where the stem nodes are. The lateral bud can be produced on it, and the new bulb gradually enlarges from the lateral buds, forming an alternation within a year, such as gladioluses, freesia, crocus, water chestnuts, and so on. After planting for a year, if the gladiolus is dug up, it can be seen that the old bulbs have shriveled up and died, attached to the new bulbs.

A bulb is the modification of the leaf and a corm is the modification of the stem.

③ Tuber type. Its underground stems are tuberous-like, with an irregular appearance, but possess bud eyes. This type includes calla lily, gloxinia, cyclamen, bulbous begonia, *Caladium bicolor* (Ait.) Vent., bletilla, etc. Tubers are modifications of stems. Tubers of most species are perennials and have little ability to split naturally.

④ Rhizome type. When the underground stem becomes hypertrophic, it forms long and thick rhizomes. There are obvious nodes and internodes on it. Each node can form a lateral bud, especially at the top section of the rhizome. When the lateral bud develops and forms more clusters, the old stem gradually dies. This type of plant includes canna, iris, water lily, lotus, convallaria, *Calla palustris* L., reed, etc.

⑤ Root tuber type. It's formed by the expansion of the roots, in which a large amount of nutrients accumulates. Root tuber is a modification of the root, without internodes or bud eyes. Flower species with root tubers include *Dahlia*, *Ranunculus asiaticus* L., Winter aconite (Ranunculaceae family, *Eranthis* genus), Yellow *Alstroemeria hybrida* (Alstroemeriaceae family, *Alstroemeria* genus), etc.

Dahlias sprout at the neck of the root, so when digging roots, cut off the old stem 10 cm above the tuber and plant it separately.

(2) They can be divided into two types according to the planting season.

① Spring planting bulbs. This kind of plant is planted in spring and blooms in summer and fall. Then, the overground part dies off, and the plant goes into dormancy. Native to the tropics or subtropics, high temperatures are required in the growing season and are weak in cold resistance, so it is necessary to dig out their bulbs in winter and store them in a place free from cold. Such flowers include gladiolus, canna, dahlia, tuberose, etc.

② Autumn planting bulbs. Planted in autumn, they start to grow and develop under cool temperatures in fall, and bloom from spring to early summer. However, in the cold area of northern China, although the root system grows fully in autumn of that year, the top bud is less extended, and it only winters under the soil surface, showing a semi-dormancy state. It will grow again when the temperature gets warm in spring, and the overground part will die and the underground part will lie dormant after flowering. They are usually planted from September to November, such as daffodils, hyacinths, tulips, *Muscari botryoides*, fritillaria, etc. Most of them are native to temperate zones and have strong cold resistance.

The stage of flower bud differentiation of bulbous flowers, whether for autumn or spring planting bulbs, takes place at high temperatures in summer.

After entering the summer, all the overground parts of the autumn planting bulbs die and enter the dormancy stage, but flower bud differentiation occurs during the dormancy stage. The optimum temperature is 11–18°C, and if the temperature exceeds 20°C, the flower bud differentiation will be blocked.

2. Ecological Habits

(1) Requirements for light. Most of these plants are light-loving, and require sufficient sunlight, while a small portion of them prefer the shade, such as *Lycoris radiata*, lilies, *Convallaria majalis* L., etc.

(2) Requirements for soil. Generally, the sandy loam with good drainage and much humus is the best, while for daffodil, tuberose, hyacinth, lily, *Lycoris radiata*, and tulip, the clayey loam is more suitable. Among them, lily prefers acidic soil, while others prefer neutral soil.

(3) Requirements for moisture. Bulbous flowers are morphologically drought-resistant plants, so the soil should not be waterlogged and should be properly irrigated during the growth period.

(4) Requirements for temperature. Bulbous flowers that originated from temperate zones do not completely stay dormant in winter, and as soon as the climate is warmer, they will bolt and bloom, such as hyacinths, tulips, daffodils, etc.

Bulbous flowers that originated in the subtropical or tropical regions have a fixed dormant period where cultivation should not happen.

(II) Reproduction of Seed Bulbs

1. Reproduction Methods

Bulbous flowers are generally reproduced with vegetative organs because it takes a long time to flower after sowing. Bulbous iris and tulips take four to five years, *Lycoris radiata* takes five to six years, and there are many species that do not bear fruit in most parts of China, such as tuberose and canna. Dahlias and cyclamen can also be reproduced by sowing reproduction.

(1) Cuttage method.

① Bud or branch cuttage. Use the bud on the bulb, the root, or the modified stem for cuttage. For dahlias, for example, foot buds or branches can be used for cuttage. Dahlia only has buds at the root neck of the root tuber. The stored root tubers are usually taken out during the period from February to March every year, and the pregermination is done first. The temperature is kept at 20–22°C during the day and 15–18°C at night. When the bud has three leaves that are 6–10 cm long, cut it for cuttage.

Branches removed during pruning can also be used. Cuttage should be carried out in the open field with branches from April to June before the rainy season so that the survival rate is high.

② Cuttage with lily scales. Choose large bulbs in autumn and dry them for a few days before cutting. The best substrate is perlite or river sand. Small bulbs are produced under the scales in the following year. They take three to four years to bloom. The outermost and innermost scales of plants should not be used for cuttage.

(2) Division method.

① Digging method for hyacinths. After the hyacinth is dried under the sunshine in August, dig out a quarter (the stem disc is removed) and place it in a dry and ventilated place at a temperature of 20–30°C. In October, many small bulbs, up to 100, will grow at the cut of the scale. The mother bulb with the small bulb is then planted with its bottom up, and the mother bulb does not flower anymore. The small bulbs can be separated and cultivated after one year, and they become large bulbs in five to six years. The wound method is also available. When the bulb is mature, first cut the bottom of the bulb flat, and then make two to three crossed cuts at the bottom of the bulb. The small bulbs will grow out of the wound. The number of bulbs obtained by this method is less than 30, but these small bulbs are large in comparison and can form large bulbs in three to four years.

② Wound division method for tuber and rhizome plants. They can be cut with buds and reproduced in the same way as potatoes. This can be applied to tuber plants such as bulbous begonias, *Caladium bicolor* (Ait.) Vent., calla lilies, etc., and rhizome plants such as canna, iris, etc. Dahlias are reproduced by the division of tubers. Do not damage the root neck.

(3) Reproduction with bulbil. Bulbil is fleshy and largely produced in the axils of the upper stem of some species of lily. There are blackish-purple bulbils in the leaf axils of the tiger lily. The bulbil of *Lilium souliei* is strong green. The bulbil of the Chinese yam is called Yu Ling Zi. Harvest the bulbils when they are ripe, but have not yet dropped, and plant them during the period from early November to late December. It takes three to four years for them to bloom.

2. Key Points of Cultivation Management and Bulb Reproduction

(1) Basal fertilizer. Basal fertilizer should not be directly applied to the bulb and must be fully decomposed to avoid plant decay.

(2) Planting method. Plants are to be spaced according to the size of the bulb. For crocuses and common snowdrops, the spacing should be 7 cm; for tulips, 12–13 cm; for daffodils, 20–40 cm; for lilies, 40 cm; for large-sized dahlias, 120–150 cm; for medium-sized ones, 60–100 cm; for small-sized ones, 40–60 cm.

The planting depth is generally two to three times as long as the bulb diameter, depending on the soil quality and cultivated species. The new bulb of gladiolus grows above the old bulb, and the rhizome of lily grows above the bulb. They can be planted more deeply. Some species of bulbous flowers have shrinking roots, such as tuberose, that will pull the bulb down. It is recommended to put some sand under the bulbs. For amaryllis and squill, part of the bulb should be exposed. In clay soil, the plant should be planted less deeply, and in sandy soil, the plant can be planted a little more deeply. At the same time, planting depth should be different according to different cultivation purposes. If the purpose is to reproduce seed bulbs or to harvest each year, the plant should be planted shallowly; otherwise, it should be planted deeply.

Large and small bulbs must be planted separately, and if they are planted together, the smaller bulbs will not grow well because of the lack of nutrients.

Most bulbous flowers have brittle, unbranched, fixed roots. They have poor regeneration ability, so they are generally not transplanted.

(3) Leaf protection. The number of leaves is fixed. It is necessary to protect the leaves. If there is too much leaf damage, it will affect photosynthesis and the plant

will not flower. For example, after a gladiolus grows nine leaves, a lily of the valley grows two or three, and a tulip grows three or four, they will not grow new leaves. Make cut flowers with fewer leaves, because after flowering, the plant still needs the leaves for photosynthesis to produce the nutrients needed to grow underground bulbs.

(4) Management. The bulbs grow only after flowering, so fertilization is necessary both before and after flowering. When cultivating bulbs for commercial use, buds must be removed to concentrate nutrients on growing underground bulbs. Dead shoots should be removed after flowering.

(5) Harvest. Harvesting is best done when the overground leaves are yellow but not dead so that the underground bulbs can be located without being damaged. Some species need to be dug annually. And there are some species that should not be dug annually. However, for single-purpose cultivation, it is necessary to dig every year. Benefits of digging annually: ① Prevent frostbite in spring planting plants when temperatures are low, and prevent root rot in autumn planting plants caused by the hot and rainy weather in summer; ② Facilitate the separation of the larger bulbs from the smaller ones; ③ Prevent the bulbs from squeezing into each other in the soil and becoming deficient in nutrients; ④ Some species will go through the after-ripening process; ⑤ Make good use of the land and allow other crops to be grown during its dormant period.

(6) Storage. The outermost film layer should be preserved and the diseased bulbs should be removed when storing the plants. If the number of diseased bulbs is small, and they are valuable species, these diseased bulbs can be preserved, but preservative, semi-soluble stone, or plant ash should be applied to the diseased spot. Light-tolerant bulbs, such as gladiolus and tuberose, can be dried under the sun, and bulbs not tolerant of light, such as dahlias and canna, should be dried in the shade. Dahlias can be stored in sand, gladiolus can be placed on mats, and some species can be bagged in cloth bags and hung indoors.

(7) Continuous cropping problem. Some species are more susceptible to pest and disease attacks. Because the residues of the plant body are left behind, the sporophytes of bacteria are preserved, so continuous cropping cannot be

done. Plants that cannot be continuously cropped include hyacinth, tulip, dahlia, and gladiolus, which cannot be planted with perennial flowers to form a flower border; the plants that can be used for continuous cropping include *Zephyranthes grandiflora* Lindl., *Zephyranthes candida* (Lindl.) Herb., *Oxalis corniculata* L., narcissus, *Lycoris radiata*, canna, bulbous iris, etc.

III. Seed Production of Perennial Flower

(I) Ecological Habits of Perennial Flower

1. Classification by Cold Tolerance and Dormancy Habits

(1) Cold-resistant perennial flowers have the characteristic of complete dormancy in winter. After all of its overground stems and leaves die, the lower sections go into dormancy. Most of them have strong cold resistance and can overwinter in the open field. By the following spring, the overground parts will germinate and continue to flower. These flowers are native to cold temperate regions, such as chrysanthemums and *Platycodon grandiflorus*.

(2) Evergreen perennial flowers still retain evergreen leaves in winter, but their growth is stagnant, showing a single dormancy state, and their cold resistance is weak. These types of plants are mostly native to warm regions in temperate zones, such as *Ophiopogon japonicus*(Linn. f.) Ker-Gawl., *Ophiopogon bodinieri* Levl., etc.

Rohdea Roth can overwinter in open fields in Jiangsu Province, but the leaves will get damaged.

2. Conditions Required for Perennial Flowers to Complete All Growth Stages

Perennials that bloom in spring require low temperatures in the temperature-sensitive stage and long sunshine in the light-responsive stage, such as *Phlox drummondii* Hook.

Perennials that bloom in autumn require high temperatures in the temperature-sensitive stage and short sunshine in the light-responsive stage, such as chrysanthemums.

Most perennial flowers require plenty of sunshine, but some species thrive in half shade, such as *Hosta plantaginea* (Lam.) Aschers., *Hosta ventricosa* (Salisb.) Stearn, *Convallaria majalis* Linn., *Rohdea Roth*, etc.

Those that grow well in the shade include *Bletilla striata* (Thunb. ex Murray) Rchb. F., *Aquilegia vulgaris*, *Platycodon grandiflorus*, *Hemerocallis fulva* (L.) L., etc.

(II) Cultivation Management and Seedling Reproduction

1. Reproduction Methods

(1) Seed reproduction.

① Cold-resistant perennial flowers are hardy plants, and they can be sowed in spring, summer, or autumn. Generally speaking, it is best to sow seeds as soon as they mature.

② Non-hardy perennials should be sown in spring or after the seeds mature.

③ Some seeds require low temperatures and moist conditions to complete their dormant stage, such as peony, iris, and delphinium, and they must be sown in autumn or spring after maturity.

(2) Sucker propagation.

Those that bloom in spring will grow ramets in autumn, so sucker propagation must be done when their overground parts are dormant and their roots are still active. Those that bloom in autumn will grow ramets during spring or autumn before they germinate, such as some evergreen species. "If the plants of peonies grow ramets in March, they will never bloom." Normally, peonies bloom in April.

(3) Cutting propagation.

① Stem cuttage. It can be done throughout the entire growth period. Perennial flowers can be cultivated as annual and biennial species, such as *Salvia splendens* Ker-Gawler, snapdragon, pansy, *Glandularia*×*hybrida* (Groenland & Rümpler) G.L.Nesom & Pruski, *Ageratum conyzoides* L., chrysanthemum.

② Root cuttage. It is suitable for species whose roots can produce adventitious buds in late autumn or early spring.

(4) Grafting. It applies to chrysanthemums.

2. Key Points of Seed Reproduction and Cultivation

① Since they are perennials and have stronger roots than annual and biennial flowers, it is necessary to carry out deep tillage and apply sufficient late-effect basal fertilizer.

② Loose soil rich in humus is preferred during the seedling stage, and clay soil is preferred from the second year. If the soil is not suitable, seedlings can be raised first and then transplanted.

③ Irrigation should be performed timely and every watering should be thorough.

④ Topdressing is usually carried out in spring when the new shoots emerge, N : P : K = 1 : 2 : 1.

⑤ Some plants, such as chrysanthemums, grow so tall that they require a supporting stand.

⑥ Carry out the work of pruning, removing dead and diseased branches, cutting off pedicel after flowering, and topping several times.

⑦ Most species are planted separately once every three to five years.

⑧ Collect seeds in time, and the methods and requirements of collecting seeds are the same as those of collecting annual and biennial flower seeds.

Review and Reflection Questions

(1) What are the reasons for the degradation of flower cultivars?

(2) What are the measures to prevent and deal with the degradation of flower cultivars?

(3) What are the signs of seed ripening?

(4) What are the reproduction methods for seed bulbs?

(5) What are the key points of seed reproduction and cultivation?

(6) What are the different sowing periods and sowing methods?

(7) What are the ways to promote rooting after the seeds are sown?

Learning Scenario III Fruit Tree Seed and Seedling Production

I. Breeding of Fruit Tree Rootstock Seedlings

(I) Breeding of Rootstock Seedlings of Pome Fruit Trees

Due to the different reproduction methods of all kinds of rootstock seedlings, they can be divided into the reproduction of tree rootstock seedlings, vegetative dwarfing rootstock seedlings, and dwarfing interstock seedlings.

1. Reproduction of Tree Rootstock Seedlings

(1) Identification of seed viability.

To ensure a reasonable sowing amount, the viability of the seeds should be assessed before sowing. There are three common methods. The first is the visual observation method. Generally, the testae of vigorous seeds are shiny, the grains are full, and the embryo and cotyledons are milky white. The second common method is the staining method. Soak the identified seeds in water for a day and a night, allowing the water to be fully absorbed. Peel off the testae, stain them in a 20-fold dilution of red ink for two hours, then wash them with water. Those whose embryos are colored are inviable seeds, and those whose embryos are not colored are viable seeds. The third method is the germination test. Spread moist cotton or soft paper in the vessel, put a certain number of seeds that have sucked water, and carry out the pregermination at a temperature of about 25°C. Calculate the seed germination rate.

(2) Seed stratification (sand storage) treatment.

The mature seeds of rootstocks of pome fruit trees require a certain amount of time to complete the after-ripening process before germinating under certain conditions of low temperature, humidity, and proper ventilation. Therefore, stratification treatment (or sand storage treatment) must be performed on rootstock seeds sown in spring. The specific steps of stratification treatment are as follows: For a large number of seeds, choose plots that are shady, dry, and not prone to waterlogging to dig some ditches. These ditches are 40–60 cm deep and 1–1.5 m wide, and their length depends on the number of seeds. During stratification, first lay a layer of clean sand at the bottom of the ditch. Mix the seeds with river sand at a ratio of 1 : 4–1 : 5 and spread them evenly on the clean sand, with a layer of wet sand on top. The temperature during stratification should be at around 2–5°C. If the wet sand is crumpled up by hand and scattered by touch, the moisture in the sand is optimal. If necessary, a bundle of sorghum stalks can be erected in the ditch to facilitate ventilation. Attention should be paid to inspection during the stratification period to prevent mildew deterioration and rodent damage. When the temperature is too high, the soil in the cellar should be turned in time or water can be added to the cellar to lower the temperature. The seeds can be sown when 10%–20% of them are germinated. If there is just a small number of seeds, wooden boxes or flower pots can also be used as containers for stratification.

(3) Sowing.

① Sowing period: Sowing can happen in autumn or spring. In areas with a warm climate in winter, good soil texture, and stable soil moisture, the seeds can be sown in autumn. Without the need for stratification, these seeds can complete the after-ripening stage naturally in the field. The seedlings can emerge early in the following spring with a long growth period and grow robustly. It is better to sow earlier before the soil freezes. In areas with a dry and cold climate in winter, heavy sandstorms, heavy soil, and serious damage from birds and rodents, it is appropriate to sow in spring. In the Yangtze River basin, seeds are generally sown in late February to late March; in North China and Northwest China, seeds are sown in mid-March to early April; in Northeast China, seeds are sown in April. If the seedlings are raised in hot beds or plastic tunnels, the seeds can be sown earlier to increase the increment of seedlings in the early stage.

② Sowing amount: The volume of seed used to produce a certain number of rootstock seedlings per unit area is expressed as kg/hm^2, which can be calculated according to the following formula:

$$\text{Sowing amount (kg/hm}^2\text{)} = \frac{\text{Number of planned seedlings per hm}^2}{\text{Number of seeds per kg} \times \text{Seed germination rate (\%)} \times \text{Seed cleanliness (\%)}}$$

The actual sowing amount should be higher than the calculated value because it is also necessary to consider the loss caused by factors such as sowing quality, sowing method, field management, and natural disasters.

③ Sowing methods: Before sowing, spread 75,000–150,000 kg of organic fertilizer and 30–45 kg of toxic soil mixed with a 1,000-fold dilution of 50% phoxim EC per hm^2 of the nursery ground, plow the soil about 30 cm deep, and then rake and grind the soil. Make high ridges in areas with more rainfall, and make level borders in areas with less rainfall. The orientation is north-south, with a width of one meter and a length of 10 m. Sow using the double-row strip tillage method, with the wider rows 50–60 cm in breadth and the narrower rows 20–30 cm in breadth. There should be two to four rows per ridge or border. Ditches should be furrowed according to a certain row spacing with a depth of 3–4 cm. Then, sow seeds in the furrows and cover them with the soil that is about 1 cm thick to suppress them.

④ Post-sowing management. After sowing, the ridges or borders should be covered with mulch film, grass curtains, or straw in time to preserve moisture, and there should also be windbreaks in the northern areas with less rain and strong wind in spring. When 20%–30% of seedlings emerge from the soil, the moisture-retaining mulch should be gradually removed.

When the seedlings grow 2 or 3 true leaves, they must be transplanted on a cloudy day or in the evening and their main roots should be cut off. The seedlings are susceptible to aphids, seedling blight, and powdery mildew. Aphids can be prevented and controlled by spraying a 4,000-fold dilution of 1.8% avermectin; seedling blight can be prevented by pouring a 6,000-fold dilution of 50% carbendazim wettable powder at the roots of seedlings; powdery mildew can be prevented and controlled by spraying 0.1–0.2°Bé of lime sulfur.

After transplanting and cutting off the main root, pay attention to irrigation to improve the survival rate of rootstock seedlings. After the rootstock seedlings start to grow, always keep the soil loose and grassless, and apply 75 kg of urea per hectare in combination with irrigation. The thinning should be finished before the end of June. The optimal spacing for rootstock seedlings is 60 cm×12 cm or 50 cm×14 cm. Retain about 140,000 rootstock seedlings per hectare and the transplanted graftings per hectare should be about 120,000. If there are excessive retained rootstock seedlings and transplanted graftings, the ratio of Grade I and Grade II transplanted seedlings will be less than 70%, which is not conducive to improving the quality of seedlings. In June, the leaves of rootstock seedlings can be sprayed with 1% urea twice at intervals of 15 days to promote their growth. In August, the leaves can be sprayed with 0.5% potassium dihydrogen phosphate once or twice to promote their robustness, and the buds in the range of 10 cm from the rootstock base should be removed in order to facilitate the budding in August.

2. Propagation of Vegetative Rootstock Seedlings

There are three propagation methods for vegetative dwarfing rootstock seedlings of pome fruits: cutting propagation, layering propagation, and propagation of vegetative intermediate dwarfing rootstock seedlings. Layering propagation is mainly used on apple dwarfing rootstock, while the pear is mostly propagated by cutting.

(1) Layering propagation.

Apple dwarfing rootstocks of M cultivar are commonly propagated by layering. It is a method of burying branches that have not been separated from the parent body into the soil, prompting the part pressed into the soil to grow roots with the nutrients provided by the parent body, and then pruning them away from the parent body to become independent rootstock seedlings. Vertical layering and horizontal layering are two types of this propagation method. The latter one is more commonly used because it can provide rootstock seedlings with a fast propagation speed, a large propagation coefficient, and a good root system. The specific practices of propagating dwarfing rootstock seedlings of pome fruits by horizontal layering are as follows.

① Soil preparation and planting. Before preparing the soil, spread 70% of soil manure (7.5 kg/m^2) mixed with calcium superphosphate (0.75 kg/m^2) and 1.5% dimethoate powder (1.5 g/m^2) on the soil surface, and then plow and level the soil. Dig a planting trench 30 cm in width and 30 cm in depth in a north-south orientation at a row spacing of 1.5 m, and spread the last 30% of the soil manure into the trench. Before the plant sap flows in spring, choose a dwarfing rootstock mother plant with a full stem and full bud eyes, cut the stem into 60 cm in length, dip the roots in mud, and plant the stem at an angle of 30°–45° to the ground northward into the planting trench at a row spacing of 40 cm, fill in the trench with the soil and compact it. Then, water the seedlings and cover the soil after the water seeps into it. The soil surface of the planting trench should be 3–5 cm lower than the surface after covering the soil so as to facilitate the horizontal layering of the dwarfing rootstock plants, hilling, preserving moisture, and rooting.

② Layering. After the mother plant is planted, when most of the buds of the plant sprout, press down the plant stem into the planting trench northward and fix it with a branch to prevent the middle of the plant stem from bulging. When pressing down the plant stem, remove the downward buds and excessively dense buds in time and make the distance between every two new shoots on the plant stem 3–5 cm.

③ Hilling. When the height of the new shoots germinating on the pressed horizontal plant stem is about 15 cm, start the first hilling of the horizontal plant stem and its new shoot base with the thickness and the width of the soil exceeding 5–10 cm, respectively, and do the hilling every 10 days. The last hilling happens in early and middle July. Cover the hilling ridges with mulch film to facilitate moisture preservation and promote rooting. The total thickness and width of the soil for hilling are about 25 cm and 30 cm, respectively. The hilling is required to use soft nutrient soil with good aeration and water preservation. The soil is composed of 30% of surface soil, 30% of decomposed sawdust, 30% of fine sand, and 10% of decomposed fine soil manure. It would be much better to fully use decomposed sawdust in hilling. The thickness and width of the soil and the quality of the nutrient soil and film are directly related to the quality of roots at the base of the new shoots and the difficulty of removing the soil later, which must be carried out strictly according to the requirements.

④ Management. During the growth period, the soil in the planting trench and the soil for hilling should be kept moist. In addition to the irrigation before hilling each time, it is also needed at the right time in drought. In June, spread 6–10 g of urea per m^2 before hilling in combination with water. Spray 0.5% Potassium dihydrogen phosphate twice at intervals of 10–15 days in July to promote the plants' robust growth. Pay attention to the prevention and control of pests and diseases. Leave one or two strong branches (to be used as the new plant stem for horizontal layering in the following spring) of the new shoots growing directly from the base of the mother plant and remove the other new shoots early.

⑤ Ramet cutting. This happens in spring or before the soil freezes, depending on local climatic conditions. In colder areas, it is advisable to perform ramet cutting before the soil freezes. During the ramet-cutting process, all the soil in the planting trench will be removed to reveal the horizontally pressed plant stem and the root system at the base of the annual shoots of the plant stem. Cut each rooted annual shoot at the base and leave a short stake of 1 cm to make it a dwarfing rootstock seedling. After grading the rootstock seedlings, store them in the cellar and put some wet sand at the root to help the seedlings overwinter, then plant them in the following spring. The pressed plant stems, the rooted short stakes left on the plant stems, and the annual shoots that haven't rooted on plant stems should be left instead of being cut. Give them proper soil for hilling and irrigate them to help them overwinter. When the buds germinate in the following spring, appropriately remove the soil and vaguely reveal the plant stem and short stakes to make the new shoots growing on the short stakes come out through the soil; the one or two strong branches at the base of the mother plant left a year ago should be cut short to 50 cm in length during winter pruning. When most of the buds are germinating, widen the trench and press down the new plant stem and the old plant stem parallelly at a spacing of 10 cm. When the shoots of the new and old plant stems are about 15 cm high, repeat the hilling, management, and ramet cutting work of last year. According to foreign reports, the stem of dwarfing rootstock mother plant can be used continuously for more than 15 years, and the quantity and quality of rootstock seedlings produced can be increased as the stem thickens and grows robustly. The stem of dwarfing rootstock mother plant has a high application value in the propagation of vegetative dwarfing rootstock seedlings in China.

(2) Cutting propagation.

It is often used on pear dwarfing rootstock. In spring, the soil should be deeply plowed, harrowed, and ridged at a spacing of 50 cm. Cover the ridge surface with the mulch film. Cut the rootstock branches into 15–20 cm long with the base in the shape of a horse's ear, seal the cut on the top of the branches with wax, and insert the cutting obliquely into the mulch-covered ridge surface at intervals of 15 cm. After the branches have rooted and the above-ground parts have started to grow, remove the mulch film and manage the branches according to the general raising method.

(3) Propagation of vegetative intermediate dwarfing rootstock seedlings.

The seedling rootstock is used as the basal rootstock (also known as the rooted rootstock). The branch that is grafted with dwarfing rootstock and has a certain length is used as the interstock. The interstock seedlings grafted with the pome fruits cultivar are called dwarfing interstock fruit seedlings, while the rootstock seedlings that are not grafted with the pome fruits cultivar are called dwarfing interstock seedlings. There are three propagation methods for dwarfing interstock seedlings: The conventional propagation method, the segmental grafting method, and the double stem grafting method. The better length of the apple dwarfing interstock segment is 20–35 cm, but in the same nursery, the variation of length should be less than 5 cm, that is, 20–25 cm, 25–30 cm, or 30–35 cm, respectively, so as to ensure that the trees are similar in growth after transplanting. Dwarfing interstock seedlings of pome fruits grow vigorously. The best length of the interstock segment is 30–35 cm. The dwarfing effect will be poor if the length of the interstock segment is shorter than 20 cm. The best length of the interstock of the pear is 15–20 cm.

In addition to that, pear dwarfing interstock can also increase the branch propagation coefficient by top grafting on trees that are five to six years old. Prune the mother tree heavily every year, fertilize and irrigate more to promote vigorous growth of it.

(II) Breeding of Rootstock Seedlings of Stone Fruit Trees

1. Collection and Handling of Rootstock Seeds

From July to August, pick the fruits of rootstocks of *Prunus davidiana*, *Amygdalus persica* L., *Armeniaca sibirica* (L) Lam, etc. when they are ripe. After the fruits of *Prunus davidiana* are picked, take out their seeds by the method of water washing and softening. Then dry and store them for use after roguing. The fruits of *Amygdalus persica* L. and *Armeniaca sibirica* (L) Lam can be eaten fresh or taken out of the seeds. In addition, their seeds can be processed and collected after being washed and dried.

2. Identification of Seed Viability

Refer to the method for identifying the seed viability of the seeds of pome fruits.

3. Seed Pregermination

The rootstock seeds of stone fruits must have a certain period of after-ripening to germinate after the dormancy is broken. The most commonly used method is sand storage stratification. Before sowing, the seeds without stratification in time can be soaked with warm water that is made from two parts of boiled water and one part of cold water or directly heated with boiled water and soaked for three to five days to pregerminate.

Seeds that have stratification or are being soaked should be pregerminated before sowing in spring. Put the seeds that are mixed with a small amount of wet sand in a place that is leeward, sunny, and warm. Fasten them with plastic film during the daytime and increase the mulch at night to keep the temperature at 15–20°C. Turn them once or twice a day, then sow them when 20% of the seeds germinate.

Seeds that have not been pregerminated or have not yet germinated after pregermination can be gently cracked open to eliminate seeds that can't germinate, such as the seeds without seed kernels or with seed kernels that are deformed, or blighted and small seeds. Using the seed kernels to sow will make the seedling emergence rate high.

4. Soil Preparation and Sowing

The sowing period, sowing amount, and sowing method refer to that of the pome fruits.

5. Management of Rootstock Seedlings

(1) Remove the mulch. The mulch film covered after sowing should be

checked frequently and broken in time after the seedlings emerge so that the seedlings can be exposed outside the film to prevent them from bending, etiolation or drying. For seedbeds with centralized seedlings, the mulch film should be removed in time after the seedlings emerge from the soil. Those seedlings covered with the greenhouse film should first be hardened by ventilation. The removal of the film can be performed only after the seedlings adapt to the outside environment, preferably on a cloudy day or in the evening.

(2) Thinning and final singling. Thinning happens when the seedlings grow two or three true leaves. Remove the seedlings that are densely packed, weak, and infested by diseases and pests, then promptly transplant seedlings in the place where the seedlings are lacking. Final singling happens when the seedlings have four or five true leaves at a plant spacing of 10–20 cm. For the seedbeds with centralized seedlings, seedlings can be planted in the nursery field when they grow one or two true leaves.

(3) Soil management. Thinning and final singling should be performed with intertillage to prevent the roots from being exposed and the seedlings from dying of wind leakage. To facilitate the survival of seedlings, transplanting seedlings should be promptly irrigated and hardening of seedlings should be carried out when seedlings have 5–7 true leaves. However, from May to June, since the weather is relatively dry and seedlings grow faster during this period, pay attention to irrigation. Fertilizer should be applied once or twice combined with irrigation, and each time spread 75–150 kg urea per hectare. If the seedlings are weak, spread fertilization again in early July.

Spring intertillage can improve the ground temperature, which is conducive to the growth of seedlings. Timely intertillage after the rain and irrigation can prevent soil compaction and overgrown weeds. The intertillage depth is 3–5 cm, which should be shallow in the early stage and properly deepened in the late stage.

(4) Topping and treatment of sublateral shoots. Rootstock seedlings of peach and apricot grow faster and are prone to have sublateral shoots, so topping should be performed about a month before grafting when the seedlings are 30 cm high so as to promote its thickening growth. The sublateral shoots growing on the plant

stem that is 10 cm high from the ground should be pruned early with the base leaves left to facilitate grafting, while the other sublateral shoots should all be retained to expand the leaf area and increase nutrient accumulation.

(5) Disease and pest control. In spring, seedlings are prone to have seedling blight and damping-off, especially in the case of low temperature and high humidity, which can cause a large number of dead seedlings. The prevention and control method is to disinfect the soil after seedlings emerge from the ground by spreading powder or spraying, and then hoe the soil shallowly. If the blight occurs, remove the diseased plants in time, make shallow trenches on both sides of the seedling ridge, and irrigate the roots with a 200-fold dilution of ferrous sulfate or 500-fold dilution of 65% Zineb wettable powders.

II. Production of Fruit Tree Seedlings

(I) Grafting Method and Post-grafting Management for Seedlings of Pome Fruit Trees

1. Preparation Before Grafting

(1) The selection of rootstock. There are many kinds of rootstocks of pome fruit trees, which are suitable for different places. The selection of rootstock should consider the following conditions: ① It should have good grafting compatibility with the cultivated cultivar and a good influence on the growth of the scion; ② It should have good adaptability to the environmental conditions of the cultivation area and strong resistance to pests and diseases; ③ The rootstock seedlings should have abundant sources and can propagate easily; ④ It should have special needed traits, such as standard rootstock, dwarfing rootstock, etc.; ⑤ It should have a well-developed root system and a good ground fixation.

(2) The selection of scions. The scions that are used as the mother trees should be collected from those adult plants with pure cultivars, robust growth, high and stable yields, known quality, and performance free of quarantine objects and virus diseases. Generally, the strong and clean vegetative shoots full of buds that grow around the periphery of the canopy or the fruiting mother branches are used as the

scions, preferably in the middle and upper parts of the branches. Annual branches are often used in spring grafting; mature shoots in the same year are used in summer grafting; spring shoots with full growth in the same year are selected for scions in the autumn grafting. If there is no mother plantation, it should be taken from the mature trees of the identified superior cultivars. The annual branches used in spring grafting should be pruned in the dormant stage, while the scions used in summer and autumn grafting can be pruned when they are needed. After the collection of the scion, prune the segments without full bud eyes on the upper and lower ends of the branch. Then, make every 50–100 of them into a bundle, mark them with the cultivars, and store them for later use. The leaves of the scions in the growth period should be pruned immediately after collection. A small section of the petiole connecting to the bud should be left to be wrapped with a wet cloth to preserve moisture. The scions to be stored should be put in a place with a low temperature of 4–13°C, 80%–90% relative humidity, and appropriate air permeability to prevent the scions from mildewing and losing water. In some cold areas in northern China, the scion is often stored in cellars or buried in sand or soil indoors. Generally, the scion can be stored for two months by checking once every 7–10 days to remove rotten branches. The scions to be transported can be wrapped with wet paper, and then covered with plastic film, the ends of which should have some gaps for ventilation and remove excess moisture. Then the scions can be transported after packing. Great attention should be paid to cooling, moisture preservation, and transporting speed in the transportation of scions in summer and autumn to prevent them from rotting. After being transported to the destination, open the package immediately and bury the scions with wet sand in a cold place. If the number of the scions is small, just spray them with water and then wrap them with a wet cloth or soak their base in water.

(3) Rootstock treatment before grafting. The base branches of the rootstock used for grafting should be pruned, and the left rootstock should be sprayed with a 1,500-fold dilution of 80% trichlorfon wettable powders once to kill pests of Limacodidae to facilitate grafting. It should be irrigated once every three or four days and have intertillage once before grafting to facilitate the departure of skin and

healing.

2. Grafting Period

(1) Spring grafting. It happens in March and April when the rootstock begins to germinate and its skin is just peelable. The branches and bud slices with xylem can be used to cultivate qualified graftings in that year. Collect the scions that have not germinated for the grafting, and the grafting process should be finished before a large number of rootstocks germinate.

(2) Early summer grafting. The budding happens from mid-May to early June when the cortex of rootstocks and scions is stripped. After the grating is finished successively, cut the rootstocks and cultivate them into seedlings in that year. The rootstocks of the seedlings sown and cultivated that year cannot be cut immediately after budding. It is better to break the upper part of the rootstocks to avoid the death of rootstock seedlings.

(3) Summer and autumn grafting. The budding happens from July to August with an average daily temperature higher than 15°C. It can last until mid-and-late September in central and northern China. No germination occurs during the year of budding, and the graftings will be cultivated after cutting the rootstock in the following spring.

3. Grafting Method

(1) Budding. Budding is divided into two categories, including budding with xylem and budding without xylem. In the period when the cortex can be stripped, the bud slices without xylem or with a little xylem should be used for grafting, while in the period when the cortex is not easy to be stripped, only chip budding with xylem can be used.

① T-form budding: Transect a scion branch through the cortex with a budding knife at 0.5 cm above the bud, and then obliquely cut into the xylem along the direction of the branch at 1.5–2 cm below the bud. The length of the oblique cut should be longer than the transverse cut. Then, pinch the bud slice with two fingers to take it. Transect the rootstock at the smoother part, and then make a longitudinal cut in the middle of the transverse cut to make a T-form cut. Split open the interface with the tip of the knife handle, and insert the bud slice gently from the top to the bottom so that the upper edge of the bud slice and the transverse cut of the rootstock can be closely connected (There is also a method known as "Dot-and-Dash" budding method that is splitting open the cortex of the rootstock in the middle of the transverse cut with the tip of the knife and then gently inserting the bud slice from top to bottom), and finally tie them tightly with a strip of plastic film, of which the petiole can be exposed or wrapped tightly.

② Chip budding: When cutting the scion bud, hold the scion upside down and first cut obliquely downward at the point 1.5 cm above the bud. The scion should be about 2 cm in length with a little xylem. Then, cut obliquely again in place 0.5 cm above the bud at an angle of 45° to the branch and make the second cut meet the bottom of the first one. After that, take the bud slice. The method of cutting the rootstock is the same as cutting the scion, but the incision is a little longer than that of the scion bud. Then, insert the bud slice into the interface with the cortex about a line in width revealed at the upper end of the bud slice. The rootstock should be aligned with the cambium of the scion bud, or be aligned with one side of the cambium when the rootstock is thick. At last, wrap the rootstock tightly with a strip of plastic film.

(2) Stem grafting. Stem grafting is usually performed with scions and needs more difficult operation techniques than budding. Stem grafting mainly happens during the spring when the sap begins to flow and the buds have not yet germinated. Stem grafting has many methods, such as cut grafting, lap grafting, side grafting, tongue grafting, rind grafting, wedge grafting, etc. There are four methods commonly used in production.

① Cut grafting: The scion used for grafting is about 6–8 cm long with 2–4 buds. The lower part of the scion should be cut into two opposing bevels with a long-cut surface of about 3 cm and a short-cut surface of about 1 cm. The surfaces should be smooth and the top of the scion should be cut into a small bevel. Prune the rootstock at the smooth part of the base that is to be grafted, flatten the section, and cut a short bevel on one side of the section, make a longitudinal cut downward from here with a little xylem. The length of the cut is the same as the long-cut surface of the scion. Then, insert the long cut surface of the scion into the rootstock interface, which should be aligned with both sides or one side of the cambium, leaving the upper end of the long cut surface slightly exposed, and bind it tightly with plastic strips. If there are few scions, cut grafting with single-bud scion can also be applied.

As for the single bud cut grafting method, please see Fig. 4-18.

Fig. 4-18 Single Bud Cut Grafting Method

A. Cut the scion. Put the prismatic side of the scion upward with the flat side down close to the index finger and cut forward obliquely in place about 1 cm below the budding point at an angle of 45°. Then, overturn the flat side of the scion to cut the cortex forward from the place near the budding point to the cambium with a smooth cut surface, and the white part of the xylem is exposed slightly. Later, turn the scion to make the budding point upward, then cut off the scion about 0.2 cm above the budding point and put it into a basin of clean water.

B. Cut the rootstock. Select the smooth side of the rootstock at 6–8 cm high from the ground, and cut the upper part of the rootstock obliquely at an angle of 45°. Then, make a longitudinal cut along the junction of the cortex and the xylem below the bevel. The white xylem of the cut surface should be slightly exposed. The length of the cut surface depends on the length of the scion. If the scion is cut above the budding point, the cut surface would be better to be as long as the bud, while the other normal cut surfaces of the single-bud scions should be about 2 cm shorter than the buds.

C. Insert buds and wrap. The scion bud used and the rootstock cut surface should be of the same size and length so that it can be inserted into the rootstock interface to make its base tightly connected with the bottom of the rootstock interface and make the cambiums of the rootstock and scion interface closely connect. If the rootstock is thicker than the scion, the cambium of the rootstock should be aligned with one side of the cambium of the scion to make the buds live. Finally, wrap it with plastic film tape and seal it with wax.

② Wedge grafting: This method is often used when the rootstock is thick. The scion should have 2–4 buds or even more. Cut the base of the scion into two opposing bevels that are 3 cm long with one side of the bevel a little thicker. Saw off the upper end of the rootstock in a smooth place at a certain height from the ground, flatten the sawn surface, use a knife to split a vertical incision in the middle of the rootstock with a depth of more than 3 cm, pry open the incision and insert the scion with the thicker side of scion outward to be aligned with one side of the rootstock cambium. With the scion slightly exposed, wrap it tightly. When the rootstock is thick, two scions can be grafted at the same time (Fig. 4-19).

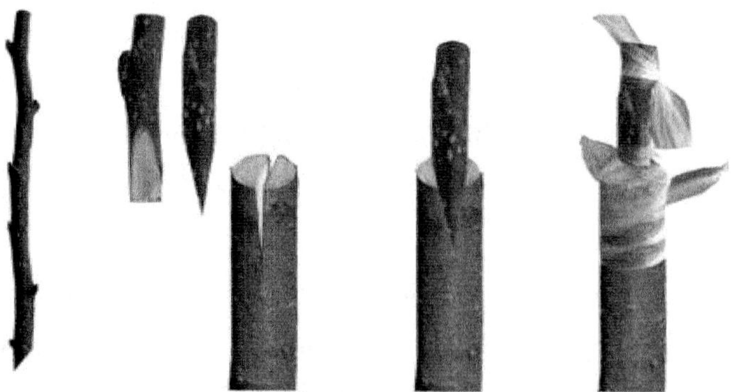

Fig. 4-19 Wedge Grafting Method

③ Side grafting: Perform stem grafting on the side of the rootstock. Cut about 3 cm in length at the base of the scion and 1.5 cm in length opposite it. Cut downward obliquely at the smooth grafting part of the rootstock at an angle of 20°–30° with a depth of more than 3 cm, and the cut should not exceed half the thickness of the rootstock. Insert the scion into the rootstock with its long cut surface facing inward and the cambiums should align. With the cutting edge of the scion bevel slightly exposed, wrap it tightly. Then, prune off the rootstock at about 1 cm above the interface with the cut slightly tilting outward.

A. Small bud side grafting method (Fig. 4-20).

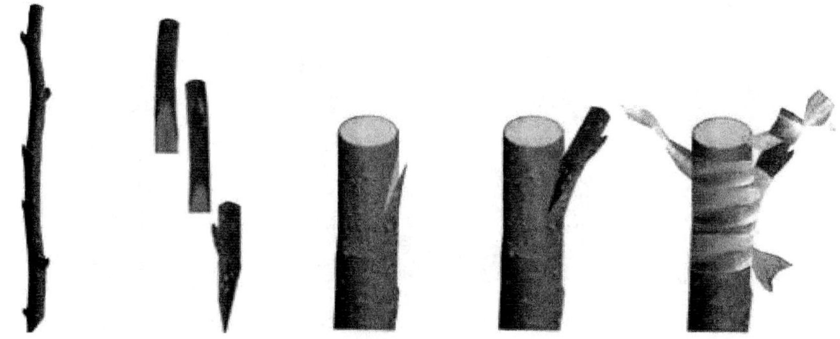

Fig. 4-20 Small Bud Side Grafting Method

a. Cut the scion. With the budding point of the scion facing upwards, cut obliquely at an angle of 45° at about 1 cm below the bud, and then make a flat cut forward at about 0.5 cm above the bud to cut off the shoot with xylem and put it into a basin with clean water. With a length of 1.2–1.5 cm, the long small buds are thinner and longer than the normal small buds used for budding.

b. Cut the rootstock. Cut the cortex on the smooth side of the trunk at 10-20 cm from the ground without cutting off the upper part of the rootstock, which requires a smooth cut surface. The cut should be deep enough to reach the cambium, and the cut surface should be long enough to allow the buds to be put in. Then, cut off 1/3 of the cortex that was cut before facilitating the wrapping and bud germination. The leaves on the rootstock at the upper and lower part of the interface should be retained as much as possible to improve the survival rate.

c. Insert buds and wrap. Insert the bud slice cut before into the rootstock incision, make the bases of the rootstock and scion closely attached, and make the cut surfaces align. Wrap it with plastic film tape with the bud eye exposed.

B. Single bud side grafting method (Fig. 4-21).

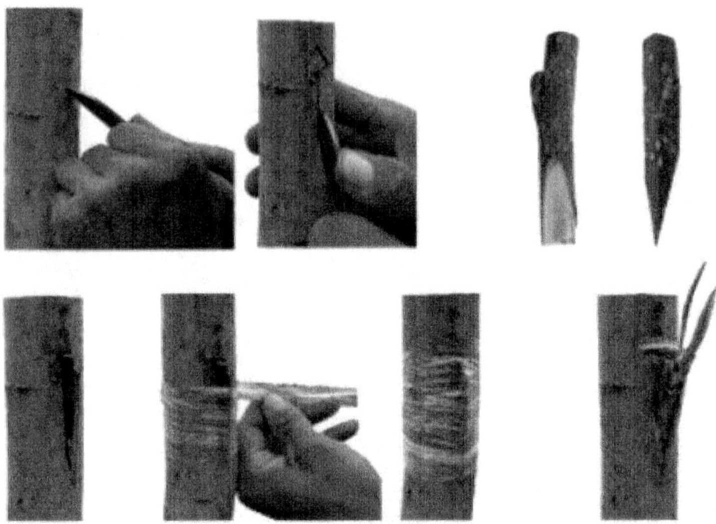

Fig. 4-21 Single Bud Side Grafting Method

The way of cutting the scion in this method is cutting the scion above the budding point, the same as that in the single bud cut grafting method. The ways of cutting rootstock, inserting buds, and wrapping refer to the small bud side grafting method.

④ Rind grafting. It is also known as bark grafting. Rind grafting is applied when the rootstock is thick and the bark of the rootstock slips. It is usually used for top grafting of pome fruits to achieve cultivar replacement. Cut or saw the rootstock with a smooth cut surface at the grafting part. Make a long cut surface that is 2–4 cm in length on the scion from the opposite side of the terminal bud, and then cut two shorter cut surfaces that are 0.5–1 cm in length behind the long one. Insert the scion between the cortex and xylem of the rootstock with its longer cut surface turned inwards. The upper end of the cut surface should be slightly exposed, about 0.5 cm in length, then wrap it tightly with plastic strips.

4. Management of Graftings

(1) Check the survival and perform re-grafting. About 15 days after budding, the survival can be checked and the tape or wax can be removed. If the bud pieces on the rootstock are fresh and the petiole drops at a touch, it means the grafting has survived, while the grafting with dry, shrinking, and brown scion buds is dead. Those grafted in spring can be re-grafted in summer and autumn, while those grafted in summer and autumn can be re-grafted immediately. The germination of branches grafted by stem grafting means that they have survived. Those failing to survive can be re-grafted in the next year or be grafted by budding as a remedy. Gradually remove the tape or wax after the grafting sprouts. The scion grafted by stem grafting, especially by rind grafting, is susceptible to wind breakage when it grows vigorously, so a pillar should be set up to support it.

(2) Cut rootstock and remove sprout. The seedlings that survive through budding should be pruned their rootstock parts above the scion buds without leaving any residual stakes. However, in windy areas, about 10 cm of live stakes can be left to tie the new shoots, and then be cut after the lignification of the base of the shoots. The cut surface of the cutting rootstock should be smooth and slightly oblique in the opposite direction of the scion bud. After cutting rootstock, the sprouts and sprout tillers should be removed as early as possible to ensure the healthy growth of the graftings. If the scion of the stem grafting germinates more than one shoot, leave one of them and remove the others.

(3) Field management. Weed and loosen the soil in time to increase the ground temperature and promote the development of the root system of seedlings. From May to June, apply urea of about 150 kg per hectare and then irrigate promptly; after July, spraying foliar fertilizers is recommended. Spray 200-fold dilution of potassium dihydrogen phosphate once at intervals of 10–15 days, twice or three times in total, to promote the fullness and strength of the seedlings. The seedlings should be irritated from May to July when they grow rapidly. After August, water should be controlled to avoid unfavorable-delayed senescence and excessive growth. Pay attention to the prevention and control of leaf rollers, aphids, and early leaf disease. Diseased plants that suffer mosaic disease and scar skin disease should

be removed and destroyed.

(II) Grafting Method and Post-grafting Management for Seedlings of Stone Fruit Trees

1. Collect Scions

The scions should be collected from the mother trees with pure cultivar, robust growth, high and stable yields, performance free of quarantine objects, severe diseases, and pests. The scions used in spring stem grafting can be collected according to the winter pruning, while the scions used in summer budding are preferably collected as they are needed.

2. Grafting Techniques

Spring grafting often happens in March and April when the sap of rootstock flows and the germination hasn't begun. In this situation, stem grafting methods like cut grafting, wedge grafting, side grafting, etc. can be used. If the rootstock germinates while the scion doesn't, chip budding can be used.

Budding in summer is better performed from late July to mid-August. The scion buds will germinate easily and can hardly overwinter if the grafting happens earlier. Meanwhile, the rootstock seedlings of peach and apricot become thick and grow rapidly in August and September, thus the scion buds grafted earlier without germination are easy to be pressed from both sides of rootstock layers, which will make it hard for them to germinate next year after cutting rootstock and will make the rootstock seedlings of peach stop germination early. However, if the grafting happens too late, when those rootstock seedlings have stopped growing, they can hardly heal their cuttings and will have excessive gummosis. It will also cause a low survival rate of grafting. The graftings will not grow vigorously even if they survive after grafting. T-form budding is usually used in summer grafting, but the T-form cut should be a little larger because the cortex of the peach rootstock is soft. The method of budding with xylem can be used when the bark of the scion and rootstock doesn't slip. The specific methods can refer to the grafting methods of the pome fruit trees.

3. Management of Graftings

The management of fertilizer and water of graftings can refer to that of rootstock seedlings. Great attention should be paid to watering the sprouts promptly after spring grafting. Waterproofing and surface drainage should be done well in the rainy season. After August, control the application of the nitrogen fertilizer and water the seedlings less to make them strong so that they can overwinter easily. As for the methods of removing the film or wax, cutting rootstock, bud picking, and prop setting can refer to those of the pome fruit trees.

The shoots of stone fruit graftings grow rapidly, with the sublateral shoots growing twice to four times a year. Therefore, trimming in the nursery is an important measure in raising those kinds of fruit tree seedlings. When the shoots grow to about 80 cm in height, leave 60–70 cm for topping and determining the erect main trunk. At the same time, prune all the sublateral shoots within the height of 30 cm from the ground and leave other sublateral shoots to grow. From late August to early September, select three or four sublateral shoots with appropriate orientation and robust growth at the 40–60 cm height of the trunk to be cultured as the main branch, and adjust its base at an angle of 60°–70°. As for other sublateral shoots, deal them with branch softening and nipping, increase their angle, and remove them if they are too dense.

If the sublateral shoots are used for trimming in the nursery, the planting spacing of rootstock seedlings should be increased with a row spacing of more than 60 cm and planting spacing of more than 30 cm.

The seedlings are susceptible to being damaged by aphids and leaf rollers. From mid-April to late May, the seedlings can be sprayed with the 1,000-fold dilution of 75% phoxim or the 1,000–1,500-fold dilution of 50% Sumithion Emulsifiable Concentrate. In the area with oriental fruit moths that affect the peach shoots, a 1,000–1,500-fold dilution of 40% phoxim emulsifiable concentrate should be sprayed once every 10–15 days and be sprayed successive two or three times during the occurrence of the moths. The affected peach shoots should be removed in time and burned together.

Review and Reflection Questions

(1) What are the methods of identifying the viability of the seeds?

(2) What are the specific practices of propagating dwarfing rootstock seedlings of pome fruits by horizontal layering?

(3) How to perform the seed stratification (sand storage)?

(4) What are the grafting methods?

(5) What are the key points in the management of rootstock seedlings of pome fruit trees?

(6) What are the key points in the management of rootstock seedlings of stone fruit trees?

Module 5 Horticultural Plant Seed Testing

Learning Objectives

(1) Make the students master the principles and techniques of seed quality testing.

(2) Be able to understand the concept and the main contents of seed standardization.

(3) Be able to perform the seed testing correctly according to standard steps and methods.

(4) Be able to perform correct seed grading of the tested seeds and fill in the report on seed testing accurately according to the results of the seed testing and the relevant national standards.

Learning Scenario I Testing Rules of Horticultural Plant Seeds

I. Meaning and Functions of Seed Testing

(I) Meaning of Seed Testing

Seed testing refers to a technical operating procedure of testing or providing services for one or more quality characteristics of crop seeds according to the standard procedures and methods of seed testing and comparing the results with the requirements.

China has been performing seed testing in accordance with the standards of the International Seed Testing Association (ISTA). The seed testing procedures or methods are called testing rules, such as *Rules for Agricultural Seed Testing* (GB/T3543.1—3543.7—1995) issued by the State Bureau of Quality and Technical Supervision. The quality index consists of test items and test values. The object of seed testing is the crop seed. There are 124 cultivars (varieties) stipulated in China, mainly including botanical seeds (such as soybeans, cotton, etc.) and botanical fruits (such as rice, wheat, etc.).

(II) Functions of Seed Testing

High-quality seeds are the basis of high yields and harvest in agriculture. The purpose of seed testing is to choose high-quality seeds for agricultural production so that farmers won't be cheated by fake and unqualified seeds and can have standard seeds.

There are various functions of seed testing. On one hand, it is one of the effective methods for seed enterprises to control seed quality. As the main link in the production technology of seeds used for field production, seed testing plays an important role in seed selection and seed retention, as well as the processing, transportation, storage, sales analysis, pricing of seeds, etc. Besides being an important part of the seed enterprise quality management system, it is also a necessary condition for seed enterprises to get a license for seed production and operation. On the other hand, it is not only a very effective means of supervising the market and providing technical services but also an important basis and technical support for seed administrative law enforcement and dispute settlement on seed quality. Specifically, seed testing mainly has the following functions.

(1) Improve seed quality. Through seed testing, the testing agencies can continuously obsolete unqualified seeds to improve the seed quality. In the critical period of seed production, the testing agencies will organize inspectors to perform field inspection in the seed production field, and then the weeds and unqualified plants can be found and removed timely, the production conditions like field isolation can be further inspected and affirmed, and the seed production can be in

line with the rules and requirements so as to improve seed quality.

(2) Checking. Through the strict test and identification of seed quilty, the quality of the seeds for storage can be ensured; on the other hand, the quality of the seed for delivery can be ensured. It can help prevent unqualified seeds from being sold in the market. What's more, it can help check if the supervision of seed quality is done well so that farmers won't be cheated by fake and unqualified seeds.

(3) Prevention. In seed production, through the control of cultivars like parents, the re-testing of purchased seeds, the testing of seeds during storage and transportation, etc., unqualified seeds can be prevented from being sold in the market and farmers will not get unqualified seeds.

(4) Supervision. Seed testing is an important technical support for seed management. By means of supervision and random testing of seed quality, agricultural administrative law-enforcement agencies will heavily punish people who produce and sell fake and unqualified seeds and can ensure that the seeds used for production in agriculture are qualified.

(5) Providing the testing reports. After the seed testing, the testing agency will issue a normative report of seed analysis, according to which the agricultural administrative law enforcement agency can deal with the disputes on seed quality and punish people who produce and sell unqualified seeds. In addition, it is also an authorized certification in the domestic and international seed trade.

(6) Seed testing. Apart from maintaining high-quality seeds and eliminating unqualified seeds, seed quality testing can effectively protect the legitimate rights and interests of seed producers, operators, and users. This is beneficial for resolving disputes over seed quality for building a harmonious and orderly society.

II. Contents and Procedures of Seed Testing

(I) Contents of Seed Testing

Rules for Agricultural Seed Testing is currently implemented in China, which has specifically stipulated the contents and methods of seed testing. There are many contents of seed testing, mainly divided into three parts, including sampling, testing, and analyzing the report. Sampling is the first and critical step of seed testing, consisting of the sampling procedures of seed lot, normatively filling in the sampling list, sample dividing procedures in the laboratory (including taking the test samples and backing up samples), and sample preservation. The testing part includes the cleanliness analysis, germination test, variety authenticity and purity identification, moisture determination, biochemical determination of seed viability, weight determination, seed health determination, etc., among them, cleanliness analysis, germination test, variety authenticity and purity identification, and moisture determination are items that must be tested, but other items like biochemical determination are not. Analyzing the report includes the allowable error analysis of the determination value, the summary of results, filling and submission, issuing, and sending.

(II) Procedures and Methods of Seed Testing

The seed testing procedure consists of three steps, including sampling, testing, and analyzing the report.

1. Sampling

It consists of steps like sampling procedures for seed lots, sample dividing procedures in the laboratory, sample preservation, etc.

2. Testing

It includes four items that must be tested: cleanliness analysis, germination test, variety authenticity and purity identification, and moisture determination. Some items that don't need to be tested, like biochemical determination of viability, weight determination, seed health determination, coated seeds testing, etc.

3. Analyzing the Report

This step includes checking the allowable error, reviewing the conditions for issuing seed testing results, and issuing the seed testing report.

The specific contents and methods are as follows.

(1) Cleanliness analysis.

It refers to inferring the composition of the seed lot according to the weight percentage determination of different components of the tested samples and the characteristics of the sample mixture.

The test sample will be divided into three components: pure seeds, other seeds, and inert matter. Then, determine the weight percentage of each component. Identify all kinds of seeds and inert matter in the samples as much as possible.

$$\text{Seed cleanliness}=\frac{\text{Pure seed weight of the crop}}{\text{Total seed weight of the sample}}\times 100\%$$

(2) Germination test.

A germination test is used to determine the maximum germination potential of the seed lot. The quality of different seed lots can be compared through it, which can also help determine the sowing value.

Only the pure seeds identified in the cleanliness analysis can be used in the germination test, which should be carried out under suitable moisture and specified germination technical conditions. After the seedlings are suitable to be evaluated, check those repeated seedlings and calculate the proportion of the seedlings of different kinds according to the result report.

It should be noted that the new regulation on the seed germination standard is different from the original one. The original regulation stipulates that those seeds germinating with their young roots being as long as the seeds, young buds being as 1/2 long as the seeds, or young roots and young buds being as long as the diameter of the seeds are all normal germinating seeds; while the new one stipulates that the germinating seed must grow into a normal seedling with a root system, seedling axis, terminal bud, cotyledons, and complete coleoptile.

The germination test result can be expressed in terms of germination potential and germination rate.

$$\text{Germination potential}=\frac{\begin{array}{c}\text{Number of normal seedlings in the early}\\ \text{stage of the germination test (under the}\\ \text{specified conditions and dates)}\end{array}}{\text{Number of tested seeds}}\times 100\%$$

$$\text{Germination rate} = \frac{\text{Number of normal seedlings in the late stage of germination test (under the specified conditions and dates)}}{\text{Number of tested seeds}} \times 100\%$$

III. Testing Report

The seed testing report is a certificated sheet with the testing result on it, which is obtained from the seed testing agency through sampling and testing according to the national standard for agricultural seed testing.

(I) Conditions for Issuing Seed Testing Reports

(1) The agency issuing the seed testing reports should be qualified and currently engaged in the testing work. The agency issuing the testing reports must have inspectors who have got licenses and specialize in seed testing. There must be specialized testing instruments and testing sites in the agency. Besides, the agency must be currently engaged in the testing work. It should have passed the professional assessment and obtained the green certification mark presented by CASL (China Accredited Seed Laboratory).

(2) The tested seed must belong to one of the cultivars listed in the seed testing regulation, which stipulates the scope of the tested crop seeds. Only the testing reports of the seeds among the 124 species listed in the testing regulation can be issued. Reports of those seeds beyond the scope can't be issued.

(3) The seed lot should conform to the regulation. Each container of the seed lot must be sealed with a batch number, and the seed lot must be divided as required. Only in this way can the testing report be linked to the seed lot. Once there are problems with the seed quality, the unqualified seeds can be traced. This makes the seeds traceable.

(4) The samples sent for testing are selected and processed as required. The results of the testing report can only be obtained from the same seed lot and the same test sample. The samples for moisture content determination must be sealed, packaged with moisture-proof measures, and inspected promptly.

(5) The test must be carried out in accordance with the methods stipulated in the national standard agricultural seed testing regulations. The test must adopt the methods stipulated in the regulations. For example, moisture determination must adopt a method such as the low-constant-temperature oven drying method, high-temperature oven drying method, or high moisture pre-drying method. The rapid determination method cannot be adopted.

After testing, the results should be calculated in accordance with the regulations and the seed testing result report should be filled out. If some items are not tested, "N" or "Not Tested" shall be filled in the corresponding blanks. If the sampling and testing are carried out or supervised by the same testing agency, a seed lot testing report should be issued; if the sampling is carried out by another testing agency or individual, that is, the sampling and testing are not carried out by the same agency, a seed sample testing report should be issued and the statement that the report is only responsible for the samples sent for testing shall be indicated.

(II) Contents and Requirements of the Testing Report

1. Contents of the Testing Report

The content of a complete testing report shall include the following items:

(1) Title;

(2) Name and address of the testing agency;

(3) Name and address of the tested unit;

(4) Name of the sampling unit and the sealing unit;

(5) Species and cultivar name of the crop;

(6) Unique reference number of the testing report and the CASL mark;

(7) Seed lot reference number, sample condition, and sample number;

(8) Sample quality, representative seed lot weight;

(9) Sampling date, date of sample reception, and test date;

(10) Production unit and production date of the seeds;

(11) Test items and results;

(12) Test categories, test basis, determination basis;

(13) Test conclusion;

(14) Report issuer, verifier, and compiler.

2. Requirements of the Testing Report

The testing report shall satisfy the following requirements:

(1) The testing report adopts a unified format and is printed out.

(2) The testing report shall not show signs of addition, modification, substitution, or alteration.

(3) The testing report is usually in duplicate, one for the entrusting party or the enterprise under random inspection, and the other for filing together with the original records.

(4) The testing report shall ensure that the data, results, and conclusions are accurate, objective, and true. The report shall bear the CASL mark, with the signature or seal of the authorized signatory, and the seal of the testing agency.

(5) The testing report shall be sent and registered as required. If the testing results are transmitted by telephone, fax, or e-mail, the identity of the entrusting party shall be confirmed and recorded.

(6) Ensure that only one valid report exists for the same sample within a certain period.

(7) The testing report shall be stored without damage, deterioration, or loss. The report shall be kept confidential for the user, and the storage period shall not be less than six years.

The testing results shall be calculated and shown in accordance with the methods specified in the regulations and shall be filled in according to the requirements of the testing report. If there is an urgent need to know the result of a certain measurement item before the end of the testing, an interim testing report can be issued, with the statement "The final testing report will be issued at the end of the testing." on the interim report.

(III) The Terminology and Filling of the Testing Report

1. Filling the Cover and Relevant Information on the Header and Footer of the Testing Report

(1) "Testing Report" shall be marked in a prominent position on the cover

of the testing report and the number of the report shall be filled in the upper right corner. The name and address of the unit issuing the testing report shall be the full name.

(2) The date filled shall be in the format ccyy-mm-dd. For example, 2011-08-16.

(3) The species epithet and scientific name shall be consistent with the relevant provisions of the testing regulations (GB/T3543.2–1995 Table 1). If the species epithet cannot be determined, the genus epithet can be used.

2. Filling the Content of the Test Items

(1) Cleanliness analysis: The results of the cleanliness analysis are expressed as the weight percentage of three components: pure seeds, seeds of other plants, and inert matter. The requirements for filling in the results are as follows: ① The weight percentage of the three components shall be kept to one decimal place, and the sum of the percentages of each component is 100%. ② If the weight percentage of the component is lower than 0.05%, "IR" or "Trace" shall be filled in. If the result of one component is zero, it must also be filled in, and "-0.0-" or "NIL" shall be used. ③ When the weight percentage of a certain inert matter or a certain seed of another plant reaches or exceeds 1%, the result shall be filled in and indicated on the result report. ④ The scientific names of seeds of other plants and the types of inert matter shall be filled in the report.

When measuring the number of the seeds of other plants, the actual weights, scientific names, and the number of seeds of each species found in the weight result shall be filled in the result report. Besides, the method applied to the full test, limited test, or reduced test shall be indicated. The results shall be expressed as the number of seeds detected in the actual determination sample. However, for the convenience of comparison, the number of seeds per kg or the number of seeds per unit weight shall be used and the species epithets shall adopt scientific names.

(2) Germination test: ① The results of the germination test shall be expressed as percentages of the grain number and be filled in with the nearest whole number. ② Normal seedlings, hard seeds, fresh ungerminated seeds, abnormal seedlings, and dead seeds shall be classified and filled in as percentages, with

the sum of 100%. If any of the results is zero, "-0-" shall be filled in the column. ③ Generally, additional instructions in the form shall be filled in with germination bed, temperature, date of seed placement, test duration, treatment method, allowable error, actual error, species of abnormal seedlings, etc.

(3) Moisture: Moisture shall be expressed as the arithmetic means of two replicates, but the difference between the two replicates shall not exceed 0.2%; otherwise, two replicates shall be re-measured. The results of moisture determination shall be filled in the blanks specified in the result report with an accuracy of 0.1%.

(4) Variety purity identification: This shall include purity percentage, testing methods, the number of identified plants, etc. The purity percentage shall be kept to one decimal place.

(5) Weight determination (Thousand-seed weight): Weight determination shall be filled in the following format: "Weight (Thousand-seed) ____g".

(6) Seed viability tetrazolium test: This shall be filled in the following format: "Tetrazolium Test:____% viable seeds". The result shall be kept as an integer.

(7) Seed health determination: This shall be expressed as the weight percentage of infected seeds or the number of pathogens in the test sample, and the scientific names of the pathogen shall be filled in the report. Information on the determination method shall also be filled in.

(8) Coated seeds: Types of coated seeds shall be indicated after the species epithets, and the format "Coating Pass Rate: ____%" shall be used in the report.

IV. Evaluation of Seed Quality

With the development of the seed industry in the 1980s, China formulated and promulgated a number of national standards related to the quality of seeds, such as the *Rules for Agricultural Seed Testing*, *Rules for Forage Seed Testing*, *Quality Standards for Agricultural Seeds*, and *Standards for Seed Packaging of Main Agricultural Crops*. The seed quality index system includes the content of seed quality and indexes of seed quality grading standards.

(I) Content of Seed Quality

Seed quality is the main indicator that determines whether the seed quality is good or bad. It is a concept synthesized by different characteristics of seeds, including both variety quality and sowing quality. Variety quality, also known as internal quality, is the quality related to the genetics of the seed and can be summarized in two words: authentic and pure. Sowing quality, also known as the external quality, is related to the seedling emergence after sowing and can be summarized in five words: clean, strong, full, healthy, and dry. Therefore, the content of seed quality should include the following aspects.

1. Authentic

It refers to the degree of authenticity and reliability of seeds, which can be expressed as authenticity. If the seeds lose their authenticity and are not the original superior cultivars, it can cause serious yield reduction.

2. Pure

It refers to the degree of consistency of typical characteristics of a cultivar, which can be expressed as variety purity. Seeds with high variety purity can obtain a bumper harvest due to the excellent characteristics of the cultivar, while seeds with low variety purity have obvious yield reduction due to mixing, degradation, and lack of uniformity.

3. Clean

It refers to the degree to which the seeds are clean, which can be expressed as cleanliness. High seed cleanliness indicates the amount of inert matter in the seeds is low and the number of pure seeds is high, which is conducive to storage and neat field emergence.

4. Strong

It refers to the degree of seed germination and seedling emergence, which can be expressed by viability, vigor, and germination rate. The seeds with high viability and germination rate germinate and emerge neatly, while the seeds with high vigor have a high field emergence rate and the seedlings are strong. At the same time, the sowing amount per unit area can be appropriately reduced.

5. Full

It refers to the fullness of seeds, which can be expressed as the thousand-seed weight (or bulk density). The fullness of the seeds indicates that the seeds are rich in storage material, which is conducive to seed germination and seedling growth.

6. Healthy

It refers to the degree of seed health, which can be expressed by the infection rate of pests and diseases. Pest and disease damage directly affects the germination rate and field emergence rate.

7. Dry

It refers to the degree of seed drying and storability, which can be expressed by seed moisture. Low seed moisture is conducive to safe seed storage and maintenance of seed germination rate and vigor.

(II) Seed Quality Grading Standards

The seed quality grading standards are a variety of grading regulations for seed quality based on the content of the seed quality. They are the criteria for the correct grading of seeds in the process of circulation. At present, China's seed quality grading standards are as follows: Conventional seeds, inbred line parents, and "three-line" parents that can be divided into original seeds and good-quality seeds; hybrids are divided into Grade I good-quality seeds and Grade II good-quality seeds. The seeds of different grades are divided according to four main indexes: variety purity, cleanliness, germination rate, and moisture. The minimum grading principle is adopted; that is, seeds that do not meet the specified standards of any indexes cannot be regarded as accredited seeds of the corresponding grade.

(1) The national standards are jointly formulated and issued by the Ministry of Agriculture and the Standardization Administration of China, which are used for the testing and grading of domestic seed purchase, sales, and allocation.

(2) The industry standards refer to the superior variety breeding regulations, field inspection regulations, seed testing regulations, and relevant seed quality standards of crop seeds promulgated by national industry authorities (agricultural authorities) as required.

(3) The local standards are the seed grading standards jointly formulated and issued by the competent agricultural departments of the provinces, municipalities, and autonomous regions in conjunction with the provincial standardization administrations. The local standards are mainly a supplementary provision for cultivars that do not have standards formulated by national industry authorities or agricultural authorities, serving as the basis for seed testing and grading of regional purchase, sales, and allocation.

(4) The enterprise standards are the internal standards of the enterprises based on the type of seeds they produce and operate.

The Seed Law of the People's Republic of China states that the quality management of seeds shall implement the label authenticity system based on national and industrial standards. When enterprises conduct seed testing to determine whether the seeds are qualified or not, the seeds must first meet the needs of national and industrial standards (except for those without national or industry standards), followed by local standards, and then enterprise standards. With the seed quality grading standards, the quality status of various crop seeds can be graded through uniformly stipulated quality standards. Seed producers and operators can also formulate target plans according to the seed quality standards to continuously improve the quality of seeds.

(III) Evaluation of Seed Quality

The variety purity evaluation of conventional cultivars is based on the results of the field inspection and indoor purity test. When the results of the two inspections are inconsistent, the lower result shall prevail; when the results of the two inspections are difficult to determine, the identification of variety purity in the field plot shall prevail. When evaluating the purity of hybrids, the field inspection results (including the rate of father hybrid, the rate of mother hybrid, the rate of dispersed pollen of father hybrid, the rate of dispersed pollen of mother hybrid, etc.) shall be put in the first place and see if the results meet the requirements. If they meet the requirements, the evaluation method is the same as that of conventional seeds based on the results of indoor testing or identification of variety purity in the field plot.

Among the indexes of sowing quality, cleanliness, germination rate, and moisture are the grading indexes of seed quality, and other indexes are used only as the content of evaluation. For example, the thousand-seed weight must meet the standards of the cultivar; the seed health must meet the requirements that there are no quarantine seeds infected with pests and diseases, the relative infection is mild, etc.

Review and Reflection Questions

(1) What is seed testing?
(2) What are the main functions of seed testing?
(3) What are the procedures for seed testing?
(4) What are the conditions for issuing testing reports?
(5) What are the contents and requirements of the testing report?
(6) What aspects does the seed quality include?
(7) What are the standards for seed quality grading?

Learning Scenario II Field Inspection of Horticultural Plant Seeds

I. Content of Field Inspection of Horticultural Plant

(I) Concept of Field Inspection

Field inspection refers to verifying variety authenticity, identifying variety purity, and investigating the crops' growth status, other crops, weeds, etc. in the seed production field during the production process. The purpose of field inspection is to determine whether the seed production process is in line with the requirements of the *Technological Specification of Seed Production*. Variety authenticity described in the field inspection refers to whether the cultivars for inspection are consistent with the documented records (e.g. labels, cultivar descriptions, etc.). Variety purity refers to the degree to which cultivars are typically consistent in terms of characteristics and features. It is worth noting that, instead of being a product inspection, the essence of verifying variety authenticity and variety purity in the field inspection is process control, especially for the seed production fields for hybrid seed production. Therefore, the Organization for Economic Co-operation and Development (OECD) states in the *Guidelines for Control Plot Tests and Field Inspection of Seed Crops* that the purposes of the field inspection are to check if the seeds in the seed production field show the characteristics and features of the cultivar that it claims to be (variety authenticity) and to ensure there are no circumstances that might be prejudicial to the quality of the seed to be harvested (variety purity).

(II) Purposes and Functions of Field Inspection

1. Purposes of Field Inspection

The purposes of field inspection are to check if the seeds in the seed production field show the characteristics and features of the cultivar that it claims to be and check various conditions influencing the quality of the seed to be harvested, so as to take corresponding measures to reduce residual genetic segregation, natural mutation, out-pollination, mechanical mixing, and other unforeseen factors affecting seed quality and to ensure compliance with the specified requirements at harvest.

2. Functions of Field Inspection

(1) To check the isolation of seed production fields to prevent purity reduction caused by out-pollination.

(2) To check the implementation of seed production technologies, especially roguing and emasculation. Carry out roguing strictly to prevent the influence of variant and impure plants (including foreign plants produced due to residual genetic segregation and natural mutation) on the purity of produced seeds.

(3) To check the growth situation in the fields, especially the anthesis synchronization. Through field inspection, measures for anthesis adjustment are put forward in time to prevent yield and quality reduction due to anthesis asynchronism. At the same time, harmful weeds and other crops can be promptly removed.

(4) To check the variety authenticity and variety purity and judge whether the seeds produced by the seed production fields meet the requirements of seed quality. To scrap unqualified seed production fields and prevent the impact of low-purity seeds on agricultural production.

(5) To provide a basis for seed quality certification through field inspection.

(III) Field Inspection Items

Field inspection items vary because of the different types of crop seed production fields. Generally, seed production fields are divided into conventional seed production fields and hybrid seed production fields.

In conventional seed production fields, inspection items are as follows: previous crops, isolation conditions, variety authenticity, rate of impure plants, rate of other plants, and overall conditions of the seed production field (lodging, health condition, etc.).

The success of hybrid production depends on the effectiveness of the male sterility system (including mechanical and artificial emasculation, self-incompatibility, cytoplasmic male sterility, genic male sterility, and chemical emasculation agents) and the fertile capability of the hybrid cultivar. The variety purity of hybrids can only be identified through post-control of the harvested seeds produced by the plots specified in GB/T 3543.5–1995. However, in order to ensure that the specified requirements are met and the variety purity of the seeds is maintained at the highest level, checks must be conducted on the following items: isolation conditions, suitable conditions for pollen dispersal, degree of male sterility, degree of pollen mixing, variety authenticity and variety purity of parents. Father (or mother) shall be harvested at an appropriate time.

(IV) Field Inspection Period

It is best to carry out field inspections in different growth stages (seedling stage, anthesis, maturity, etc.) several times in the seed production fields in order to obtain accurate seed quality information. When there is no condition to carry out the inspection many times, it is also acceptable to carry out an inspection in the period when the characteristics and features of the cultivar are mostly shown (such as the anthesis of hybrid seed production fields and the maturity stage of edible organs of vegetable crops). Seed production fields can be inspected several times during the growing season, but at least one inspection should be conducted during the period when the characteristics and features of the cultivar are most fully and obviously shown in order to evaluate the variety authenticity and variety purity. The most suitable period for field inspection of many crops is at the anthesis or shortly before the anther dehiscence, while some crops also require inspection of vegetative organs, and some crops must be observed before full maturity. Conventional seeds must be inspected at least once at the maturity stage. Hybrid seeds such as hybrid rice, hybrid corn, hybrid sorghum, and hybrid rape must be inspected 2–3 times during the anthesis. Vegetable crops at the maturity stage of product organs (such as leafy vegetables at the leafy head maturity stage, fruit vegetables at the fruit maturity stage, root vegetables at the taproot, rhizome, tuber, and bulb maturity stage) must have an additional inspection.

(V) Content of the Field Inspection and Requirements for Field Inspectors

1. Content of Field Inspection

(1) Conventional seeds.

Conventional seed field inspection should include the following items:

① Confirm that the seed production field meets the requirements for the production of seeds of this species.

② Check if the sown seed lots are the same as the labels claim.

③ Check if the seeds belong to the inspected horticultural cultivar as a whole (variety authenticity) and check the variety purity.

④ Identify weeds and seeds of other plants, especially those difficult to separate by processing.

⑤ Check if the isolation conditions meet the requirements.

⑥ Check the overall conditions of the seed production field (lodging, pest and disease infection, etc.).

(2) Hybrid seeds.

Although the variety purity of hybrids can only be identified through post-control of the harvested seeds produced by the plots, maximum assurance of variety purity can be achieved through the following items of field inspection:

① Ensure suitable isolation distance from sources of pollen contamination.

② Ensure a high degree of male sterility or self-incompatibility.

③ Ensure ideal conditions for transferring the father pollen to the mother plant.

④ Ensure high variety (inbred lines) purity of each combination (parent plants).

⑤ Ensure that the father is harvested before the mother.

2. Requirements for Field Inspectors

In order to ensure the accuracy of the inspection results, inspectors must be trained. They are required to pass the written assessment of field inspection principles and procedures, and the practical assessment of field inspection skills. Inspectors must be familiar with field inspection methods, field standards, characteristics and features of cultivars, methods, and procedures of seed production, as well as other aspects of knowledge. Inspectors must have the ability to confirm the variety authenticity based on the characteristics and features and to identify and quantify the impure plants in the seed production fields. Inspectors must have extensive knowledge of the cultivars for inspection, and be familiar with the differences between characteristics and features of cultivars for inspection.

II. Procedures of Field Inspection of Horticultural Plant

(I) General Procedures of Field Inspection

1. Understand the Situation and Check Seed Labels

Through interviews with producers and inspections, have a comprehensive understanding of information, such as the number of the seed production field, name of the applicant, crop, cultivar, category (grade), the name and contact method of the farmer, site of the seed production field, field number, area, details of previous crops (file), seed lot reference number, etc.

In order to confirm the variety authenticity, it is necessary to check the labels for details of the seed source. Therefore, the producer shall keep at least two labels of the seed lot, one erected in the field and the other kept for inspection; producers of hybrid seeds shall also keep the seed labels of the parents and the field distribution maps.

2. Inspect Isolation and Overall Conditions of the Seed Production Field

(1) Inspection of isolation conditions.

Based on the maps of the seed production field and surrounding fields provided by the producer, inspectors shall walk around the outer circle of the seed production field to inspect the isolation conditions (including mechanical mixing isolation during harvest and isolation from other fields that have been infected with seed-borne diseases). The isolation distance shall reach the minimum distance specified. In addition, inspectors shall observe volunteers, other crops, or weeds in the seed production field and surrounding fields because they might be sources of pollen contamination.

(2) Inspection of seed production field condition.

An overall evaluation of the conditions of the seed production field will determine whether a detailed inspection of variety purity is necessary. When inspecting the overall conditions of the seed production field, inspectors shall check conditions in the field and surrounding areas in detail. If inspectors find the seed production field where growth is blocked or poor due to different seeds sown, contaminated seeds, severe lodging, overgrown weeds, pests, diseases, or other reasons, the seed production field shall be eliminated and will be unqualified for the evaluation of variety purity.

3. Inspect Variety Authenticity and Variety Purity

(1) Variety authenticity.

Usually, after walking around the seed production field, inspectors shall inspect no less than 100 plants to ensure that they are consistent with the characteristics and features described in the given cultivar.

(2) Sampling.

In order to evaluate the variety purity, the sampling procedures must be followed, i.e. concentrating on a small area of the seed production field (sample area) for detailed inspection. In general, all connected fields of the same cultivar, the same source, the same propagation generation, and the same cultivation conditions are regarded as inspection areas. The sampling point shall cover the whole seed production field, and it shall be determined randomly according to the shape and size of the field and the characteristics and features of each crop species. The sampling points shall be determined evenly. Common methods are diagonal sampling, quincuncial sampling, checkerboard sampling, V-shaped ridge (bed) sampling, etc.

4. Analysis Check and Calculation

Field inspection shall not be conducted in strong sunlight, bad weather, or heavy rain. Inspectors shall identify every plant (ear) at sampling points. The number of this cultivar, other cultivars, other crops, weeds, and plants infected with pests and diseases shall be recorded, respectively. For hybrid seed production fields, the number of father hybrid, mother hybrid, and mother plant with dispersed pollen shall also be recorded. The following formula shall be adopted to calculate the percentage of each item. If there are sporadic quarantine weeds or plants infected with pests and diseases outside the inspection points, they shall be recorded separately.

$$\text{Variety purity} = \frac{\text{Number of plants of this cultivar}}{\text{Total number of plants of the tested crop}} \times 100\%$$

$$\text{Other crops} = \frac{\text{Number of plants of other crops}}{\text{Total number of plants of the tested crop} + \text{Number of plants of other crops}} \times 100\%$$

$$\text{Weeds} = \frac{\text{Number of weeds}}{\text{Total number of plants of the tested crop} + \text{Number of weeds}} \times 100\%$$

$$\text{Pest and disease infection} = \frac{\text{Number of plants infected with pests and diseases}}{\text{Total number of plants of the tested crop}} \times 100\%$$

$$\text{Other cultivars} = \frac{\text{Number of plants of other cultivars}}{\text{Total number of plants of the tested crop}} \times 100\%$$

$$\text{Mother plants that have dispersed pollen} = \frac{\text{Number of mother plants that have dispersed pollen}}{\text{Total number of the tested mother plants}} \times 100\%$$

$$\text{Father (mother) hybrid plants that have dispersed pollen} = \frac{\text{Number of father (mother) hybrid plants that have dispersed pollen}}{\text{Total number of the tested father (mother) plants}} \times 100\%$$

(II) Field Inspection Report

After completing the field inspection, inspectors shall fill in the field inspection result sheets in accordance with the prescribed format. Generally, inspection results during the seedling stages are for reference, and the results during the anthesis and the maturity stage are used as the basis for grading. If the inspection results during the anthesis and the maturity stage are different, the lower one shall be used for grading. The field inspection result sheets shall be filled in triplicate.

In the field inspection, inspectors shall sign rectification proposals if some requirements (variety purity, etc.) are not met in accordance with the standards and can be met through rectification measures (such as roguing). If the standards cannot be met through rectification, for example, the isolation conditions do not meet the requirements, or there are severe lodgings, etc., inspectors shall recommend the elimination of the inspected seed production fields.

Review and Reflection Questions

(1) What is the concept of field inspection?

(2) What are the purposes of field inspection?

(3) What are the functions of field inspection?

(4) What are the contents of field inspection?

(5) What are the general procedures of field inspection?

Module 6 Processing, Storage, and Transportation of Horticultural Plant Seeds

Learning Objectives

(1) Be able to enumerate the significance of processing horticultural plant seeds.

(2) Be able to propose the main methods of seed cleaning and selection based on the basic principles of processing horticultural plant seeds.

(3) Be able to propose the main methods and operating principles of seed packaging, seed coating, and seed drying according to the basic principles of processing horticultural plant seeds.

(4) Be able to properly control the environmental conditions in seed storage for different plant species.

(5) Be able to take appropriate measures to control pests and diseases, control microorganisms, and prevent and deal with moisture condensation and heat during seed storage.

(6) Be able to propose the main measures for proper transportation according to the basic requirements of transporting horticultural plants.

Learning Scenario I Horticultural Plant Seed Processing

I. Overview of Seed Processing

Seed processing refers to the various treatments of seeds from seed collection to sowing, including seed cleaning, drying, selection and grading, seed coating, and processing procedures of quantitative or fixed number packaging. That is, the processing procedures of turning newly collected seeds into commercial seeds. Through seed processing, the seed cleanliness, germinability, variety purity, and viability can be improved, and the moisture content of seeds can be lowered, which can improve the seed storability, stress resistance, seed utilizing efficiency, value, and commodity characteristics. Seed processing is an important means of improving seed quality, and it is also a key part of seed commercialization.

(I) Purposes of Seed Processing

First, to improve the cleanliness and germination rate of seeds; second, to make the seeds uniform in appearance; third, to prevent pests and diseases from infecting seeds; fourth, to promote seed germination and the growth and development of crops; fifth, to improve seed storability and facilitate transportation and marketing.

Processed seeds have significant advantages in production. After processing, the seed cleanliness will be increased by 2%–3% and the germination rate will be increased by 5%–10%. Thus, the seed quality is improved obviously and the sowing amount can be reduced, lowering the cost of agricultural production. Seeds processed have neat seedlings that are more, stronger, and have more tillers and ears. The yield can be increased by 5%–10%, which significantly improves the crop yield per unit area. Processed seeds are divided into different grades according to various uses and sales markets and will be sold in standardized packages. Therefore, the commodity characteristics of seeds can be improved, and the circulation and sale of counterfeit and shoddy seeds can be effectively prevented. After processing, seeds will be full and of uniform size, conducive to simultaneous growth and consistent maturity stages of crops, which benefits mechanized sowing and harvesting and improves labor efficiency. Most of the seeds infected with pests and diseases are removed, and the rest of the seeds are coated during processing, allowing pesticides to be released slowly, which not only reduces the application amount of chemical fertilizers and pesticides but also is conducive to environmental protection because the application way of pesticides has been changed from open to conceal. The seeds processed are clean and dry, which can improve the stability of seed storage and extend the storage period. As a result, the normal commodity circulation of seeds is ensured.

(II) Significance of Seed Processing

The seed processing is the most economical and effective measure to increase crop yields. Processed seeds have neat and strong seedlings, which is conducive to a yield increase of 5%–10%. The sowing amount of seeds can be reduced and the grains can be conserved. Seeds processed are suitable for mechanized operations and have a relatively less severe infection with pests and diseases, which is conducive to a good growth environment for crops during the seedling stage. Seed processing is an important part of achieving the standardization of seed quality, which can be divided into different grades according to various uses and sales markets, and the processed seeds will be sold in standardized packages. Therefore, the commodity characteristics of seeds can be improved. In this way, farmers can avoid the mixing of inferior seeds. It is conducive to mechanized operations in the fields, thus improving labor efficiency and reducing labor intensity. It can also reduce the pollution of pesticides and fertilizers, promoting the sustainable development of agriculture. It plays a significant role in the safe storage and transportation of seeds, as well as maintaining their relatively high viability and vigor.

(III) Seed Processing Techniques

Since there is a wide variety of seeds, they have different initial statuses (impurity content, moisture content, etc.) with varying requirements and content of processing. However, the basic content of seed processing should include threshing, preliminary cleaning, drying, selection and grading, chemical treatment, measurement and packaging, and other operational processes.

1. Threshing

Threshing is the preliminary preparation for processing seeds and is also a critical process that affects seed quality. To minimize seed damage, the moisture content of the seeds must be reduced to a certain level when threshing. Otherwise, the breakage rate will increase. The seeds shall be put in the threshers uniformly and the number of seeds shall be close to the maximum capacity of the thresher. The speed at which the seeds are put in shall be reduced to ensure a high threshing rate.

2. Preliminary Cleaning

In general, seeds shall go through preliminary cleaning before they are dried and selected. During the process, large debris and light inert matter, such as broken stems, leaves, ears, etc., are removed to improve the fluidity of the seeds. Therefore, preliminary cleaning can ensure the work efficiency and performance of dryers and separators, and reduce the heat consumption of the machinery. There are many different types of machines for preliminary cleaning, generally performed with airflow by vibrating screens or rotary screens.

II. Seed Cleaning and Selection

Seed cleaning will separate impurities mixed in the seeds, such as the stems, leaves, ears, damaged seeds, weed seeds, silt, stones, hollow seeds, etc., through technical measures to improve the cleanliness of seeds and prepare for safe drying, packaging, and storage of seeds. Seed selection will remove the seeds of other crops or other cultivars and seeds that are imperfect, moth-eaten, or deteriorated to improve accuracy in grading, utilization efficiency, purity, germination rate, and vigor of seeds.

(I) Basic Methods of Seed Cleaning and Selection

Seed cleaning and selection are mainly based on differences in seed size, seed specific gravity, aerodynamics characteristics, surface characteristics of seeds, seed color, and electrostatic properties of seeds.

1. Seed Cleaning Based on the Shape of Seeds

During seed cleaning, seeds and debris are separated based on three basic shape elements: length, width, and thickness. They are to be separated with different screens and methods. At present, screens commonly used for seed cleaning can be divided into several kinds, such as perforated screens, woven screens, and chaffer (Fig. 6-1).

Fig. 6-1 Crop Chaffer

2. Cleaning and Pneumatic Separating Seed Based on Aerodynamic Principles

The floating characteristics of seeds and various debris in the airflow are different. The main influencing factors are the weight of seeds and the windward area of seeds. Based on aerodynamics, many methods can be adopted in seed cleaning.

3. Seed Cleaning Based on Surface Characteristics of Seeds

Seed cleaning can be carried out by taking advantage of the different surface shapes and smoothness of seeds and debris, as well as the different frictional resistances of seeds and debris on the inclined plane. The most commonly used separators based on the surface characteristics of seeds are inclined draper seed separators (Fig. 6-2).

4. Seed Cleaning Based on the Seed Color

It is based on the color or light-dark characteristics of seeds. Let the seeds be separated, and pass through a brightly illuminated viewing area. The lights reflected from the seeds are compared with a standard light color chosen in advance on the background. When the lights reflected from the seeds are different from the standard light color, a signal is generated. The seed will then be ejected from the mixture and fall into a different duct.

Fig. 6-2 Inclined Draper Seed Separators

5. Seed Cleaning Based on Seed Density

Due to different species, fullness, moisture content, and the degree of infection with pests and diseases, seed density varies a lot. The more significant the difference in density is, the more efficient the separation will be. At present, the most commonly used method for this kind of seed cleaning is to use the different buoyancy of seeds in liquids. When the seed density is greater than the liquid density, the seeds will sink; conversely, the seeds will float. Those floating seeds can be removed from the liquid to achieve the purpose of separating lighter seeds from heavier ones.

(II) Seed Selection and Grading

The physical differences between different seeds and inert matter are used to separate various seeds and inert matter, thus achieving the purpose of selecting and grading the seeds. The main techniques adopted are as follows.

1. Seed Selection Based on Aerodynamic Principles (Pneumatic Separators)

The seeds are selected and graded based on the difference in the critical wind speed of the different components of the seed lot.

2. Seed Selection Based on Seed Size Characteristics (Screens)

For example, indented cylinder screens can be used for separation based on seed length; round-hole screens can be used for separation based on seed width; slotted-hole screens can be used for separation based on seed thickness.

3. Seed Selection Based on Density (Gravity)

Seed density varies depending on different species, fullness, moisture, and the degree of infection with pests and diseases. Besides, there are greater density differences between the seeds and the inert matter. The greater the difference in density is, the more efficient the separation will be. Seed selection based on density can be divided into liquid selection (selection of seeds by saltwater, mud, etc.) and gravity selection.

4. Seed Selection Based on Surface Characteristics (Roughness) of Seed

This kind of selection is usually used for forage seeds. The commonly used machinery includes velvet roll separators (based on different levels of smoothness or roughness) and frictional separators (based on different friction coefficients).

(III) Machinery Commonly Used for Seed Cleaning and Selection

The most commonly used machinery for seed cleaning includes air screen cleaners, draper-type winnowing machines, etc., such as 5X-4.0 seed separator, etc. Indented cylinder, indented disc separator, density separator, inclined draper seed separator, separator of photoelectric color, electrostatic separator, etc. are often used for seed selection, such as 5XZ-3.0 positive-pressure gravity separator, 5XP-3.0 flat-screen seed grader, 5XW-3.0 indented cylinder separator, 5XY-2.0 cylinder screen separator, 5XZ-1.0 gravity separator, 5XF-1.3A multiple seed separator, etc. However, during the process of seed cleaning and selection, multiple seed separators will be used to clean and select simultaneously.

1. Air Screen Cleaners

An air screen cleaner is a seed cleaning device that takes advantage of both airflow and screens by using characteristics of aerodynamics and the size of seeds. At present, it is the most widely used cleaning machine for seed cleaning.

There are various structures, sizes, and types of air screen cleaners, ranging from small ones with only a single fan and a single screen to large ones with multiple fans, 6–8 screens, and several air chambers. The commonly used air screen cleaner has four screens, and the seeds will flow to the feeder through the feeding hopper based on the seed weight. The feeder will regularly send the mixture (that is fed to the hopper) into the airflow. The airflow will first remove light ears, and the remaining seeds are scattered on the top layer of the screen. Other large lumps of impurities will be removed through the screen. The seeds falling from the top layer of the screen will flow onto the second layer of the screen. Then the seeds can be roughly graded by size. Next, the seeds from the second layer of the screen are transferred to the third one, which will finely separate the seeds and let them reach the fourth layer for the last grading. When the seeds have completed four separations, they will go through an airflow. The heavier seeds, which mean better seeds, will fall directly, and the lighter seeds and ears will be lifted and removed.

The screens are chosen based on the seeds for cleaning and selection and the inert matter for removal. When the shape and size of the screens are decided, the actual separation of seeds depends on their size. Round-hole screens and slotted-hole screens are used to identify and separate seeds according to their width and thickness respectively.

2. 5XZ-1.0 Gravity Separator

This separator is mainly used for cleaning and grading seeds of the same shape and size but with different densities. It comprises three parts: the supporting suction blower, the feeding device, and the main body of the separator.

When using the separator, the seeds are transported to the inlet feeding tube by the feeding device and fall into the vibrating screen by pushing the spring-controlled valve with their weight. Then, the seeds will be evenly distributed on the screen surface. Since every grain weighs differently, the lighter ones will be on the upper layer, while the heavier ones will not be lifted and will stick to the screen surface and move upwards due to the vibration of the vibrating screen. The seeds separated will flow out of three rubber sleeves. Full grains will flow out from outlet A, imperfect grains will be out from outlet C, while mixed seeds will flow out from outlet B. Moderate mixed grains from outlet B will be sent back to the inlet to be separated again.

Factors influencing the selection include the vertical and horizontal inclination of the screen surface, the airflow, and the amplitude of the motor.

3. 2.5XF-1.3A Multiple Seed Separator

This separator comprises screens for selection based on seed width and thickness, circular indented cylinders based on seed length, and fans based on seed weight and aerodynamic characteristics.

(1) Screening. Perforated screens are adopted in this separator because of the accurate hole patterns and positions, as well as an efficient seed selection.

(2) Indented cylinder. The indent diameter is 5.6 mm.

(3) Pneumatic separating. Vertical airflow is used to cooperate with the screens. In this way, seeds falling from the feeding device will be transported to the screen surface. When the seeds slide down the screen surface, lighter seeds and the inert matter will be uplifted because their critical velocity is lower than the airflow velocity. In comparison, heavier seeds will slide down the screen surface. The cross-section of the upper end of the air passage expands, and the airflow velocity will gradually decrease. Thus, those uplifted light seeds and the inert matter will fall into the deposition chamber. Dust and other substances will be discharged from the outlet.

(4) Working process. When the seeds to be selected are fed into the feeding hopper, they will reach the front suction duct through the valve and octagonal rubber roller. Under the airflow of the fan, seeds will be uplifted, and when they reach the reflector plate, they will fall into the front deposition chamber. When they reach the bottom of the chamber, the weight allows them to push the valve and fall onto the surface of the upper screen. The lighter inert matter will flow to the middle deposition chamber with the airflow, while some of the heavier insert matter will fall to the bottom and be discharged through the outlet after reaching the collecting tank through valves. However, the lighter inert matter will be discharged with airflow through the fan outlet and the external duct. Since the critical velocity of the heavier inert matter is higher than the airflow velocity, it cannot be uplifted, but will fall to the ground through the entrance of the front suction duct. Those seeds falling on the screen will fall to the lower screen if their size is smaller than the mesh size of the upper screen. Those extra-large seeds and inert matter with

a size larger than the mesh size of the upper screen will go through the collection tank and be discharged through the outlet. Those seeds or inert matter falling on the lower screen will fall on the bottom of the screen box through the lower screen and be discharged through the outlet if their size is smaller than the mesh size of the lower screen. The better seeds will flow to the rear screen through the screen surface of the lower screen. The pneumatic separation will be carried out again with the cooperation of the rear screen and the rear suction duct. The rear suction duct will suck the remained inert matter and lighter immature seeds into the deposition chamber. The heavier ones will fall to the bottom of the chamber, and when there are enough of them, the valve will be pushed open. Then, they will go through to the collection tank and be discharged from the outlet. With airflow, the lighter inert matter will be discharged through the fan. There are two outlets for the seeds flowing through the screen surface of the rear screen. One is the outlet, discharging seeds that will not be further selected by length, and the other is the indented cylinder, which will carry out a further selection of seeds by length. Among seeds that need further selection, shorter ones will fall into the V-shaped groove after being uplifted to a certain height. Then, they will be discharged through the outlet. The selected seeds that remain will go through the winch and discharge chute, and be discharged from the outlets. At this point, the entire process of seed selection is completed.

(IV) Requirements for Seed Cleaning and Selection

To improve efficiency and productivity, it is necessary to understand separation purposes, as well as the composition and separation features of the selected seeds before, during, and after the seed cleaning and selection. It is necessary to choose proper separation machinery, adjust machine operation properly, and inspect the separation in time.

1. To Clarify Separation Purposes

Before the seeds are sent for separation, the purpose shall be clarified first to decide whether they are to be cleaned or selected. The grades and standards that selected seeds are required to meet also need to be clarified so that separators can be chosen properly.

2. To Understand the Composition of the Selected Seeds

Before choosing a separator, seed characteristics such as size, density, color, etc., and features of impurities shall be analyzed. Besides, the requirements for the seeds to be selected shall be clarified so that the proper machinery for cleaning and selection can be chosen. Moreover, technical parameters, such as the size of screens for cleaning and the diameter of the indent on the indented cylinder, should be determined properly.

3. To be Familiar with Mechanical Properties

Adjust operating parameters properly because different cleaning and selection machines are designed for different separation purposes. The optimum results can only be obtained through a complete command of mechanical properties, the proper choice of machine parts, and the proper adjustment of operating parameters.

4. To Inspect the Separation in Time

During the separation, the results of cleaning and selection shall be understood in time to improve and adjust the operating parameters for obtaining the optimum separation results.

III. Seed Drying

The moisture of newly collected seeds is relatively high, which can be up to 25%–45% in general. Thus, the respiration intensity of seeds is high, and a high amount of heat and moisture is released. As a result, seeds are prone to heat and mildew, or alcohol intoxication may appear because anaerobic respiration occurs when the oxygen in the seed pile is quickly exhausted, or seeds may die from freezing injury when the temperature is below 0°C. Therefore, seeds must be dried in time to reduce the moisture to the level of safe packaging and storage. In this way, vigorous germinability and vitality of the seeds can be maintained, and the storage period of seeds can be improved. The main methods of seed drying are as follows.

(I) Natural Drying

At present, natural drying is one of the main methods of seed drying in China.

With the help of sunlight, wind, and other natural forces, the moisture content of seeds can be reduced. The advantages of natural drying are that it is simple, economical, and safe. The seeds do not tend to lose their viability under normal circumstances. What is more, the ultraviolet rays in the sunlight can also play a role in sterilizing and killing insects. The limits of natural drying are that it is limited by site conditions and weather. Besides, labor intensity is high for natural drying, especially in areas with a humid climate and much rain, which influences the drying efficiency.

Natural drying can be carried out before and after threshing. For example, the seeds of cabbage, kale, carrot, celery, etc. can be dried before threshing. They can be dried in the field, or spread or hung in a pergola trellis built after harvesting. However, drying seeds after threshing is often carried out on sunning grounds or cement fields. The following points shall be noted when drying seeds on a sunning ground.

1. Cleaning and Pre-heating the Floor

A day when the weather is sunny shall be chosen. Cleaning the floor first as "dry the ground before you dry seeds". The time for drying shall be after 9:00. If the seeds start drying too early, the seeds on the floor may condense (because the floor's temperature is too low) and water stratification may occur, which affects the drying efficiency.

2. Thinly Spreading Seeds on the Floor and Turning Them over Regularly

When the seeds are spread out for drying, the layer of seeds shall not be too thick. In general, the thickness of layers of small seeds shall not exceed 5 cm and that of layers of medium and large seeds shall not exceed 10–25 cm. Besides, the seeds shall be turned over at regular intervals to improve the drying efficiency.

3. Timely Warehousing

Seeds after exposure to sunlight need to be cooled before warehousing. Otherwise, when stored in the warehouse, heated seeds will easily condense at the bottom as they meet the cool floor, which is not beneficial for seed storage.

(II) Mechanical Drying Through Ventilation

For newly collected seeds with high moisture, when the weather is cloudy and rainy, or when there is no hot-air drying machine, cool and dry air from the outside can be blown to the seed pile through an air blower. The water vapor and heat produced by seed respiration in the gaps of the seed pile can be taken away continuously to dry seeds and cool them down. This is a way that temporarily prevents damp seeds from heating up and deteriorating, and inhibits the growth of microorganisms.

Drying through ventilation takes the outside air as a drying medium. Therefore, the reduction of seed moisture content is affected by the air's relative humidity outside. In general, drying through ventilation is the most economical and effective way only when the external relative humidity is lower than 70%. However, in humid areas in the South or when it is raining in the North, the atmospheric humidity will not be very low, with which seed moisture cannot be reduced to a balanced level of the atmospheric relative humidity. When the water-holding capacity of seeds is balanced with the water absorption ability of air, seeds will not give off water or absorb it from the air. Assuming that the seed moisture is 17%, the air with a relative humidity of 78% at a temperature of 4.5°C is in equilibrium with the seed moisture. If the air relative humidity exceeds 78% at this time, seed drying cannot be carried out.

In addition, the relative humidity at which equilibrium is reached is lower as the seed moisture decreases. Therefore, when the seed moisture is 15%, the relative humidity of air must be lower than 68%. Otherwise, the seeds cannot be dried.

In general, the equilibrium relative humidity will increase with the rise of temperature. Therefore, seeds with a moisture content of 16% cannot be dried in the air with a relative humidity of 73% at a temperature of 4.5°C.

The equilibrium between seed moisture and relative air humidity can indicate why drying with natural air must be supplemented by artificial heating. When drying with natural air, air blowing can be suspended when the moisture content of seeds has been reduced to about 15%. Blow again when the relative humidity of air is lower than 70% to dry the seeds further. At the usual temperature for drying with natural air, the relative humidity of air is 70%, at which equilibrium is reached with seeds with a moisture of 15%. If the relative humidity exceeds 70%, instead of drying the seeds further, the air blower will make the seeds absorb water from the air. Therefore, this method can only be used for ventilation and drying of newly collected wet seeds when temporary safe storage is required.

Seed drying through ventilation is relatively simple because it can be carried out as long as there is an air blower. Generally, it is believed that when the airflow is greater than 9 m^3/min, only power consumption instead of the speed of seed drying will increase, which is not economical because the thickness of seed layers has resistance to the airflow. Therefore, the efficiency of seed drying through ventilation is also related to the thickness of the seed pile and the air volume blown to the seed pile. If the seed pile is high and the thickness is low, the air volume will be large. As a result, the seed drying will be efficient and fast; Otherwise, the seed drying will be slow.

(III) Seed Drying Through Heating

This is a way that heated air is used as a drying medium to directly go through the seed layer, vaporizing the seed moisture and thus drying the seed. It is adopted in tropical and subtropical regions that are warm and humid, especially in units that produce seeds on a large scale or store vegetable seeds for a long time.

When drying through heating, the air is heated and the relative humidity of air is reduced to improve its water retention ability. Besides, heated air can act as a heat carrier to provide the seeds with the heat needed to evaporate moisture. According to the heating temperature and the speed of drying, drying through heating can be divided into the following two kinds.

1. Slow Drying with Low Temperature

The airflow temperature used is generally at most 8°C higher than the atmospheric temperature. Besides, a low airflow is utilized. For example, an airflow lower than 6 m^3/min is often adopted for seeds of an area of 1 m^3. The time needed for slow drying with low temperatures is relatively long, and it is mostly used for drying in the warehouse.

2. Fast Drying with High Temperature

A higher temperature and greater airflow are used when drying. It can be divided into two types: one is for static seed layers and the other is for dynamic seed layers.

The heated air temperature should not be too high when drying the static seed layers by convection. According to the types of dryers, the initial moisture content of seeds, and different seasons, the temperature is generally higher than the atmospheric temperature by 11–25°C, and the maximum temperature shall not exceed 43°C. This type of dryer includes bag dryers, box dryers, and hot airflow drying chambers that are often used today. The heated air can heat the seeds evenly by drying the dynamic seed layers, improving productivity and saving energy. Wet seeds are fed to the dryer continuously, and then continuously discharged after drying. Therefore, the process is also called continuous drying. For example, when a dryer with an air trough is drying seeds with a moisture content of less than 20%, the temperature of the heated air is preferably 43–60°C and the temperature of the seeds discharged is 38–40°C.

In addition, there are drying methods where far-infrared rays and solar energy are used as heating sources.

(IV) Freeze-drying

Freeze-drying, also called lyophilization, freezes seeds at temperatures below freezing. In this way, the seeds can be dried through sublimation.

1. Freeze-drying Equipment

According to different drying scales and requirements, the freeze-drying equipment can be divided into large and small facilities. A small freeze-drying facility is composed of the following parts.

(1) Drying room. It is used to place seeds for drying. Its lower part is a heater with a duct at the base leading to the vacuum system. The temperature is usually kept at -10– -30°C, and the pressure is about 133.3 Pa.

(2) Vacuum pump system. Since the pressure in the system must be maintained at about 1.333 Pa during the process of freeze-drying, there must be a vacuum pump and pipes for air exhaust. Oil-sealed rotary vane pumps are commonly adopted.

(3) Low-temperature sealed device for water collecting. In order to collect the vapor produced during the freeze-drying, a sealed device for water collection shall be set. In general, a low-temperature sealed device for water collecting is used.

Besides, a freezer with a cooling capacity below -40°C is required.

(4) Accessory apparatus. In the freeze drying facility system, vacuum gauges, thermometers, flow meters, and other related instruments must be included.

2. Freeze-drying Method

There are usually two freeze-drying methods. One is conventional freeze drying, during which seeds are placed in a Teflon-coated aluminum box with a volume of 254 mm×38 mm×25 mm; then the aluminum box with the seeds will be placed on a freezer rack that has been cooled to -10 – -20°C. The other method is rapid freeze-drying, during which the seeds shall first be frozen in liquid nitrogen. Then, the seeds will be placed in a tray on a rack at a temperature of -10 – -20°C, and the pressure inside the box will be reduced to about 40 Pa. After that, raise the temperature of the rack to 25–30°C to slightly heat the seeds. Since the pressure is reduced, the ice inside the seeds will gradually become less through sublimation. Because sublimation is an endothermic process, it requires a small supply of heat. If the pressure inside the box is maintained below the vapor pressure of ice, the water vapor sublimed will freeze and hinder the melting of the ice in the seeds. As the ice in the seeds reduces, the sublimation effect decreases, and the temperature of the seed pile will gradually increase to the same temperature as the rack.

(V) Other Drying Methods

1. Desiccant Drying

Seeds and desiccant will be sealed in a certain proportion in an airtight container. Use the moisture-absorbing ability of desiccant to continuously absorb the moisture diffused from the seeds and dry the seeds until reaching the equilibrium moisture. The desiccants currently used include lithium chloride, self-indicating silica gel, calcium chloride, activated alumina, quicklime, phosphorus pentoxide, etc.

2. Radiation Drying

Radiation energy will be transferred to the wet seeds by radiation elements or invisible rays. When the seeds absorb the energy, which will be converted into heat, the temperature of the seeds will rise and vaporization occurs so as to achieve seed drying. Solar energy drying, infrared drying, and far-infrared drying belong to radiation drying.

3. High-Frequency Drying

High-frequency induction heating equipment with a current frequency of 1-10 MHz will be used to rapidly change the polarization directions of the polar molecules in the seeds under the high-frequency electric field. As a result, thermal motion similar to friction occurs, increasing the temperature of the seeds and the moisture evaporates rapidly.

4. Microwave Drying and Resistance Drying

By adopting these two drying methods, seeds will be dried quickly and uniformly. Besides, the growth and reproduction of pests can be inhibited.

5. Vacuum Drying

According to the principle that the boiling point of water can be greatly reduced under vacuum conditions, mechanical means are used to draw the air out of the drying room with a vacuum pump to form a low-pressure space. As a result, the boiling point temperature of moisture is lower than the limited temperature, at which the seeds can be dried. Under the premise that the viability of seeds will not be affected, the moisture inside the seeds vaporizes rapidly when it reaches the boiling point while the seeds are dried rapidly and effectively.

IV. Seed Coating

Seed coating applies adhesives or film formers to wrap non-seed materials such as fungicides, pesticides, micro-fertilizers, plant growth regulators, colorants, or fillers around the seeds to make the seeds spherical or to substantially maintain their initial shapes. It is a new seed technology that improves stress resistance and disease resistance of seeds, accelerates germination, promotes seedlings, increases yield, and improves seed quality.

(I) Classification of Seed Coating Methods

1. Seed Pelleting

Seed pelleting refers to the method by which adhesives are used to wrap non-seed materials such as fungicides, pesticides, dyes, fillers, etc. around the seeds. Usually, seeds are made into spherical individual seed cells that do not differ significantly in size and shape. This coating method is mainly applicable to small seeds of vegetables such as carrots, onions, cabbage, kale, beets, etc., which are used for precision sowing. Since inert materials such as fillers (e.g. talc) are added during the seed pelleting process, the volume and weight of the seeds increase, and the thousand-seed weight also increases.

2. Film Coating

Film coating refers to the method by which film formers are used to wrap non-seed materials, such as fungicides, pesticides, micro-fertilizers, dyes, etc. around the seeds to form a thin film. After film coating, seeds are basically seed cells that resemble the initial seed shapes. However, the range of changes in size and weight varies depending on the type of seed coating agents. In general, film coating is suitable for large and medium seeds.

(II) Types of Seed Coating Agents and Their Properties

1. Pesticide Type

This type is mainly used for the prevention and control of seed and soil-borne diseases. The seed coating agents are mainly composed of pesticides. Applying this type of seed coating agents in a large amount could contaminate the soil and cause poisoning to humans and animals. Therefore, try to choose pesticides of high efficiency and low toxicity to add to seed coating agents.

2. Compound Type

This type is designed for multiple purposes such as disease prevention, resistance improvement, growth promotion, etc. Therefore, seed coating agents of this type include chemical components such as pesticides, micro-fertilizers, plant growth regulators, resistance substances, etc. Many currently used seed coating agents belong to this type.

3. Biological Type

It is a newly developed seed coating agent worldwide. Based on the principle of antagonism between biological fungi, beneficial antagonistic rhizobacteria are selected to resist the multiplication and attack of harmful pathogens, achieving disease prevention. For example, biological seed coating agents are used to prevent and control black rot of Cruciferous seeds, seed-borne diseases of celeries, tomato diseases, and pepper diseases. From the perspective of environmental protection, it is a trend to develop natural, non-toxic, and non-polluting biological seed coating agents.

4. Special Type

This type of seed coating agent is specially designed for different crops and purposes.

(III) Requirements for Coating Machinery

Seed coating operation is the process of putting seeds into the coating machine and evenly wrapping the seed coating agents around the surface of the seeds by mechanical action.

It is a continuous batch production, in which seeds are measured one bucket at a time, and the liquid drug is measured one spoon at a time. When the measured seeds and liquid drug fall simultaneously, the liquid drug atomized in the atomizing device will be sprayed on the seeds, achieving seed pelleting or film coating. In the end, the seeds will be stirred and discharged.

During seed coating, the requirements for the machine are as follows.

1. Ensure Sealing

The seed coating machinery must be sealed completely during operation to ensure that the chemical agents do not harm operators. When mixing powder drugs, the powder cannot be scattered into the air or thrown on the ground. When mixing liquid drugs, the liquid cannot be dripped outside the container so as not to contaminate the operation environment.

2. Ensure Uniform Mixing and Coating

The machinery used for seed coating shall be able to apply powder and liquid drugs or to apply powder and liquid drugs at the same time. Ensure that the seeds and the coating agents can be mixed in proportion so that the seeds can be coated evenly. The proportion shall be adjusted as needed and the adjustment methods shall be simple and easy to learn. When coating, the solution should be able to adhere evenly to the seed surface or pelletize the seed.

3. With High Economic Value

The machinery should be efficient, simply constructed, and low-cost in production. The parts in contact with drugs should use anti-corrosion materials or take anti-corrosion measures to extend the service life of the machinery.

(IV) Preparation Before Seed Coating

The machinery, agents, and seeds should be prepared before the coating operation starts.

1. Selection of Coating Machine

The coating machine is selected appropriately according to the seed species and the coating method.

2. Preparation of Machinery

First, the technical state of the coating machine should be checked to see if it is in good condition, like whether the installation is stable and horizontal, whether the fastening bolts are loose, whether the rotating parts are stuck, and whether there are tools or foreign objects left in the machinery. Second, the trial operation should be carried out to check whether the motor rotates are in the right direction and whether the rotation of each part is stable. Move the dosage hopper shaft to swing and observe whether the powder and liquid supply devices can work normally. Listen carefully to find any abnormal sounds during the trial operation. When problems are found, they should be carefully sorted out one by one and properly handled. Once it is confirmed that the technical condition of the machine is good, it can be put into operation.

3. Preparation of Agents

Seed coating agents should be selected according to the various requirements of different seeds, and sufficient amounts of drugs should also be prepared in line with the number and proportion of the seeds to be processed.

For the preparation of liquid drugs, the main thing is to prepare a mixture according to the different requirements of different drugs. The general instructions for using liquid drugs will specify the mixing ratio of drugs to water, and the mixture can be prepared according to that ratio. Be sure to stir when mixing so that the solution is well blended.

4. Preparation of Seeds

All seeds for coating must be the seeds selected after being processed, and the seed moisture should be within the range of the safe storage moisture. The coating is the last process of the seed processing line, and all coating machinery should be placed at the end of the line. According to China's current production habits, the coating operation is carried out before sowing, i.e. the processed seeds will be first stored for overwintering and then coated when being sown the next spring. Before coating is carried out, the seeds should be inspected to confirm that the cleanliness, germination rate, and moisture of the seeds meet the requirements.

5. Germination Test

Any crop seed must be subjected to a germination test before mechanical coating treatment with seed coating agents. Only seeds with a high germination rate can be treated with seed coating agents. Each batch of seeds treated with seed coating agents should also be tested for germination to check the coated seeds germination rate. The following methods can be used to perform germination tests on coated seeds.

(1) Wet sand and petri dish method. Add 17% of the sand weight of water into the sifted fine sand, mix them well, and then put them into a large Petri dish with a diameter of 15 cm. The mixture's thickness is half the depth of the Petri dish. Then flatten it. After sowing a certain number of seeds, they are placed in an incubator at a specified temperature to measure their germination potential and germination rate.

(2) Large sandbox method. The sand used in this method is the same as that in the wet sand and Petri dish method. After sowing, add water to flatten the soil, and then put the sandbox into the specified greenhouse with a lid on top to check the germination potential and germination rate.

(V) Methods of Seed Coating

The two main methods of coating seeds with seed coating agents are mechanical coating method and artificial coating method. These methods are performed in seed processing plants and require the mastery of specially trained technicians.

1. Mechanical Coating Method

The coating of good-quality seeds should be performed under the centralized arrangement of seed companies. At present, the main coating machines applied in China include the 5BY-5A type, 5BY-LX type, 5BYF-5 type, etc.

2. Artificial Coating Method

Artificial methods can be adopted if there is no coating machine, which includes the following methods.

(1) Round-bottomed cauldron coating. Fix the round-bottomed cauldron, weigh the seeds, and put them into the cauldron. Weigh the seed coating agents proportionally and pour them into the cauldron to mix with the seeds. Then, quickly stir them with a large prepared spatula. Let them mix well and leave for sowing.

(2) Large bottle or small iron drum coating. Prepare a large bottle or a small iron drum with a lid that can hold 5 kg of seeds. Weigh 2.5 kg of seeds into the bottle or drum and a certain amount of seed coating agents in proportion to the seed. Then, pour the mixture into the bottle or drum with seeds in it. Seal the lid, shake it quickly, mix well, and pour it out for sowing.

(3) Plastic bag coating. When using plastic bags to coat seeds, prepare two plastic bags of different sizes first, pack the two bags together, weigh a certain amount of seeds in proportion to the seed coating agent, and then pour the seeds with the agents into the inner plastic bag. Tie the bag up and rub the bag quickly with your hands. Mix them well and pour them out for sowing.

V. Seed Packaging

(I) Purpose and Requirements of Seed Packaging

Reasonable packaging of cleaned, dried, and selectively processed seeds can prevent seed mixing, pest and disease infection, moisture absorption and moisture regain, and seed deterioration. Besides, it can improve the commodity characteristics of seeds, maintain their viability, and ensure the safe storage, transportation, and easy sale of seeds. The requirements for seed packaging are as follows.

1. Requirements of the Seeds

The seeds in moisture-proof packaging must meet the standards of seed moisture and cleanliness required by packaging to ensure that the seeds will not deteriorate during storage and transportation and maintain their original quality and viability.

2. Requirements of the Container

The packaging container must be moisture-proof, clean, non-toxic, not easy to break, and light. The seed is a living organism. If moisture-proof packaging is not used, the seeds will absorb moisture and regain moisture under high-temperature conditions, and the toxic gases produced will damage the seeds and cause them to lose viability.

3. Determine the Packaging Quantity

The suitable packaging quantity should be determined according to the species of seeds, the sowing amount, the production areas, and other factors to facilitate use or sales.

4. Requirements for Longer Preservation

Longer preservation time requires lower moisture in the packaged seeds and better packaging materials.

5. The Storage Conditions of Packaged Seeds

The storage conditions are less demanding in areas with low humidity and dry climates, while it is strict in humid and warm areas.

(II) Types, Characteristics, and Selection of Packaging Materials

1. Types and Characteristics of Packaging Materials

At present, the common packaging materials are mainly jute bags, multi-layer paper bags, tin cans, polyethylene aluminum foil composite bags, and polyethylene bags.

(1) Jute bags: Good strength, but easily permeable to moisture. Poor moisture, insects, and rodent resistance.

(2) Metal cans: High strength, high moisture-proof, lighttight, flood-proof,

harmful fumes-proof, insect-proof, and rodent-proof performance. Suitable for fast automatic packaging and sealing. One of the most suitable seed packaging containers.

(3) Polyethylene aluminum foil composite bags: Appropriate strength, very low moisture permeability, the most suitable moisture-proof material. The composite bag has several layers. Because the aluminum foil has tiny pores, the innermost and outermost polyethylene film has a sufficient moisture-proof effect. It is generally believed that the moisture of the seed in it will not change within a year.

(4) Porous plastics like polyethylene and polyvinyl chloride: Not completely moisture-proof. Dry seeds sealed inside the bags and containers, which are made of this material, will slowly absorb moisture. Therefore, its thickness must be above 0.1mm. This moisture-proof packaging is only valid for about one year.

(5) Polyethylene film: It is the most widely used thermoplastic film. Usually, it can be divided into low-density type (density of 0.914–0.925 g/cm^3), medium-density type (density of 0.930–0.940 g/cm^3), and high-density type (density of 0.950–0.960 g/cm^3). All three types of polyethylene films are microporous materials, and their permeability to water and other gases varies depending on their density.

(6) Aluminum foil (Thickness less than 0.038 mm): Though having many micro-pores, the water vapor permeability is still very low.

(7) Aluminum foil with polyethylene film composite products: With better moisture-proof and breakage-proof performance, they are ideal seed packaging materials, including aluminum foil/cellophane/aluminum foil/heat-sealing varnish, aluminum foil/tissue paper/polyethylene film, and kraft paper/polyethylene film/aluminum foil/polyethylene film.

(8) Paper bags: Mostly made of bleached sulfite paper or kraft paper, with its surface covered with a layer of white clay for printing. Many paper seed bags are multi-layered structures made of several glossy or crepe paper layers. Multi-layer paper bags have different structures depending on their purposes. Ordinary multi-layer paper bags have poor resistance to breakage, moisture, insects, and rodents,

and will dry out and break easily when they are under dry conditions, which cannot protect the viability of seeds.

(9) Cartons and cardboard cans (or fibre core): These are widely used for seed packaging. Multi-layer kraft paper can protect most of the seeds' physical qualities of the seeds and is suitable for automatic packaging and sealing equipment.

2. Selection of Packaging Materials and Containers

Packaging containers should be selected according to the seed species, seed characteristics, seed moisture, shelf life, storage conditions, seed usage, transportation distances, and regions.

(1) Porous paper bags or knitted bags are generally used for seeds that require good ventilation (such as beans), or for a large number of wholesale seeds stored in dry and low-temperature places with a short shelf life.

(2) Small paper bags, polyethylene bags, aluminum foil composite bags, and tin cans are commonly used to package retail seeds.

(3) Containers such as steel cans, aluminum boxes, plastic bottles, glass bottles, and polyethylene aluminum foil composite bags are usually used for the long-term conservation of a small number of seeds or seeds at high prices, or the conservation of a variety of resources.

(4) In tropical and subtropical regions with high temperatures and humidity, seeds should be packaged in tightly packed and moisture-proof containers. The seeds should be dry enough to meet the moisture requirements for safe packaging, and then they should be sealed into moisture-proof containers to prevent the loss of seed viability.

(III) Safe Moisture of Packaged Seeds

According to the principles of safe packaging and storage, the seed life can be extended when the seed moisture content is reduced to the amount that is in equilibrium with 25% relative humidity, which will help maintain the seed viability. However, this moisture content varies depending on the seed species. Failure to achieve a certain degree of dryness will accelerate the deterioration and death of the seeds, as seeds with high moisture will quickly run out of oxygen and accumulate carbon dioxide due to respiration in sealed containers, eventually leading to anaerobic respiration and death of seeds because of poisoning. Therefore, seeds packaged in moisture-proof sealed containers must be dried to reach the moisture content of safe packaging to maintain the original viability of the seeds.

(IV) Packaging Machinery and Methods

1. Seed Packaging Quantity

Seed packaging mainly includes the packaging according to seed weight and seed number. The former one is generally adopted. The weight of each package differs by the scale of production, the sown area, and the amount of seed used. For example, vegetable seeds are often packaged in different weights, such as 4 g, 8 g, 20 g, 100 g, or 200 g per bag. The need to improve the quality of seeds and precision sowing requires packaging according to seed numbers for more expensive vegetable and flower seeds, such as 100 or 200 seeds per bag. Therefore, to adapt to quantitative packaging and fixed number packaging, seed packaging machinery also has two corresponding types.

2. Seed Packaging Process and Machinery

(1) Seed packaging process: Transporting seeds from the silo warehouse to the feed box→weighing or counting→bagging (or putting into the container)→sealing (or seaming)→sticking (or hanging) labels.

(2) Seed quantitative packaging machine: Seed packaging in some areas of China has largely realized automated or semi-automated operations. The seeds are transported from the silo warehouse using gravity or air lifters, belt conveyors, lifts, and other machinery to the feed box, and then into the weighing equipment. When the predetermined weight or volume is reached, the seed flow is automatically cut off, and then the seed enters the packaging machine. Open the mouth of the packaging container and let the seeds flow into it. After seaming or sealing the seed bags (or containers) and attaching (or pre-printing) the labels, the packaging operation is completed.

(3) Seed fixed number packaging machine: With the advanced seed fixed

number packaging machine, as long as the selected seeds are put into the hopper, they will run through the fixed number photoelectric counter, flow into the packaging line, and be automatically sealed and moved to the exit. After workers put them into the customized cartons, the packaging process is completed.

(V) Preservation of Packaged Seeds

Although packaged seeds have certain characteristics such as moisture-proof, insect-proof, and rodent-proof, they will still be affected by high temperatures and humid environments, which will accelerate deterioration. Therefore, packaged seeds must be stored in moisture-proof, insect-proof, rodent-proof warehouses or places with dry and low temperatures. Seed bags of different crop species and cultivars should be stacked separately. To facilitate proper ventilation, there should be appropriate space between seed bag stacks. In addition, it is also necessary to do management work well, like fire prevention and routine inspection, to ensure the safe preservation of packaged seeds and truly exert the advantages of seed packaging.

Review and Reflection Questions

(1) What are the purposes and significance of seed processing?
(2) What is seed cleaning and selection?
(3) What are the basic methods of seed cleaning and selection?
(4) What are the processes for seed selection and grading?
(5) What are the main methods of seed drying?
(6) What are the types of seed coating methods?
(7) What are the preparations before seed coating?
(8) What are the methods of artificial coating?
(9) What are the precautions for using the seed coating agent to coat seeds?
(10) What are the requirements of seed packaging?

Learning Scenario II Storage and Transportation of Horticultural Plant Seeds

I. Construction of Seed Warehouse

Seeds need to undergo a long or short storage phase from harvest to sowing. The task of seed storage is to use reasonable storage equipment and advanced scientific technology to artificially control storage conditions, reduce changes in seed quality to the minimum, and maintain vigorous germinability and viability of the seeds, thus ensuring their sowing value.

(I) Seed Storage Conditions

The environmental conditions that will affect seed storage mainly include relative humidity, warehouse temperature, and ventilation conditions.

1. Relative Humidity

The change in the moisture of seeds during storage is mainly determined by the relative humidity in the air. When the relative humidity of the warehouse is greater than that of the equilibrium moisture of the seed, the seed will absorb moisture from the air, making its internal moisture gradually increase. Its vital activities will also change from weak to strong with the increase of moisture. In the opposite case, the seed will release moisture into the air and become progressively drier. Its vital activities will be further inhibited. Therefore, it is essential to keep the air dry during the seed storage.

Maintaining a low relative humidity is based on actual needs and possibilities. Germplasm resources require a low relative humidity owing to their long-term preservation time and dry seeds, generally at about 30%. The storage period for the seeds used in production is relatively short, so the required relative humidity is not very low. It only needs to reach a balanced humidity with the safe moisture of the seed, roughly within the range of 60%–70%. Considering the safe moisture standard of seeds and the current situation, the relative humidity of the warehouse is generally controlled below 65%.

2. Warehouse Temperature

The seed temperature will change due to the influence of warehouse temperature, which will change by the air. But there is often a gap between these three temperatures. In the season when the temperature rises, the air temperature is higher than the temperature of the warehouse and seeds; In the season when the temperature drops, the air temperature is lower than the temperature of the warehouse and seeds. The temperature of the warehouse will not only make the seed temperature change but also sometimes cause the transfer of moisture in the seed pile or even condensation because of temperature differences. Especially in spring and autumn, this phenomenon occurs more when the temperature changes dramatically. For example, if the seeds are stored in the high-temperature season, due to the decline of temperature in autumn, which causes the temperature of warehouse walls to decrease, the seed temperature and warehouse temperature near the walls will reduce. The density of the air in this part increases and free convection occurs. The air near the wall will form an airflow downward. Because the center of the seed pile is less affected by the temperature, the seed temperature remains high, forming an upward airflow. So, the downward airflow passes through the bottom layer, reaches the center of the seed pile, and turns into the upward airflow. Then, it passes through the central layer where the seed temperature is high, and then reaches the colder part of the top center. After that, it leaves the surface of the seed pile, forming a loop with the surrounding descending airflows. In this airflow circulation loop, the air continuously absorbs moisture from the seed pile and keeps flowing with the airflow. If the water vapor meets the cold air, it will condense at 35–75 cm below the upper surface layer. If measures are not taken in time, the top seed layer will deteriorate.

Another scenario is that when the temperature rises in spring, the airflow in the seed pile will be the opposite of the above situation. At this time, the temperature in the seed pile is low. The seed temperature around the warehouse wall is increased by the air temperature. The air falls in the center of the seed pile and then rises along the nearby walls. Therefore, the moisture in the airflow condenses at the bottom of the warehouse. Therefore, the influence of temperature in spring will not only cause condensation on the surface of the seed pile but also cause the seeds at the bottom to gain moisture easily. Over time, it will also cause seed deterioration. In order to avoid the disparity between the seed temperature and the air temperature, heat insulation can be adopted inside the warehouse to keep the seed temperature constant. When the temperature is low, ventilation methods can be used to change the seed temperature to the air temperature.

Generally, the increase in temperature of the warehouse will accelerate the respiration of seeds and cause pest and mold infestation. Therefore, the seeds are most prone to decay and spoilage during the summer, late spring, and early fall. Low temperatures can reduce the vital activities of seeds and inhibit the damage of molds. Germplasm resources have a long preservation time, often at very low temperatures, such as 0°C, -10°C, or even -18°C. The number of seeds used in production fields is relatively large. For practical considerations, the storage temperature is generally controlled at 15°C.

3. Ventilation Conditions

Besides water vapor and heat, the air also contains various gases like nitrogen, oxygen, and carbon dioxide. If the seeds are stored under ventilation conditions for a long time, moisture absorption and increasing temperature will cause their vital activities to change from weak to strong, soon losing viability. It is more favorable to store dry seeds under airtight conditions, which are designed to isolate oxygen, inhibit the vital activity of the seed, reduce material consumption, and maintain the seed's potential ability. It also prevents the outside water vapor and heat from entering the warehouse. But it is not absolute. When the temperature and humidity inside the warehouse are higher than outside, doors and windows should be opened for ventilation. If necessary, mechanical ventilation can be used to accelerate the air circulation, so that the temperature and humidity inside the warehouse can be reduced as soon as possible.

In addition to this, the warehouse should be kept clean. If seeds are infected with pests and microorganisms, the water vapor and heat produced by their propagation and activities will deteriorate the storage conditions and harm the seeds directly and indirectly. The life activities of pests and microorganisms require a specific environment. If the warehouse is kept dry, at a low temperature, and airtight, then it can inhibit them.

(II) Basic Types of Seed Warehouses

At present, the commonly used warehouses in China can be divided into five types: simple warehouses, house-type warehouses, cylindrical silos, mechanized cylindrical silos, and low-temperature warehouses. Among them, the company mostly adopts house-type warehouses, mechanized cylindrical silos, and low-temperature warehouses.

1. House-type Warehouse

This type of warehouse is like common house, which can be divided into several categories according to the materials they are made of, such as wood structure, brick, and wood structure, reinforced concrete structure, etc. The wood-structure warehouse has been gradually dismantled and rebuilt because of the difficulty in obtaining materials, the poor sealing performance, and bad rodent and fire prevention. Most of the current warehouses are house-type warehouses constructed with reinforced concrete. This kind of warehouse is relatively sturdy. It has good airtight performance and can meet the requirements of rodent-proof, finch-proof, and fire prevention. There are no columns in the warehouse. With a ceiling on top, the warehouse's inner wall and floor are paved with asphalt layers to prevent moisture. This type of warehouse is suitable for storing bulk or packaged seeds.

2. Cylindrical Silo

This type of warehouse is shaped like a cylinder. Because the cylindrical shape is relatively tall, it is generally equipped with a remote-measuring thermohygrograph, in-and-out warehouse conveyor, automatic weighing machinery, and automatic cleaning machinery. The general mechanized cylindrical silo consists of more than 10 simplified arrangements, and the center of the four simplified combinations becomes the interstice silo. This kind of warehouse makes full use of space and has a large capacity but a small footprint. Generally, it is six to eight times more land-saving than house-type warehouses, but the cost is higher and the storage of seeds has more stringent requirements.

3. Low-temperature Warehouse

This kind of warehouse is built according to the basic requirements of low temperature, dryness, and airtight for the safe storage of seeds. The characteristics and structure of the warehouse are generally the same as those of the house-type warehouse, but its structure is quite tight. Moisture-proof layers cover the inner walls and the floor, while the walls and ceilings have thick heat insulation layers, respectively. There is a buffer room in the warehouse. Windows cannot be installed in low-temperature warehouses to prevent the outside temperature and humidity from passing through the window gap into the warehouse. Sometimes, the excessive temperature difference between the inside and outside of the warehouse will make water condense on the glass and drip into the seed pile. An 18 cm high breathable wooden pad frame is provided at the bottom of the stack, a 60 cm wide aisle is set between two stacks of seeds in the room, and the perimeter of the stack is 50 cm away from the wall to facilitate sampling, inspection, and moisture-proof. The warehouse is equipped with cooling and dehumidification machinery that can control the seed temperature below 15°C and the relative humidity to around 65%. It is an ideal seed storage warehouse at present.

4. Simple Warehouse

The simple warehouse is renovated from a private house. It retains the original structure of the house. By overhauling and plugging the loopholes in the house, smoothing and whitening the beams and walls with lime, and treating the ground with moisture-proof measures, the storehouse can achieve the requirements of no hole, no seam, no leakage, no dampness, and flat and smooth, which can be used for the temporary storage of seeds after being fully dried.

(III) Site Selection and Daily Maintenance of Seed Warehouses

1. The Principle of Selecting the Site of the Warehouse

The site of the warehouse should meet the following requirements:

(1) The side of the warehouse is better to face south, and the terrain should be high and dry to prevent water leakage from the warehouse floor. Especially in the south of the Yangtze River, except for mountainous and hilly areas, the groundwater

level is generally higher and there is more rain, so it is necessary to choose a site higher than the flood level or heighten the foundation of the warehouse according to the local hydrological data and people's experience.

(2) The soil at the site of the warehouse must be solid and stable. It is not advisable to build warehouses in areas that are likely to collapse. General seed warehouses require soil solidity to withstand more than 10 t of pressure per m^2. If this requirement cannot be met, the foundation of the four corners of the warehouse and brick piers should be reinforced. Otherwise, the foundation of the house will sink, or the ground will rupture and cause unnecessary losses.

(3) The site of the warehouse should be as close as possible to the railway, road, or waterway transport lines to facilitate the transportation of seeds.

(4) The site of the warehouse should be as close as possible to the seed breeding and production base to reduce the cost of seed transportation.

2. Standards of Warehouse Building

(1) The warehouse should be firm. It should be able to withstand the pressure of seeds on the ground and warehouse walls, as well as the effects of wind and adverse weather. From the roof and body of the warehouse to the wall base and flooring, building materials should be heat-insulated and moisture-proof to facilitate safe seed storage.

(2) The warehouse should be airtight and ventilated. The purpose of sealing is to insulate the seeds from adverse weather conditions such as rain, humidity, or high temperatures, and to achieve the desired effect of killing pests by using pharmaceutical fumigation. The purpose of ventilation is to dissipate water vapor and heat inside the warehouse, so as to prevent the seeds from being exposed to high temperatures and high humidity for a long time, which will affect their viability. At present, natural ventilation is more common since mechanical ventilation equipment has not yet been popularized. Natural ventilation is based on the principle of air convection, so doors and windows should be set symmetrically. Windows are best in the form of a flip-up and are airtight when shut. The height of windows should be appropriate, as too high will hinder air convection and ventilation, and too low will reduce the utilization rate of the warehouse.

(3) The warehouse should be insect-proof, impurity-proof, rodent-proof, and finch-proof. The warehouse should have a ceiling and its inner walls should be flat and painted white with lime, which can help check insect traces. Leaving no gaps in the warehouse not only helps eliminate the habitat of pests but also facilitates the cleaning of seeds to prevent mixing. The warehouse door should be installed with rat guards, and the windows should be equipped with barbed wire to prevent rodents and finches from entering.

(4) Buildings such as the sunning ground, storerooms, and testing rooms should be set up near the warehouse. The sunning ground is used for drying or processing the seeds before they enter the warehouse. Its area depends on the size of the warehouse, generally equivalent to 1.5–2 times the area of the warehouse is appropriate. The storeroom is especially used for storing warehouse equipment and tools. Its size can be determined by the actual needs of the warehouse and the number of equipment. The testing room should be located in a quiet area with sufficient light. Meanwhile, a fumigation room can be set up as needed for the fumigation of seeds and the packaging supplies such as jute bags.

3. Maintenance of the Warehouse

The warehouse must be thoroughly inspected and maintained before the seeds are put into storage to ensure their safety during storage. The first thing to inspect is the warehouse's overall structure. Carefully observe whether the warehouse has signs of sinking, tilting, etc. If there is a possibility of collapse, seeds cannot be stored here. Next, inspect gradually from the outside to the inside for signs such as leakage from the roof or whether the warehouse flooring is kept flat and smooth. If water seepage, cracks, and pockmarks are found on the flooring, it must be repaired to remain flat and smooth. Likewise, interior walls should be kept smooth and white. Gaps found on the walls should be patched, smoothed, and painted white with whitewash. No small holes should be left in the warehouse to prevent rats from sneaking in. The new warehouse should conduct short-term trial storage to observe its reliability. After the trial, it should be overhauled according to the warehouse standards. When its safety and reliability are determined, the seeds can be stored for a long time.

4. Commonly Used Equipment in Seed Warehouses

In order to improve the technical level and work efficiency, and reduce the labor intensity of the management, the seed warehouse should be equipped with the following equipment.

(1) Testing equipment. In order to correctly grasp the dynamic changes of seeds during storage and the quality of seeds in and out of the warehouse, the seeds must be tested. The testing equipment should be set according to the required measurement items, such as thermodetector, moisture meter, oven, germination box, volume-weight instrument, magnifier, microscope, hand screen, etc.

(2) Handling and conveying equipment. Mechanical transport is adopted when seeds enter and exit the warehouse, which can be equipped with wind suction conveyors, mobile belt conveyors, stackers, mechanical elevators, etc. If matched with various machinery, joint operations can be carried out.

(3) Mechanical ventilation equipment. When the natural wind cannot reduce the temperature and humidity in the warehouse, mechanical ventilation shall be adopted quickly. Ventilation machinery mainly includes blowers (blast and suction) and ducts (underground and aboveground). A suction ventilator is better than a blast blower in general conditions.

(4) Seed processing equipment. The processing equipment includes three major parts: cleaning, drying, and chemical treatment. Cleaning machinery is also divided into two types: roughing and sifting. In addition to the sunning ground, the drying equipment should be equipped with an artificial dryer. Chemical treatment machinery such as coating machines, disinfection machines, and chemical seed dressers can disinfect and sterilize the seeds to prevent the spread of seed diseases.

(5) Fumigation equipment and fire fighting equipment. Fumigation equipment is essential to control pests in the warehouse. Various types of gas masks, anti-poison respirators, balling guns, fumigants, etc. should also be equipped.

One of the important elements of warehouse management is fire prevention. Therefore, the warehouse should be equipped with firefighting equipment, including a variety of fire extinguishers, dry sand, fire hydrants, etc.

(6) Other equipment. To keep track of the temperature and humidity of the

various parts of the seed pile and the bagged seed stack. Now, the remote-measuring thermohygrograph is usually used to observe the temperature and humidity changes in the seed pile in time by burying its probes in different parts of the pile, so as to understand the stability of seed storage. In addition, the warehouse also needs packaging equipment such as balers, sealing machines, various specifications of jute bags, composite bags, appliances for the sunning ground, and measuring appliances like scales, electronic scales, etc.

II. Seeds Storage

(I) Standards for Seed Storage

The stability of seeds during storage varies significantly depending on the crop species, maturity degree, and harvest season. For example, under the same moisture conditions, seeds of general oil crops are less easy to preserve than seeds containing more starch or protein. The requirements for seed moisture during storage are also different. For example, the safe moisture of Indica rice seeds in the South must be below 13% to ensure that they survive the summer. The moisture of seeds containing more oil, such as rape, peanuts, sesame, and cotton, must be reduced to less than 10%. Seeds with damaged grains or poor maturity are easily damaged by pests and microorganisms and lose viability under high-moisture conditions due to their high respiratory intensity. Therefore, this type of seed must be strictly cleaned, selected, and removed. Seeds that do not meet the warehousing standards should not be put into the warehouse rashly and must be re-processed, cleaned, or dried. The seeds can only be stored in the warehouse after passing the test.

The climatic conditions in China's northern and southern provinces vary greatly, so the warehousing standards of seed cannot be forced to be consistent. The State Bureau of Quality and Technical Supervision stipulates that the variety purity index is the basis for dividing the seed quality level. If the seed purity cannot reach the original seed index, the seeds will be degraded to Grade-I good-quality seeds. Those that do not meet the index of the Grade-I good-quality seed purity are called Grade-II good-quality seeds. If the seeds cannot reach the Grade-II index, they will be defined as unqualified seeds. Indicators are set to measure cleanliness, germination rate, and moisture, and a seed that fails to reach any of the indicators is unqualified. The moisture of rice, corn, and sorghum seeds in the north of the Great Wall and in the alpine areas is allowed to be higher than 13%, but not higher than 16%. The moisture of seeds transported to the south of the Great Wall (except seeds in the alpine areas) shall not be higher than 13%.

(II) Seed Batches

Before entering the warehouse, crop seeds should not only be strictly separated according to different cultivars but also be stacked and treated according to the origin, harvest season, moisture, and cleanliness. Whether the quantity of seeds in each batch is large or small, it shall be uniform. The samples obtained from different parts are required to represent the characteristics of each batch of seeds.

There are some differences among different seeds. If the differences are significant, these seeds should be stacked separately or organized again to make them basically consistent. Otherwise, the quality of the seeds will be affected. If some seeds of low cleanliness are mixed in a seed stack of high cleanliness, the value of the stack will be reduced for use in production, and the stability of the seeds will also be affected during storage, as seeds of low purity tend to absorb moisture again. Likewise, mixing different batches of seeds with too much moisture disparity can cause the transfer of moisture within the seed stack, resulting in rotting and deterioration. Seeds should be stacked in batches when their pathogenic conditions and maturity stages are different. When there is a large number of seeds in the same batch (e.g. rice and wheat seeds more than 2.5×104 kg), the seeds should also be separated. Seed batching before warehousing should be treated with caution, as it is very important to ensure the sowing quality and long-term safe storage. When warehousing, “five separations” should be noted, that is, separating fresh seeds from aged seeds, dry seeds from wet seeds, seeds infected with pests from insect-free seeds, and seeds of different species and cleanliness, so as to improve the stability of storage.

(III) Clearance and Disinfection

Proper warehousing and disinfection are the basis for preventing cultivar mixing as well as diseases and insects, especially in warehouses that have stored seeds for a long time and are in disrepair (including renovated warehouses).

1. Clearance

The clearance work includes both cleaning the warehouse and remediating inside and outside the warehouse. Cleaning the warehouse is not only to completely remove seeds of other cultivars, inert matter, and garbage in the warehouse but also to clean the warehouse equipment, scrape the worm nests, repair the wall surface, and caulk and paint. Outside the warehouse, weeds should be regularly removed and sewage should be discharged to keep the environment outside the warehouse clean. The specific practices are as follows.

(1) Cleaning warehouse equipment. Bamboo mats, baskets, hemp bags, and other equipment frequently used in the warehouse are the places where pests like to hide. They must be cleaned by picking, knocking, striking, washing, brushing, and disinfected by exposure to the sun, fumigation with agents, and boiling and scalding in boiling water, so as to thoroughly remove residual seeds and hidden pests embedded in the warehouse.

(2) Scraping worm nests. There are many holes and gaps in the wooden warehouse, which are good places for the inhabitation and reproduction of warehouse insects. Therefore, all beams, columns, walls, and floors in the warehouse must be thoroughly scraped. The scraped seeds and other debris shall be cleaned up, and the insect carcasses shall be burned in time to prevent infection.

(3) Repairing wall surface. If plaster on walls inside and outside the warehouse has been peeling for years due to being in bad repair, the wall surfaces shall be repaired in time to prevent pests from hiding.

(4) Caulking and painting. After scraping the worm nests, the joints in the warehouse, regardless of size, shall be caulked with paper strips mixed with lime

mortar. The warehouse wall shall be fully painted before and after it is used to store the seeds. The purpose of painting is not only to present a neat style but also to expose worm marks on white walls.

2. Disinfection

Disinfection should be done regardless of whether seeds have been stored in old or new warehouses. Spraying and fumigation are two methods for disinfection. Disinfection must be carried out before repairing the wall, caulking, and painting, especially before comprehensive painting, because the newly painted lime is highly alkaline before drying, which makes it easy to decompose and invalidate the drugs.

Disinfection in empty warehouses can be treated with agents such as trichlorfon or DDVP. If trichlorfon is adopted, dilute the trichlorfon original solution to 0.5%–1% and fully mix the solution. Then, spray it evenly with a sprayer. 3 kg of 0.5%–1% aqueous solution can be applied to a warehouse over 100 m^2; The disinfection can also be achieved by soaking the sawdust in the 1% trichlorfon aqueous solution, drying the sawdust, and fumigating to kill pests. After spraying, close the doors and windows for no less than 72 hours to kill pests. After the warehouse is disinfected, it must be cleaned before storing seeds.

If DDVP is adopted, use 100–200 mg of 80% Emulsifiable Concentrate (EC) for each cubic meter of the warehouse. Application methods are as follows.

① Spraying Method: Mix 1–2 g of 80% DDVP EC with 1 L of water to synthesize 0.1%–0.2% diluent, and then spray.

② Hanging Method: Hang the wide cloth or paper strips that have been soaked in 80% DDVP EC in the air of the warehouse, with the rows spacing of about 2 m and the strip spacing of 2–3 m, and let the agent volatilize to kill the pests.

After applying the above two methods, the doors and windows must be closed tightly for 72 hours. After disinfection, the warehouse must be ventilated for 24 hours before the seeds enter the warehouse, so as to ensure the safety of operators. Aluminum phosphide can also be used for fumigation and disinfection, but safety should be noted.

(IV) Seed Stacking

Labels and cards must also be made before the seeds are put into storage. The label should indicate the crop, cultivar, grade, and the full name of the business unit, and be fastened to the outside of the seed bag. The card shall indicate the crop, cultivar, purity, germination rate, moisture, year of production, and business unit, and the card should be packed in seed bags or placed in seed piles or stacks when sealed. At the time of warehousing, the seed bags must be immediately weighed, registered, and stacked separately according to the seed category and grade to prevent mixing. The stacking can be divided into bag stacking and bulk stacking.

1. Bag Stacking

The bag stacking forms could be selected according to the warehouse conditions, storage purpose, seed quality, storage season, and temperature. Operation paths should be set for management and inspection, so the stacks should be 0.5 m from the wall and 0.6 m from each other. The stack height and width could be increased or decreased depending on the dryness and other conditions of the seeds. For seeds with higher moisture, the stack should be narrower for effective ventilation that could dissipate the moisture and heat in the seeds. Dry seeds can be stacked more widely. The direction of the stacks should be parallel to the doors and windows of the warehouse, i.e. if the doors and windows are set oppositely on the south and north walls, then the stacks should be arranged from south to north for better management and air circulation when the doors and windows are opened. During stacking, the bag mouth shall face inward to prevent the stacks from pests and collapsing. Generally, the packed seeds shall be padded with a plate at the bottom about 20 cm above the ground for ventilation. Bag stacking methods are as follows.

(1) Solid stacking. Bags should be stacked in order regularly without spacing. There can be two, four, six, or eight rows for a stack, and four bags in a row are generally used. The stack length is decided by the size of the warehouse, and sometimes it is equal to the length of the warehouse. The two ends of the stack shall

be stacked into a half "非" shape to prevent collapse. This method could achieve a high utilization rate of storage capacity but could only be used for stacking the seeds of a certain quality, such as the seeds stored at low temperatures in winter or temporarily stored.

(2) "非"-shaped or half "非"-shaped stacking method. This stacking method provides good ventilation, anti-collapse ability, and convenience to check. Stack up in a "非" shape or a half "非" shape. For example, in the "非"-shaped stacking method, two bags are placed longitudinally in the middle of the first layer, and other bags are placed horizontally on the left and right sides of the middle two bags, which forms the shape of the Chinese character "非". The second layer is transposed with two longitudinal bags on both sides and other horizontal bags in the middle, and the stacking form of the third layer is the same as that of the first layer.

(3) Ventilation stack. This stacking method with larger spacings is convenient for ventilation and moisture dissipation and is mostly used for stacking seeds of high moisture. This method is used in summer to check the conditions of the seeds bag by bag. There are many forms of ventilation stacks, such as "井"-shaped, "口"-shaped, holed coin-shaped, and "工"-shaped. If it is difficult to operate, the stacks should not be piled too high and wider than two bags. Besides, pay attention to safety.

2. Bulk Stacking

In the case of a large number of seeds, insufficient storage capacity, or lack of packaging tools, bulk stacking is mostly used. Bulk stacking is appropriate for storing seeds that are sufficiently dried and of high cleanliness. It could be operated in the following ways.

(1) Bulk stacking in the entire warehouse and bulk stacking in several compartments. A large number of seeds could be stacked by this method, which contributes to a high utilization rate of warehouse capacity. A warehouse can be divided into several compartments according to the number of seeds and the purpose of convenient management, or in the case that the seeds of multiple cultivars need to be stacked separately to avoid mixing. Generally, seeds could be stacked 2–3 m high, but they must be below the safety line. Bulk stacking in the

entire warehouse is often used to store a large number of seeds. Therefore, it is necessary to strictly follow the seed warehousing standard, strengthen management at ordinary times, and pay special attention to abnormal phenomena such as condensation on the outer layer of seeds.

(2) Bulk stacking in one large enclosure. This method may be used for warehouses with less solid walls or without moisture barriers, or for stacking seeds with large free flowability (such as soybeans and peas). Before stacking, consider the size of the warehouse, package the seeds of the same batch and cultivar into gunny bags, and stack the bags along the wall at a 0.5 m distance to the wall to form an enclosure. The stacking height should not be too high. The enclosure height can be up to about 2 m, and the seeds can be stacked in bulk within the enclosure. The seed stack inside the enclosure must be lower than the enclosure height and care must be taken to prevent collapse. When the enclosure is stacked, the bags should be close to each other and cover the seam layer by layer, so that the enclosure can be interlocked to prevent collapse. At the same time, it should be retracted by about 3 cm layer by layer from bottom to top.

(3) Bulk stacking in several small enclosures. This method is adopted for a small number of seeds of various cultivars. It can also be used as a temporary stacking measure when the seeds are of different cultivars and grades or when the seeds that do not meet the warehousing criteria cannot be handled in time. The stack shall be enclosed while stacking, and the stacking height is generally about 2m.

(V) Seed Warehouse Management System

After the seeds are stored, it is necessary to establish or improve the management system, clarify management responsibilities, and strengthen the daily management. The management system is as follows.

(1) Production post-responsibility system. Personnel with a strong sense of responsibility and professional dedication should be selected for this work. Custodians should constantly improve their professional skills and strive to improve their scientific management level. Relevant departments shall also conduct regular assessments.

(2) Security system. The warehouse should be provided with a duty system that arranges personnel to patrol to eliminate unsafe factors promptly, and with fire prevention and theft prevention measures to avoid accidents.

(3) Cleaning and sanitation system. Qualified cleaning and sanitation jobs are prerequisites for eliminating diseases and pests in the warehouse. The inside and outside of the warehouse must be cleaned and disinfected frequently to keep it clean. The six sides of the warehouse must be clean and three aspects outside the warehouse (weeds, garbage, and sewage) must not be left. When the seeds are discharged from the warehouse, the warehouse shall be cleared to prevent the mixing and infection of diseases and pests.

(4) Daily inspection. The inspection contents include air temperature, warehouse temperature, seed temperature, atmospheric humidity, warehouse humidity, seed moisture, germination rate, pests and mildew, warehouse condition, etc.

(5) Filing system. When seeds are delivered to the warehouse, the source, quantity, and quality status must be registered. The results of each inspection must be recorded and saved in detail for comparative analysis and examination. Such registration and recording are conducive to finding problems and taking timely measures for improvement.

(6) Financial accounting system. The approval procedures and weighing bookkeeping must be strictly carried out for each batch of seeds entering and leaving the warehouse. The accounts shall be clear and the surplus and shortage of seeds shall be kept in mind. Do not miss the farming time, and track down responsibilities for unreasonable additional losses.

(VI) Seed Check

1. Seed Temperature Check

Under normal conditions, the increase and decrease of seed temperature are influenced by the air temperature. The temperature change is related to the heat insulation and sealing performance of the warehouse, the size of the seed pile, and the stacking mode. Generally, if there are a small number of packaged seeds stacked in a warehouse with poor heat insulation and sealing performance, the seed temperature changes rapidly as it is greatly affected by the air temperature. On the contrary, if a small number of seeds are stacked in bulk in the warehouse with good heat insulation and sealing performance, the temperature of the seeds is less affected by the air temperature, especially of the seeds in the middle and lower layers, of which the temperature is relatively stable.

The change in seed temperature generally reflects the health condition of stored seeds and is quite sensitive and easy to check. Therefore, the method of checking the seed temperature is generally included in the guideline for storage work.

The cycle of checking the seed temperature in different seasons can be found in Table 6-1. It is better to check the temperature from 9:00–10:00 daily to avoid being affected by the external temperature. In addition to checking the seed temperature, the warehouse temperature and air temperature shall also be recorded.

Table 6-1　　Temperature Check Cycle in Different Seasons

Seed Moisture (%)	Summer and Autumn		Seed Temperature in Winter		Seed Temperature in Spring		
	Newly harvested or seeds without after-ripening	Seeds with after-ripening	Above 0°C	Below 0°C	Below 5°C	5–10°C	Above 10°C
Below Safe Moisture	Once a day	Once every three days	Once every five to seven days	Once every half a month	Once every seven to 10 days	Once every five days	Once every three days
Above Safe Moisture	Once a day	Once a day	Once every three days	Once every seven days	Once every five days	Once every three days	Once a day

2. Seed Moisture Check

In general, the seed pile moisture is mainly influenced by the relative humidity of the air. Seed moisture changes little throughout the day and only at the surface layer. The change of moisture in a year varies with seasons. The seed moisture is high in low temperatures and the rainy seasons, while the seed moisture is low in summer and autumn. The change of seed moisture at different layers is also different. The upper layer, which is most affected by air and generally around 33 cm (one foot) in depth, has a particularly prominent change in seed moisture. The change of the seed moisture in the middle and lower layers is smaller, except for the seeds about 17 cm above the ground in the lower layer, of which the moisture rises a lot sometimes due to ground influence. The above changes often occur in spring and summer in production, so it is said that "check the surface in spring and the bottom in summer".

Moisture transfer also occurs when there is a temperature difference between different parts of the seed pile. For example, the moisture of seeds with higher temperatures in the sunny part will transfer to the part with lower temperatures, so that the originally uniform seed moisture changes. The seed moisture becomes lower in the high-temperature part and higher in the low-temperature part. The cycle of checking the moisture depends on the seed temperature. When the seed temperature is below 0°C, the checking shall be carried out once a month. When it is above 0°C, the checking shall be carried out twice a month. The checking shall also be carried out after each seed sorting. If the airtightness of the warehouse is poor, the checking times can be increased as needed in rainy seasons. The moisture of seeds should also be known before the seeds enter or leave the warehouse and before killing pests in the warehouse.

(VII) Warehouse Pest Check

1. Pest Habits

The activity of pests varies with temperature. When the temperature is below 15°C, the pests move slowly and do little harm. Within the temperature range suitable for life activities, the pests become more active with the increase in temperature, and pest infestation becomes severe. Therefore, pest infestation in winter is the least severe due to the low temperature. As the temperature rises in spring, the pest infestation increases gradually. Pest infestation is the most severe in summer, especially from July to August, during which the pest activities are rampant. As the temperature drops in autumn, the pest infestation decreases gradually. The activity of the pests in the seed pile is influenced by the change in seed temperature, generally moving towards the parts with higher temperatures.

The pests would like to stay in the bottom part (below the point of 33 cm to the ground) to the South in spring and to the North in autumn, mostly on the surface of the seed pile in summer, and below the point of 1 m to the surface of the seed pile in winter. The movement ability of pests in the seed pile is related to their species. Most moth pests can only move on the surface of the seed pile and about 30 cm below the surface. Crustacean pests can move to the depth of the seed pile. Moisture-loving pests move to the part with high moisture in the seed pile, while some pests move to the parts where inert matter and broken particles are concentrated. Pests are afraid of light and generally move to shadows.

2. Check Method

Generally, it is performed by screening. Take 1 kg of seeds at one inspection point, screen the pests after a certain period of vibration screening, analyze the species and number of live pests, and then determine the control measures. If there is no sieve, spread out the inspection samples on white paper, carefully find out the pests, and count the pest species and the infection rate per kilogram of seeds (grains/kg).

In the season of low temperature, the dead and living status of pests should be correctly judged. The method is to put the detected pests on the palm or blow a breath of hot air to it, and the live pests will wake up and crawl after being heated. Generally, the screening method is not used to test moth pests, because moth pests are good at flying and will fly away after vibration, which is not easy to screen. The visual inspection and statistics of moth flies can be carried out by throwing seeds in the air, seeing the flying moth pests when the seed surface is vibrated, and then observing the pest population density, so as to decide the prevention and control measures.

3. Screening Cycle

It depends on the air temperature and seed temperature. Generally, when the temperature is below 15°C in winter, screening shall be carried out once every two to three months. When the temperature is above 15°C in spring and autumn, screening shall be carried out once a month. If the temperature exceeds 20°C, screening shall be carried out twice a month. In the case of high temperatures in summer, screening shall be carried out once a week.

(VIII) Seed Germination Rate Check

During the storage period, the germination rate of seeds changes due to different storage conditions and storage times. The seed germination rate would hardly decrease when stored under perfect conditions for a short time. For some seeds with physiological dormancy, storage over some time can increase the germination rate. Therefore, it is necessary to perform regular germination tests on seeds. According to the changing status of the germination rate, measures should be taken in a timely manner to improve storage conditions and avoid losses.

In general, the seed germination rate should be checked every four months. The checking times can be increased according to the temperature change. For example, it should be checked once after high or low temperature and before and after chemical fumigation. The last time shall be completed no later than 10 days before the seeds are discharged from the warehouse.

(IX) Mold, Mice, and Birds Check

Generally, visual and smelling inspections are used to check the mold. The parts to be checked are generally the corners of the wall, the bottom and upper layers where the seeds are vulnerable to moisture, and places along with the doors and windows where rain leakage may occur. Checking mice and birds can be done by observing whether there are any feces or signs of their activities in the warehouse. At ordinary times, the surface of the seed pile should be leveled to find the signs. Once found, capture and elimination shall be done, and the loophole shall be blocked.

(X) Warehouse Facilities Check

Check the water seepage of warehouse floors, the rain leakage of roofs, and the falling off of ash walls, especially in the case of strong tropical storms, typhoons, and rainstorms. At the same time, check the flexibility of opening and closing of doors and windows and the firmness of bird-proof nets and mouse-proof plates.

Reasonable ventilation is necessary.

(1) Purpose of ventilation. After the seeds are put into the warehouse, they need to be ventilated at an appropriate time, regardless of their storage duration. Ventilation is an important management measure for seeds during storage. Ventilation can reduce the temperature and moisture, so as to keep the seeds dry and keep the seed temperature low for a long time, in which case the physiological activities of seeds and the harm of pests and molds could be restrained. Ventilation can also maintain the balance of temperature in the seed piles and avoid moisture transfer due to temperature differences. In addition, it can accelerate the convection current in the seed piles and eliminate the harmful substances caused by the seed catabolism and poisonous gas caused by fumigation. For seeds with fever symptoms or seeds that are mechanically dried, ventilation and heat dissipation are more necessary. In conclusion, ventilation is an essential technical measure in the management of seeds during storage.

(2) Principles of ventilation. Before ventilation, the temperature and humidity inside and outside the warehouse must be measured to determine whether ventilation can be performed, mainly in the following cases:

① Ventilation is not suitable on rainy and foggy days.

② When the outside temperature and humidity are lower than those in the warehouse, ventilation is needed. However, attention should be paid to the influence of cold currents. Preventive measures should be taken to avoid dew condensation on the outer layer of seeds due to excessive temperature differences in the seed pile.

③ Ventilation is feasible when the temperature outside the warehouse is the same as that inside the warehouse and the humidity outside the warehouse is lower than that inside the warehouse, or the humidity inside and outside the warehouse is basically the same and the temperature outside the warehouse is lower than that inside the warehouse. The former reduces humidity, and the latter reduces temperature.

④ When the temperature outside the warehouse is higher than inside and the relative humidity is lower than inside, or the temperature outside the warehouse is lower than inside and the relative humidity is higher than inside, whether ventilation is feasible depends on the absolute humidity at that time. If the absolute humidity outside the warehouse is higher than inside, ventilation is not allowed. On the contrary, ventilation is feasible.

(3) Method of ventilation. There are two ventilation methods, namely natural ventilation and mechanical ventilation, which can be selected according to the current equipment conditions and needs of the warehouse.

① Natural ventilation: It is a method of selecting the time conducive to cooling and dehumidification according to the temperature and humidity inside and outside the warehouse. Opening doors and windows for natural air exchange is a method to cool and disperse moisture in the warehouse. The effect of natural ventilation is related to temperature difference, wind speed, and the way seed stacking is done. When the temperature outside the warehouse is lower than that inside, the pressure difference between the air inside and outside the warehouse will be generated, and the air will naturally exchange. The cold air enters the warehouse and the hot air is discharged outside the warehouse. The greater the temperature difference is, the more air exchange between inside and outside is, and the better the ventilation effect is. If the wind speed increases, the wind pressure increases, the airflow also increases, and the ventilation effect becomes better. The ventilation effect of packing stacking in the warehouse is better than that of bulk stacking, and the ventilation effect of small packing and ventilation stacks is better than that of large and solid stacks.

② Mechanical Ventilation: It refers to using machinery to supply (or exhaust) air. This is a way to cool and dehumidify the seed pile through ventilation management or ventilation slots for air exchange, mostly used on bulk seeds. Because it is mechanically powered, the ventilation effect is good, with short ventilation time, fast cooling, uniform cooling, and other characteristics.

There are two types of mechanical ventilation, namely air trough mechanical ventilation and air duct mechanical ventilation. According to the different ways of air flowing in the seed pile, it can be divided into supply type and exhaust type. The way of using a fan to press dry and cold air from the duct or inner trough into the seed pile and get the hot and humid air inside the pile discharged from the surface is called supply type; The way of sucking out the hot and humid air from the duct or air trough and making the dry and cold air enter the seed pile from the surface is called exhaust type. Which of the above ventilation types to choose can be determined according to the warehouse conditions and seed stacking needs. It is best to stall the fan in a way that it can both supply and exhaust so that we can determine which type or alternate use of the two types according to the specific situation of ventilation and cooling. In order to facilitate movement and achieve multiple uses, the fan is generally installed on a cart that can be moved, so that when a warehouse is successfully ventilated and cooled, it can be moved to another warehouse for use.

(4) Notes on mechanical ventilation are as follows.

① The effect of mechanical ventilation is related to the temperature and humidity at that time. When the external temperature and humidity are low, the ventilation effect is better. On the contrary, except in special cases, ventilation is generally impossible. The effect is usually more pronounced when the temperature difference between air temperature and seed temperature is above 10°C. In order to prevent the seeds from absorbing moisture from the humid air, it is necessary to ventilate at a relative humidity lower than that of the seed equilibrium moisture.

② Before the ventilation, the surface of the seed pile should be raked to make the thickness of the seed pile even so as not to cause uneven ventilation due to the uneven thickness of the seed pile.

③ When supply ventilation is carried out, in order to prevent dew condensation on the surface of the seed pile, a layer of straw bales can be spread on the surface. The ventilation cannot be stopped before the dew disappears. When exhaust ventilation is carried out, the container should be used to collect the condensed water at the outlet of the fan. The fan cannot be stopped before the water drops stop, nor can it be stopped halfway so as to prevent the condensed water in the air pipe from flowing into the seed pile through the pipe hole. For underground air trough ventilation, it is possible to use containers to collect water or put small bags containing rice chaff in the air outlet to absorb moisture.

④ When implementing exhaust air duct ventilation, the upper and middle layers of the seed pile cool down quickly, while the lower layer cools down poorly, so the ventilation cannot be stopped when the ventilation requirements are not met. It is also possible to adopt exhaust ventilation first and then supply ventilation to improve the ventilation effect. When using a single air duct or multiple air ducts for ventilation, the joint of each duct should be tight, with no air leakage and no hose bending, etc., so as not to affect the ventilation effect. The end of the duct can not be too high from the ground; otherwise, it will reduce the ventilation effect of the bottom. Preferably, it should be 30–40 cm from the ground.

⑤ During ventilation, it is necessary to strengthen the temperature check to keep track of the seed temperature drop and possible dead space. The dead space is mainly caused by the unreasonable arrangement of ventilation pipes and ventilation slots. The remedy is to install more ventilation pipes in the dead space.

⑥ In the implementation of ventilation operations, all doors and windows should be opened to speed up the airflow. When ventilation is not carried out, the air inlet should be tightly blocked so that the seeds will not be affected by the hot and humid outside air.

III. Microbiological Management of Seeds During Storage

There are two major categories of microorganisms commonly found on seeds. One is the yellow herbaceous sporeless bacterium and *Pseudomonas fluorescens* attached to fresh and healthy seeds, which do no harm to seed storage and have an antagonistic effect on mold; the other is some molds in fungi, which are the most harmful microorganisms to the safe storage of seeds. In seed storage, the main ways to control mold are as follows.

1. Improve the Quality of Seeds

High-quality seeds are more resistant to microorganisms. In order to improve the viability of the seeds, it is necessary to harvest, thresh and dry the seeds in time when they are mature, and perform the seed cleaning carefully, including removing sundries, broken grains, and grains that are not full. When storing, pay attention to separate storage of new and old seeds, dry and wet seeds, seeds with pests and seeds without pests, and seeds of different species and cleanliness to improve the stability of the stored seeds.

2. Keep Dry Conditions

When seed moisture and indoor relative humidity are lower than the minimum water required for microbial growth, microbial activity can be inhibited. To this end, firstly, the seed warehouse should be able to prevent moisture and humidity, and have good ventilation and airtightness condition. Secondly, the seeds should be fully dried before storage so that the moisture content is kept below the safe moisture limit balanced with 65% relative humidity. In the process of seed storage, dry and airtight storage methods can be used to prevent the seeds from absorbing moisture again. In the season with temperature change, it is necessary to control the temperature difference and prevent condensation, and it is necessary to ventilate and reduce humidity in time after the seeds with high moisture are stored.

3. Maintain Low-temperature Conditions

Controlling the temperature of seed storage below the temperature suitable for mold growth can inhibit the activity of microorganisms. If the seed temperature is kept below 15°C and the relative humidity of the warehouse is below 65%, the purpose of pest and mildew prevention and safe storage can be achieved. This is also the temperature and humidity limit of the so-called "low-temperature storage".

Natural low temperatures can be used to maintain low-temperature conditions, such as the temperature in northern China. Low-temperature storage can also be performed by mechanical refrigeration. When carrying out storage, the seed moisture should also be reduced to below-safe moisture to prevent the activity of some cryogenic microorganisms under high moisture conditions.

4. Use Chemical Agents

The chemical agent commonly used is aluminum phosphide. Phosphine produced by aluminum phosphide has a good bacteriostatic and mildew-proof effect. Because it is also a pesticide, the amount of its pesticide composition is sufficient to inhibit bacteria, so as long as fumigation is carried out once in use, the purpose of pest killing and bacteriostasis can be achieved at the same time.

IV. Seed Storage Management at Room Temperature

(I) Management of Seeds During Storage

Management during seed storage is very important. It is necessary to establish various systems, put forward measures, and conduct strict inspections according to the specific situation so as to find and solve problems in time and avoid seed storage loss.

(1) Warehouse post responsibility system. Select responsible and career-minded people for this job. The custodians should constantly improve their professional skills and strive to improve their scientific management level, and relevant departments should assess them regularly.

(2) Security system. The warehouse should establish an on-duty system, organize personnel patrols, and eliminate unsafe factors in time. It should also do a good job of fire and theft prevention to ensure that there are no accidents.

(3) Cleaning and sanitation system. Cleaning and sanitation are prerequisites for eliminating pests and diseases in the warehouse. The inside and outside of the warehouse must be cleaned and disinfected frequently to keep it clean.

(4) Inspection system. The contents of the inspection include air temperature, warehouse temperature, seed temperature, ralative humidity, humidity inside the warehouse, seed moisture, germination rate, pest and mold situation, etc.

(5) Archival system. When each batch of seeds is put into storage, it is necessary to register their source, quantity, and quality condition item by item, and the results after each inspection must be recorded and kept in detail for comparative analysis and investigation, which is conducive to finding problems and taking timely measures to improve the work.

(II) Management Measures

1. Prevent Mixing

When the seeds go in and out of the warehouse, the varieties are easily mixed, so special attention should be paid. There should be labels inside and outside the seed package, and they should be checked repeatedly when they are in and out of storage.

2. Reasonable Ventilation

The ventilation methods (natural ventilation and mechanical ventilation) can be selected according to the current equipment conditions and needs of the warehouse, which have been mentioned before.

3. Temperature Check

When checking the seed temperature, the whole pile of seeds can be divided into three layers, namely upper, middle, and lower layers, with five check points in each layer. It can also be increased or decreased according to the size of the seed pile. If the size of the seed pile exceeds 100 m^3 the number of check points should be increased accordingly, and additional auxiliary check points should be added in areas that are usually suspected, such as the points near the wall, corner, window, or rain leakage area, so as to grasp the situation of the seed pile comprehensively. Check the seeds every three days within half a month after storage (The number of inspections can be reduced in northern China, while the number of inspections should be appropriately increased in southern China.), and then every 7–10 days. In the second and third quarters, check the temperature once a month.

4. Moisture Check

When checking moisture, the seed pile is also divided into three layers with five check points in each layer. Mix the samples taken at each check point, and then take the samples for tests. The sampling must be representative, and samples taken in suspicious areas can be tested separately. The period of checking moisture depends on the seed temperature and it should be checked once a quarter in the first and fourth quarters, once a month in the second and third quarters. The check should also be performed after each seed sorting.

5. Germination Rate Check

The seed germination rate is generally checked every four months. However, an additional check should be added according to temperature changes, after high or low temperatures, and after fumigation. The last check must be done 10 days before the seeds are discharged from the warehouse.

6. Pest, Mold, Rodent, Bird Check

Pests are generally checked by screening, that is, by counting the number of live pests per kilogram after a certain period of vibratory screening. The method of grain scattering is used to check moths, and the statistics are made by visual observation.

7. Warehouse Facilities Check

Check the water seepage of warehouse floors, the rain leakage of roofs, and the falling off of ash walls, especially in the case of strong tropical storms, typhoons, and rainstorms. At the same time, check the flexibility of the doors and windows' opening and closing mechanisms and the firmness of bird-proof nets and mouse-proof plates.

(III) Characteristics and Management of Seed Storage in Low-temperature Warehouses

In order to adapt to the needs of the seed trade, preserve the temporary stock of seeds, and maintain a higher level of seed viability, many elite seed enterprises in China have built many low-temperature and low-humidity seed banks. The aim is to achieve safe storage of seeds and ensure the high viability of seeds by controlling the two main factors of temperature and humidity in seed storage.

In the low-temperature warehouses, the mechanical cooling method is adopted to keep the temperature below 15°C, and the relative humidity is controlled at about 65%. In terms of management, in addition to the requirements of the general seed warehouse, the following points should also be noted.

(1) The bottom of the seed stack must be equipped with a breathable wooden or plastic matting frame. A certain spacing should be retained between two stacks and between the stacks and the walls.

(2) Arrange the position of seed stacks reasonably, use the warehouse space scientifically and improve the utilization rate.

(3) Try to open the sealed door of the warehouse as little as possible. Even if

it's necessary to check the warehouse, a number of matters should be carried out as a whole to reduce the number of times opening this door.

(4) Strictly control the temperature and humidity in the warehouse. Generally, the temperature in the warehouse should be controlled below 15°C, the relative humidity should be controlled below 70%, and the temperature and humidity should be kept stable.

V. Rational Transportation of Seeds

1. Standards of Rational Seed Transportation

Seeds should be transported quickly and over short distances, and they should be transported in a straight line to reduce the cost. Therefore, time and transportation price are the standards for evaluating transportation programs and checking transportation work.

2. Organizational Measures for Rational Seed Transportation

First, according to the distribution of seed production and marketing and transportation conditions, determine the reasonable flow of seeds. Show the connections of seed production, supply, transportation, and sale with a flow chart, choose the transportation way that is reasonable, and make seed transportation rationalized and institutionalized. Second, use direct, straight-line transportation. In seed transportation, unnecessary procedures and links should be reduced as far as possible, and seeds should be transported directly from the place of origin or shipment to the place of sale and buyers. Third, when selecting the means of transport, the most reasonable means of transport shall be selected, and the transportation shall be organized in accordance with the principle of "timely, safe, accurate, and economical" by comparing and calculating the transport time, distance, link, loss, and cost.

3. Notes for Seed Transportation

(1) Each batch of seeds must be accompanied by a "Seed Test Certificate" and a "Plant Quarantine Certificate", which must be marked with the detailed address and postal code of the unit or individual that sends and receives the goods.

(2) The means of transportation must be clean and dry, equipped with

rainproof equipment, and protected from oil stains.

(3) The seeds need to be bagged for transportation and marked layer by layer.

Review and Reflection Questions

(1) What are the environmental conditions for seed storage?

(2) What are the basic types of seed warehouses?

(3) What types of equipment are commonly used in seed warehouses?

(4) What are the different ways of seed stacking?

(5) What are the principles of seed warehouse ventilation?

(6) What are the main ways to control mold in seed storage?

(7) Why should the seed storage warehouse be properly ventilated?

(8) What are the characteristics of seed transportation?